Rikugun: Guide to Japanese Ground Forces 1937–1945 is the first nuts-and-bolts handbook to utilize both the voluminous raw allied intelligence documents and post-war Japanese documentation as primary sources. This first volume covers the tactical organization of Army and Navy ground forces during the 1937–45 war. Using the wartime IJA mobilization plans, and the Unit Organization Tables, Unit Strength Tables and Unit History Tables compiled by the War Ministry and the 1st Demobilization Bureau during and after the war, a complete picture of IJA ground forces through the war is presented.

The evolution of the Japanese force structure is examined, including infantry, armor, cavalry, artillery and naval ground combat units from battalion to division level, each thoroughly discussed and illustrated with tables of organization and equipment and mobilization data. This forms the framework for any discussion of the IJA's capabilities and intentions.

Leland Ness has been conducting and supervising defense analysis and writing military history for over 40 years. He served as director of special projects at DMS/Jane's, published a newsletter on ground ordnance for the defense industry, and has been an editor at Jane's for the last ten years (for Jane's Ammunition Handbook and Jane's Infantry Weapons). He is also the co-author of the classic *Red Army Handbook*, and the author of the HarperCollins *WWII Tanks and Combat Vehicles*. He has been particularly interested in Asian military history since graduating with a degree in Oriental Studies and Language. During 28 years in the Army Reserve, he served with Headquarters, US Army Intelligence Agency on active duty during Desert Storm.

RIKUGUN

GUIDE TO JAPANESE GROUND FORCES 1937–1945

Volume 1

Tactical Organization of Imperial Japanese Army & Navy Ground Forces

Leland Ness

Helion & Company Ltd

Helion & Company Limited
26 Willow Road
Solihull
West Midlands
B91 1UE
England
Tel. 0121 705 3393
Fax 0121 711 4075
email: info@helion.co.uk
website: www.helion.co.uk
Twitter: @helionbooks
Visit our blog http://blog.helion.co.uk

Published by Helion & Company 2014
Designed and typeset by Farr out Publications, Wokingham, Berkshire
Cover designed by Euan Carter, Leicester (www.euancarter.com)
Printed by Lightning Source, Milton Keynes, Buckinghamshire

Front cover: A 75mm infantry regimental gun in action in China early in the war. Rear cover: A cavalry lineman with the pole that allowed him to run wire above ground in trees.

ISBN 978-1-909982-00-0

British Library Cataloguing-in-Publication Data
A catalogue record for this book is available from the British Library

Contents

List of Illustrations

List of Tables

List of Acronyms and Abbreviations

AA	Anti-Aircraft
AP	Armor Piercing
Art	Artillery
AT	Anti-Tank
Att	Attached
Bde	Brigade
BG	Border Garrison
Bty	Battery
Cav	Cavalry
CD	Coast Defense
Co Off	Company Grade Officer
DP	Dual Purpose
EU	Expeditionary Unit
FIU	Fortress Infantry Unit
Fld Off	Field Grade Officer
GL	Grenade Launcher
HE	High Explosive
HQ	Headquarters
Hvy Art Bn	Heavy Artillery Battalion
IGC	Independent Garrison Command
IGdU	Independent Guard Unit
IGU	Independent Garrison Unit
IMB	Independent Mixed Brigade
IMR	Independent Mixed Regiment
Ind Garr Inf Bn	Independent Garrison Infantry Battalion
Ind Inf Bn	Independent Infantry Battalion
Inf	Infantry
LMG	Light Machine Gun
LoC	Line of Communications
MG	Machine Gun
MSG	Master Sergeant
Mtn	Mountain
n/a	not available
NCO	Non-Commissioned Officer
Off	Officer
PVT	Private
Recon	Reconnaissance
Regt	Regiment
SGT	Sergeant
SMG	Submachine Gun
SNLF	Special Naval Landing Force

SP	Self-Propelled
SSD	South Seas Detachment
TO&E	Table of Organization & Equipment
WO	Warrant Officer

Preface

When I first started researching the Japanese ground forces in World War Two some forty-plus years ago I found there was very little that had not been lifted verbatim from a few wartime US intelligence manuals. As I dug into the archives in Washington, DC I realized that there was a huge gold mine of raw intelligence information that been overlooked or misinterpreted in the preparation of those manuals. Having served in military intelligence, and having participated in the preparation of similar manuals during war, I was not surprised to find that things are clearer in hindsight than they were at the time. I was shocked, however, to find that the Army's Military Intelligence Division either did not get or did not use the large volume of excellent information coming from the Navy's command in the Pacific.

Twenty years later, while still waiting for someone to produce a decent guide to Japanese ground forces, I began using the Japanese resources in the US National Archives and Library of Congress. These included both wartime documents seized during or at the end of the war, and documents created in the first five years after the war by the Japanese authorities to aid in demobilization and provide background for US military histories. Most of these have, over the years, been returned to Japan, fortunately after microfilming.

About eight years ago I gave up waiting for someone else to produce a useful guide. The Japanese scholars have produced some very useful works, but are hampered by the destruction of most of the original source material at the end of the war and an aversion to using US sources. Western historians, on the other hand, have been intimidated by the completely alien Japanese language. The few excellent models, Alvin Coox and Edward Drea, for example, concentrated on operational and institutional histories rather than the more mundane nuts-and-bolts reference approach that I had been looking for.

Thus I launched my own effort to produce two reference sources for the war in the Far East, one covering the Japanese ground forces and the other the Chinese. Both have been challenging, for different reasons. The documentation for the Japanese forces is voluminous and fairly reliable, although with some large and frustrating gaps. Here it is often a matter of what to leave out, as well as how to bridge those gaps.

The absence of historians engaged in serious research on the subject in the West has made this necessarily a rather solitary endeavor, but I have been fortunate to receive help and guidance from two of the premier Japanese authorities, Akira Takizawa and Tadashi Yoda. Both have pointed me towards new sources and have corrected such errors as they saw. Errors inevitably remain, and those, of course, are solely my responsibility.

1

General

The Japanese victory over Russia in 1905 spawned a number of "lessons learned," some accurate, some erroneous, that would inform Japanese doctrine and organization for the next forty years. Among those, some were contradictory, leading to rancorous feuds within the Army leadership that continued into the Second World War.

Certainly there was an initial feeling that an opening offensive, if sufficiently devastating and awe-inspiring, could cause a much larger opponent's will to falter and lead to a quick victory. This had clearly happened in the war, but there seems to have been little recognition that the Russian social and government structure at the time was already rotten to the core and about to implode. There was also popular, although not universal, sentiment that the war had been won largely as a result of the intangible samurai ethic unique to Japan.

Japan's role in World War One was minor, but many Army officers followed the fighting on the Western Front closely. As a result, where previously there had been various disagreements among the Army leadership on various issues, the different cliques now tended to coalesce into two main schools of thought. One, the traditionalists, continued to believe that Japan possessed a unique advantage in the warrior ethos of its people, and that doctrine should therefore be built around ferocious, close-quarters infantry combat, preferably featuring the bayonet. The reformers tended to look at the carnage of the Western Front and see how modern technology, particularly when integrated into a team, had slaughtered unsupported infantry. Similarly, the traditionalists continued to believe that an opening offensive of spectacularly overwhelming violence and speed could quickly break the will of an opponent, and to that end they favored the retention of the large square divisions that could absorb casualties and keep fighting, at least in the short term. The reformers tended to believe the next significant conflict could degenerate into a long war of attrition, for which Japan would not be suited without industrial modernization. They also advocated reorganization of the Army into smaller triangular divisions, providing a better firepower ratio and more flexibility in employment.

A swirling and bewildering array of cliques, associations and informal groups, many of which loathed each other, muddied the waters of Japanese doctrine from World War One through the mid-1930s. Conservatives of varying flavors, however, were the dominant force until 1936. In February of that year hot-headed young officers of the 1st Division launched a coup in Tokyo in the name of traditional martial values. When the coup collapsed after four days the reformers took advantage of the opportunity to discredit many of their opponents.

Nevertheless, the dominance of the conservatives through the mid-1930s resulted in considerable organizational stability. Through the early 1920s the Imperial Japanese Army (IJA) was built upon eighteen regional districts, each of which maintained allotted forces through conscription, training and maintenance of reserves, plus the Imperial Guards district in Tokyo, which recruited from throughout Japan. Two further districts (19th and 20th) were formed in Korea in 1915 to administer the Japanese now resident there as well as Korean volunteers. In 1924, however, four divisional districts (13th, 15th, 17th and 18th) were abolished in their entirety as an economy measure.

Each district maintained a division and other smaller elements at peace strength through conscription of local male residents in their twentieth year for a two-year term of service. On completion of this active duty a soldier was released to the "first reserve" where he remained liable for recall for another 15 years. Those males not needed to fill out the district's units and those of a lower physical quality were given some initial training (not to exceed 180 days) and immediately released to the "conscript reserve" where they remained liable for call-up for 17 years.

The division resident in a particular district carried the same numerical designation as that district. It was normally maintained at peacetime strength, which was generally about two-thirds of its wartime strength. In order to facilitate expansion the officer corps was usually kept at a higher level and the number of privates at a lower level. Examples of peace and wartime authorized strengths for 1937 show this clearly:

	Officers	Warrant Officers	Master Sergeant	Sergeant & Corporal	Private
Infantry Regt (Peace)	85	38	45	154	1995
Infantry Regt (War)	88	16	45	299	3319
Divisional Cavalry Regt (Peace)	12	3	5	26	186
Divisional Cavalry Regt (War)	18	2	2	54	341
Field Artillery Regt (Peace)	69	12	20	74	846
Field Artillery Regt (War)	85	9	23	139	1813

In addition to the division, most districts also maintained a portion of smaller units required by the army as a whole, such as medium artillery regiments, cavalry brigades, anti-aircraft artillery regiments, etc., as well as administering the local coast defense ("heavy artillery") regiments.

When a division (or other unit) was needed for service outside the district area two options were available. The quickest method was to bring the existing unit up to full (war) strength through the recall of reservists or an expanded intake of conscripts. The division would then depart, leaving a small cadre behind in the district to operate a "depot division" that would carry on the normal territorial functions, such as conscription, training, etc. Although quick, this method caused considerable disruption to the district operations. The preferred method, if time was available, was to create an entirely new field division in the district through conscription and recall, leaving the original division behind to carry on its territorial functions.

While the traditionalists in the Army hierarchy placed most of their faith in the perceived warrior ethos of Japan, that did not mean they disregarded the advantages of firepower altogether. Indeed, they keenly recognized the need for infantry weapons and light artillery to support the offensive. The infantry received three new weapons that significantly enhanced their small-unit firepower in 1922, the most important of which was a light machine gun that could accompany the rifle platoons and provide a base of mobile firepower. The other two provided high-explosive fire support at the battalion level, a 70mm mortar for area targets and those in defilade, and a 37mm gun for point-type targets. Although the Type (Taisho) 11 light machine gun was later to prove less reliable than most, the IJA's infantry in the 1920s was generally as well armed as any in the world.

Through the 1920s the IJA infantry closely resembled its western counterparts, but maneuvers highlighted the need for even more effective close support of the infantry if they were to execute their vaunted maneuver and assault role. The result was the development of two uniquely Japanese weapons. The Type 89 grenade discharger was adopted in 1929 and could fire hand grenades and special 50mm projectiles from a hand-held tube. Three years later the Type 92 battalion gun was introduced to replace both the 37mm flat-trajectory gun and the 70mm mortar.

The artillery enjoyed its own modernization program in the late 1920s, launched largely to take advantage of technology newly developed by the Schneider firm. Not surprisingly the light artillery was accorded priority and the first results were the Type 90 (1930) 75mm field gun and the Type 91 (1931) 105mm field howitzer. A 105mm field gun followed in 1932 and finally a 150mm howitzer in 1936. The 75mm gun was a modified version of a Schneider weapon and was placed in local production immediately, while the 105mm howitzer was a direct copy of the French original and initial lots were purchased from France until indigenous production could begin. The 105mm gun and 150mm howitzer were placed in production as well, but only at very low rates.

Manchuria Beckons

The Army's expansion plans were far from academic. Beginning in 1907 the IJA had rotated each of its divisions through a 2-year tour as the garrison division in Port Arthur, providing the "muscle" behind the Japanese claims to Manchuria, and in 1919 the headquarters supervising this area was named the Kwantung Army. The garrison division led a rather sedentary life, since the Japanese did not officially "own" Manchuria, but only the railways running through it. To guard their property, a right extracted from China in 1915, the IJA formed six independent garrison battalions of railway guards, reduced to four in 1925 but restored to its former strength in 1929. These railway guards quickly developed a reputation for mobility and initiative in conducting hard-hitting operations to support or chastise various Chinese warlord factions.

Through the 1920s the Kwantung Army had become increasingly involved in local Chinese politics, only rarely bothering to tell Tokyo what it was doing. It used its forces, mostly the garrison battalions, to shield some favored warlords from their enemies, and assassinated others. By 1931, however, several factors converged to push the Kwantung Army to even greater stretches of its authority. The wide open, often fertile, spaces of Manchuria beckoned to many Japanese both in a romantic sense and as a source for future Japanese autarky. The Japanese had had wide latitude in Manchuria due to the fluid situation created by feuding Chinese warlords, but looming on the horizon was the prospect of a unified China under Chiang Kai-Shek. Warlord armies the Japanese held in contempt as corrupt and inefficient, but a unified China under a dynamic leader might be a different story. If Japan was to secure Manchuria it would have to act quickly. The government in Tokyo was reluctant, but the Kwantung Army had no such doubts.

The forces available to the Kwantung Army were scanty. The 2nd Division had rotated in in April 1931, but was actually a brigade in strength. Since it was not deploying, Tokyo thought, to an active theater it was on the peacetime strength organization, and on top of that had left over a third of its men behind to form the 2nd Depot Division. Its four infantry regiments had an average strength of 40 officers and 832 enlisted, the cavalry regiment had only 117 men, the field artillery regiment 542 men (with four 75mm batteries), only one company of the engineer battalion was present, and the signal unit was platoon strength, for a total division strength of only 4,350. Far more potent was the Independent Garrison Unit of 4,000 hard men in six garrison battalions totaling 24 companies. Also present was the 279-man Port Arthur Heavy Artillery Regiment with eight 150mm howitzers, 500 military police and a few armored trains.

Table 1.1: Divisions Activated 1881–1939

Activation	Div	Type	Inf Regts	Cav/Recon	Art Regt	Fate
14 May 1881	1	A	1, (3), 49, 57	1 Cav	1 Fld	destroyed Leyte 1/45
14 May 1881	2	A	4, 16, 29, (30)	2 Cav	2 Fld	ended war in Indochina
14 May 1881	3	A	6, (18), 34, 68	3 Cav	3 Fld	ended war in China
14 May 1881	4	A	8, 34, 61, (70)	4 Cav	4 Fld	ended war in Thailand
14 May 1881	5	A	11, 21, (41), 42	5 Cav	5 Fld	ended war in East Indies
14 May 1881	6	A	13, 23, 45, (47)	6 Cav	6 Fld	ended war on Bougainville
19 Oct 1886	7	A	(25), 26, 27, 28	7 Cav	7 Fld	ended war in northern Japan
14 Dec 1893	1 Gds	A	1, 2, (3), (4) G	G Cav	G Fld	split into 1st & 2nd Guards Divs
1 Oct 1898	8	A	5, 17, 31, (32)	8 Cav	8 Fld	destroyed Luzon 6/45
1 Oct 1898	9	A	7, 19, 35 (36)	9 Cav	9 Mtn	ended war on Formosa
1 Oct 1898	10	A	10, 39, (40), 63	10 Cav	10 Fld	destroyed Luzon 6/45
1 Oct 1898	11	A	12, (22), 43, 44	11Cav	11 Mtn	ended war in Japan
1 Oct 1898	12	A	(14), 24, 46, 48	12 Cav	12 Fld	ended war on Formosa
6 Jul 1905	14	A	2, 15, (50), 59	18 Cav	20 Fld	ended war on Palau
18 Jul 1905	16	A	9, 20, 33, (38)	20 Cav	22 Fld	destroyed Leyte 12/44
24 Dec 1915	19	A	73, (74), 75, 76	27 Cav	25 Fld	destroyed Luzon 6/45
24 Dec 1915	20	A	(77), 78, 79, 80	28 Cav	26 Fld	ended war in New Guinea
9 Sep 1937	13	S	(58), 65, 104, 116	17 Cav	19 Mtn	ended war in China
9 Sep 1937	18	S	55, 56, 114, (124)	22 Cav	18 Mtn	ended war in Burma
11 Sep 1937	101	S	101, 103, 149, 157	101 Cav	101 Fld	disbanded 6/40
11 Sep 1937	108	S	52, 105, 117, 132	108 Cav	108 Fld	disbanded 6/40
11 Sep 1937	109	S	69, 107, 119, 136	109 Cav	109 Mtn	disbanded 6/40
30 Sep 1937	26	S	11, 12, 13 Ind	none	11 Ind Fld	destroyed Leyte 12/44
4 Apr 1938	15	S	51, 60, 67	15 Recon	21 Fld	ended war in Burma
4 Apr 1938	17	S	53, 54, 81	17 Recon	23 Fld	ended war at Rabaul

Activation	Div	Type	Inf Regts	Cav/Recon	Art Regt	Fate
4 Apr 1938	21	S	62, 82, 83	21 Recon	51 Mtn	ended war in Indochina
4 Apr 1938	22	S	84, 85, 86	22 Recon	52 Mtn	ended war in Indochina
15 May 1938	116	S	109, 120, 133, (138)	120 Cav	122 Fld	ended war in China
20 May 1938	106	S	113, 123, 145, 147	106 Cav	106 Fld	disbanded 6/40
15 Jun 1938	104	S	108, 137, 161, (170)	104 Cav	104 Fld	ended war in China
16 Jun 1938	110	S	110, 139, (140), 163	110 Cav	110 Fld	ended war in China
21 Jun 1938	27	S	1, 2, 3 China	27 Recon	27 Mtn	ended war in China
11 Jul 1938	23	A	64, 71, 72	23 Recon	13 Fld	destroyed Luzon 6/45
7 Feb 1939	32	S	210, 211, 212	32 Recon	32 Fld	ended war on Halmahera
7 Feb 1939	34	S	216, 217, 218	34 Recon	34 Fld	ended war in China
7 Feb 1939	35	S	219, 220, 221	35 Recon	35 Fld	ended war on New Guinea
7 Feb 1939	36	S	222, 223, 224	36 Cav	36 Mtn	ended war on New Guinea
7 Feb 1939	37	S	225, 226, 227	37 Recon	37 Mtn	ended war in Indochina
25 Mar 1939	33	S	213, 214, 215	3 Cav	33 Mtn	ended war in Burma
30 Jun 1939	38	S	228, 229, 230	38 Cav	38 Mtn	ended war at Rabaul
30 Jun 1939	39	S	231, 232, 233	39 Recon	39 Fld	destroyed Manchuria 8/45
30 Jun 1939	40	S	234, 235, 236	40 Cav	40 Mtn	ended war in China
30 Jun 1939	41	S	237, 238, 239	41 Cav	41 Mtn	ended war in New Guinea
7 Oct 1939	24	A	22, 32, 89	24 Recon	24 Fld	destroyed Okinawa 4/45
Note: infantry regiments in (parenthesis) are those lost during triangularization.						

Against this, the local warlord, Chang Hsüeh-liang, had an estimated 250,000 regulars in 33 infantry and 10 cavalry brigades, plus 80,000 irregulars. Nevertheless, following the suspicious derailment of a train on the Japanese railway in September 1931 an offensive by the Kwantung Army was quickly, albeit painfully, successful, with Chang's forces abandoning Manchuria and pulling back into China itself. The Kwantung Army had finally succeeded in getting in over its head.

Tokyo now had no choice but to reinforce its wayward cousin in Manchuria, but devised a new structure for the reinforcement. They did not want to mobilize divisions to war strength and ship them off, probably out of concern for both domestic and international perception, but the organization of the 2nd Division as fielded was clumsy. Its regiments were actually battalions in strength, and its battalions that of companies. The compromise was to direct selected districts to raise temporary groupings called mixed brigades for dispatch overseas. By tasking each specified division to provide two full-strength infantry regiments and the bulk of its artillery and support units a more manageable force was created.

In early December two mixed brigades, a tank unit and artillery were dispatched to Manchuria, followed by the full 20th Division and one mixed brigade from the 19th Division in Korea. To distract attention from that theater the Japanese provoked a riot in Shanghai in January 1932 that quickly grew out of hand and the General Staff had to dispatch three divisions and a mixed brigade to that city to "protect Japanese interests." The build-up in Manchuria continued through 1932, with the full 10th Division arriving in April and the 14th Division in May. In October 1933 the Japanese recognized a puppet government in Manchuria that split it off from China and placed it firmly in the Japanese camp.

Even before the docile government of Manchuria, now known as Manchukuo, was recognized the Kwantung Army had begun the process of building a Manchurian Army to complement their efforts at security. Created in 1932, by 1935 the Manchukuoan Army consisted of 26 infantry brigades and eight cavalry brigades.

By the mid-1930s, however, it was becoming apparent that the Kwantung Army was having to direct its attentions in two directions. To the south they wanted a buffer zone, usually a province wide, between "their" Manchukuo and a restless, resentful China. To the north they noted a Soviet build-up more rapid than they had thought possible, from an estimated six rifle divisions in 1931 to fourteen in 1935, plus three cavalry divisions and two mechanized brigades. Nevertheless, the Kwantung Army was dependent on the General Staff in Tokyo for its field formations, and its strength remained at three divisions during 1933–36.

Partly to relieve the Kwantung Army of its responsibility for north China, and partly to ensure more direct control from Tokyo the China Garrison Army, an old but previously tiny command, was expanded and given responsibility for Japanese military interests in the northern five provinces. It was expanded to ten rifle companies in 1935 and in April 1936 was made directly subordinate to the General Staff. In May the bulk of the combat forces were reorganized into the China Garrison Infantry Regiment, later expanded to two such regiments, then three.

Into the China Morass

In spite of the build-up by the Soviets the Japanese continued to be fascinated by the land beckoning to the south. In April 1937 the Japanese government demanded that the Chinese provinces of Hopei, Shansi, Shantung, Chahar, and Suiyan be split off from China to create a second Manchukuo-style puppet regime for the benefit of Japan. The nationalist government of China, of course, refused and on 7 July 1937 the Japanese manufactured an incident at the Marco Polo Bridge outside Peking.

To follow this Japan planned a massive pincer attack on the Chinese capital. The China Garrison Army, guarding the enclave at Tientsin, was to drive northwest, while another force was to drive south from Manchuria. For this operation the China Garrison Army deployed the 20th Division plus three brigades loaned from the Kwantung Army.[1] The northern arm consisted of the 1st Independent Mixed Brigade (IMB) and the 11th Independent Infantry Brigade. Peking quickly fell.

The independent mixed brigade was a new type of formation, not to be confused with the mixed brigade used earlier. Here each infantry regiment and the artillery regiment in a district was tasked to provide a full-strength battalion, each battalion of engineers and transport a company, and the divisional HQ a command element. This, with a little extra infantry, created an expeditionary unit of five infantry battalions, an artillery battalion,

1 The China Garrison Army was reinforced by the 5th, 6th and 10th Divisions in mid-August.

one company each of signals, transport and engineers and a medical detachment. This device was to remain in service through to the end of the war, albeit later adopted for its speed and minimal disruptive effect rather than any concern for appearances.

With the need to both guard against the perceived Soviet threat and to chastise the Chinese with a major punitive expedition the force structure of the IJA had been stretched beyond its peacetime limits. Plans were made for a doubling of the number of divisions by directing each divisional district to prepare to activate a second division of reservists or conscripts. Four of the divisions would carry the designations lost in the 1922 reduction in force, raised by the districts that had taken over their areas (e.g., 13th Division in the 2nd District, 15th Division in the 3rd District, 17th Division in the 10th District, and 18th Division in the 12th District). The remaining divisions would take the number of their parent district plus a hundred (i.e., 104th Division from the 4th District).

In July orders went out to create and activate the 13th, 18th, 101st, 108th and 109th Divisions, these being known as "special divisions" since they were filled with reservists and not officially incorporated into the regular army. One additional division, the 26th was formed by expansion of an anomalous early mixed brigade.[2] The six divisions were officially activated during 9–30 September 1937. All were "square" units of two brigades each of two regiments, except the 26th Division, which was activated with simply three infantry regiments.

As a further measure towards a war footing the Imperial General Headquarters (IGHQ) was activated on 17 November.[3]

In the meantime, the expansion of the war in China was proceeding apace. China was threatening Shanghai and in August the 3rd and 11th Divisions were landed in there, followed shortly by the 9th, 13th, 16th and 101st Divisions. Initial attempts to advance west of the city were repulsed by Chiang's elite German-trained divisions with heavy losses, but an amphibious flanking maneuver by the new Tenth Army in November broke the stalemate and on 1 December the Tenth Army was ordered to advance to the new nationalist capital of Nanking. By the end of 1937 the army had 16 divisions and 600,000 men committed to the China theater.

In early 1938 the Japanese forces in China were placed under two new commands, the North China Area Army and the Central China Expeditionary Army. To fill these out, as well as strengthen the Kwantung Army further against a now restless Soviet Union, further force expansion was necessary. On 4 April 1938 two more special divisions were activated, the 15th and 17th. These were followed by two more divisions in May (106th and 116th) and a further two in June (104th and 110th). In addition, increased conscription allowed the creation of three additional divisions, the 21st and 22nd in April and the 27th in June.[4] These three, plus the 26th Division formed earlier, were referred to as "new divisions," distinct from the reservist divisions although still not considered part of the regular army force structure. All the non-regular divisions could be considered "wartime" divisions, to

---

2 The division was made up of the 11th, 12th and 13th Independent Infantry Regiments and the 11th Independent Field Artillery Regiment. These are not to be confused with similarly-numbered units elsewhere, for instance the 11th Infantry Regiment of the 5th Division.

3 Both services resisted efforts to provide IGHQ any real powers, and it served mainly as discussion venue rather than a command element through the war.

4 The 27th Division was activated on 21 June by expansion of the China Garrison Brigade, this being the former guard unit held in the international settlement area of Shanghai.

include the special, new and later types, such as the security divisions.

As the Japanese were finding out, China is a huge country and striking deep into the countryside to force a decisive battle also brought into the Japanese fold large areas that had to be garrisoned. For this role they expanded the use of the independent mixed brigades with a higher ratio of infantry to combat support and service support than was found in the divisions. Six of these formations were activated in February 1938, along with combat support and service units. This would be accomplished by increasing the number of draftees. While the number of young men entering their 20th year remained relatively constant, the percentage chosen for induction into the Army rose from 22.9 in 1937 to 44.4 the next year, or in absolute terms, from 170,000 to 320,000.

The conscript divisions were notable beyond simply expanding the force structure. As early as 1936, with the reformers now in the ascendancy, the General Staff had agreed that a triangular division configuration, that is, with three infantry regiments under a single HQ, was preferable to the current binary configuration of two infantry brigades each with two regiments. Every other nation had reached the same conclusion, but it was not until the 26th, 15th, 17th, 21st and 22nd Divisions were raised that this new organization was actually used. Apparently the IGHQ now awaited field experience with these divisions before embarking on a radical reorganization of the army, for few further moves in this direction were forthcoming for some time.

By mid-1938 all but four (Guards, 7th, 11th and 19th) of the regular peacetime divisions were mobilized and serving under overseas commands:

Command / Formation	Units
Kwantung Army	
six divisions	1, 2, 4, 8, 12 & 26
Cavalry Group	1st & 4th Cavalry Brigades
3rd Cavalry Brigade	
three independent garrison units	1, 2 & 3
North China Area Army	
two divisions	5 & 114
three mixed brigades	3, 5 & 13
1st Army	
four divisions	14, 20, 108 & 109
two independent mixed brigades	3 & 4
Central China Expeditionary Army	
five divisions	15, 17, 18, 22 & 116
2nd Army	
four divisions	3, 10, 13 & 16
11th Army	
five divisions	6, 9, 27, 101 & 106
China Garrison Brigade	

In addition to this, the 19th Division, nominally at home in Korea, was in fact responsible for the southern end of the border between Manchukuo and the Soviet Maritime Province near Korea, this a result of the large number of Koreans living in that area. Further, eight "border garrisons," equivalent to battalions and regiments, were formed by the Kwantung Army in March 1938 to provide covering forces for the border areas with the Soviets.

The 23rd Division, activated on 11 July, was unique on two counts, first by being regarded as an active division to be added to the regular force structure, and second by being optimized for conditions in the far west of Manchuria.

In July and August 1938 the 19th Division engaged in combat with the Soviets around Lake Khasan, suffering 1,440 casualties. Settled diplomatically, the battle convinced the Japanese that the Soviets, while adamant about their border claims, were not prepared to attack Manchuria. This perception freed troops for adventures further south in China.

Nanking had fallen in December 1937, but Chiang had simply fallen back even further to Chungking. By March of 1938 the Japanese controlled much of northern and central China. They accomplished this by sending out fast-moving columns in divergent directions to find and hammer the ill-prepared Chinese forces as they met them. In March, however, Chiang adopted a different strategy. He would no longer attempt to meet the Japanese columns head-on, but would instead let them pass by with only nominal resistance, and then come down on their long, exposed lines of communication. At Taierhchwang this tactic was tested against the 10th Division (reinforced by a brigade from the 5th Division), which had marched light and fast deep into China. Swooping down from all sides the Chinese cut off the division's supplies and launched massive hard-fought attacks that mangled the division.[5] The Japanese became notably more conservative after this, and such victories were thereafter rare for the hard-pressed and ill-equipped Chinese.

The quick war to chastise China and persuade it to give up the northern provinces had failed. Seeking to find Chiang's vulnerable points the IGHQ hit on the idea of cutting off his supplies. The bulk of these were flowing in through ports in south China and in October of 1938 the new 21st Army, commanding the 5th, 18th and 104th Divisions with supporting units, invaded the south and occupied Canton. Still Chiang refused to give in.

Yet further force expansion was clearly needed and in early 1939 ten more "wartime" divisions were ordered raised from scratch. The first five (numbered 32 and 34–37) were activated on 7 February 1939, one more (33) on 25 March, and the final four (38, 39, 40 and 41) on 30 June. These were all raised as triangular divisions, consisting of three infantry regiments each, along with an artillery regiment, a cavalry or reconnaissance regiment, an engineer regiment, transport regiment, signal battalion, and service elements. In mid-1939 ten more IMBs were raised for duty in China to garrison the expanding Japanese footprint there.

If IGHQ thought that they could now devote their full attention and energies to the problem of China, however, they were sadly mistaken. In May 1939 a small force of Soviet-Allied Outer Mongolian cavalry entered the small village of Nomonhan in a disputed area of remote southwestern Manchuria. The new 23rd Division, specially raised for this area, was told to see the intruders off, and this they initially did. When their 64th Infantry Regiment returned a week later, however, it was surrounded and mauled by Mongolian and Soviet troops. On 1 July the full division entered the fray, supported by the 3rd and

5 There is considerable dispute over the number of Japanese deaths in the battle. At the time the Chinese trumpeted to the world that they had killed all but 2,000 of the 16,000 Japanese engaged. Post-war Japanese accounts place casualties much lower. In any event it was widely perceived as a significant (if uncommon) Chinese victory.

4th Tank Regiments and part of the 7th Division. In August the Soviets launched a major counterattack, using large quantities of tanks and artillery, and crushed the Japanese force, inflicting 17,000 casualties. A cease-fire was declared in mid-September, but the Japanese had learned not to be complacent about their Manchurian holdings. Presumably a lesson learned should have been the need for a heavier force structure, but in fact most of the Army leadership blamed poor fighting spirit of the troops.

A brief hiatus in fighting in China in late 1938 and early 1939 provided the opportunity to replace some of the regular divisions with "wartime" divisions. During 1939 ten divisions were pulled out, with the 9th, 10th, 14th and 16th Divisions being brought back to Japan and reduced to peacetime status; the 20th Division returning to Korea, where it was also reduced to peace strength; and the 101st, 106th, 108th, 109th and 114th Divisions ordered back to Japan to be deactivated and the men demobilized.[6] The pause in fighting had been brief and Japanese strength tied down in China remained about a million. By September 1939 the order of battle for Japanese forces in China was as follows:

North China Area Army	
1st Army (Shansi province)	
four divisions	20, 36, 37 & 108
three independent mixed brigades	3, 4 & 9
12th Army (Shantung province)	
two divisions	21 & 32
three independent mixed brigades	5, 6 & 10
Mongolia Garrison Army	
one division	26
one independent mixed brigade	2
Cavalry Group	1st & 4th Cavalry Brigades
three divisions	27, 35 & 110
four independent mixed brigades	1, 7, 8 & 15
11th Army (Wuhan area)	
six divisions	3, 6, 33, 34, 101 & 106
one independent mixed brigade	14
13th Army (lower Yangtze river)	
four divisions	15, 17, 22 & 116
three independent mixed brigades	11, 12 & 13
21st Army (south China)	
two divisions	18 & 104
Formosa Brigade	
one independent infantry brigade	1

Non-divisional support elements were sparse by international standards, but still impressive

6 The return of the 106th Division was actually delayed until early 1940.

when judged from the small size of the earlier standing army: six machine gun battalions, two tank regiments (= battalions), twelve tankette companies, seven mortar battalions, four independent mountain artillery regiments, two brigades and two regiments of medium artillery, two anti-aircraft regiments, and six engineer regiments. The conquered area, however, was so vast that almost all these forces were tied up on garrison duty, with few available for reserve or maneuver roles.

Although no fewer than seventeen new divisions had been raised as triangular units during the past two years, no effort had been made to convert existing divisions to this configuration until late 1939. Finally, however, one infantry regiment was taken from each of the 8th and 11th Divisions and added to a new infantry regiment, along with reconnaissance and artillery regiments to form the new 24th Division, officially activated on 7 October 1939. The process continued slowly and haltingly. The 28th Division was activated on 10 July 1940 from infantry regiments drawn from the 1st, 2nd and 9th Divisions; and the 25th Division followed 19 days later with regiments shed by the 4th, 10th and 12th Divisions; then the 29th, activated on 1 April 1941, which drew regiments from the 3rd, 14th and 16th Divisions. The final division to be directly formed from thrown-off regiments was the 31st, in May 1943 from regiments of the 13th, 18th and 116th Divisions. All these joined the ranks of the regular army as permanent additions to the force structure.

The four new divisions were assigned to the Kwantung Army to bring that force back to respectable strength after the debacle at Nomonhan. The other square divisions gradually shed their fourth infantry regiments as well. The 5th, 19th and 20th Divisions sent a regiment each back to the depots in late 1942, where they remained until needed to form the 30th Division in 1944. The 7th Division sent its 25th Infantry Regiment back to its home territory, where it formed the basis for the Karafuto Mixed Brigade. In June 1940 the Imperial Guards formations were split into two: the Guards Division and the Guards Brigade.

Force Expansion

A far more sweeping reorganization of the army was in the works, however. In late 1939 the War Ministry had proposed a force expansion to 65 divisions by 1942, this was contingent on a reduction in forces in China to permit build up of inventories and stockpiles. The China Expeditionary Army, however, requested an increase in its strength, killing the plan before it was finally approved.

Instead, force structure transformation would take a new tack. Work had begun on the 1941 Army mobilization plan in the spring of 1940 and the new document would, for the first time, attempt to adapt the regular army force structure to the new, hostile, environment in which Japan found itself.

The mobilization plan served to define the regular army, and as such up until 1940 it had only concerned itself with the seventeen regular divisions plus additions and supporting arms. The "wartime" divisions were regarded as temporary formations, and thus not included in the calculations of the mobilization plan. As it became clear that hostilities would drag on for some time, however, it also became apparent that the definition of the army as a 17-division force would be fiction for the foreseeable future.

The new plan now divided the regular army divisions into two categories, "A" and "B." The "A" divisions were the pre-war regular divisions, plus the 23rd Division raised for Manchuria; and the 24th, 25th and 28th Divisions, all formed from infantry taken

Table 1.2: Divisions Activated 1940–1943

Activation	Div	Type	Inf Regts	Cav/ Recon	Art Regt	Fate
3 Jun 1940	2 Gds	A	3, 4, 5 G	Gds Cav	2 G Fld	ended war on Sumatra
10 Jul 1940	28	A	3, 30, 36	28 Cav	28 Mtn	ended war on Ryukus
10 Jul 1940	54	B	111, 121,154	54 Recon	54 Fld	ended war in Burma
10 Jul 1940	55	B	112, 143, 144	55 Cav	55 Mtn	ended war in Burma
10 Jul 1940	56	B	113, 146, 148	56 Recon	56 Fld	ended war in Burma
10 Jul 1940	57	B	52, 117, 132	57 Recon	57 Fld	ended war in Japan
29 Jul 1940	25	A	14, 40, 70	75 Cav	15 Mtn	ended war in Japan
1 Sep 1940	51	B	66, 102, 115	51 Recon	14 Fld	ended war on New Guinea
30 Nov 1940	48	A	1&2 Formosa, 47	48 Recon	48 Mtn	ended war in East Indies
1 Apr 1941	29	A	18, 38, 50	29 Cav	29 Mtn	destroyed on Guam
2 Feb 1942	58	S	Bdes 51 & 52	none	none*	ended war in China
2 Feb 1942	59	S	Bdes 53 & 54	none	none*	ended war in Korea
2 Feb 1942	60	S	Bdes 55 & 56	none	none*	ended war in China
2 Feb 1942	68	S	Bdes 57 & 58	none	none*	ended war in China
2 Feb 1942	69	S	Bdes 59 & 60	none	none*	ended war in China
2 Feb 1942	70	S	Bdes 61 & 62	none	none*	ended war in China
16 Apr 1942	71	S	87, 88, 140	71 Cav	71 Mtn	ended war on Formosa
13 Mar 1943	61	S	101, 149, 157	none	none*	ended war in China
22 Mar 1943	31	S	58, 124, 138	none	31 Mtn	ended war in Burma
1 May 1943	62	S	Bdes 63 & 64	none	none	destroyed Okinawa 3/45
1 May 1943	64	S	Bdes 69 & 70	none	none*	ended war in China
1 May 1943	65	S	Bdes 71 & 72	none	none*	ended war in China
4 May 1943	46	S	123, 145 & 147	46 Recon	46 Fld	ended war in East Indies
15 Jun 1943	100	S	Bdes 75 & 76	none	one bn	destroyed on Mindanao 8/44
30 Jun 1943	63	S	Bdes 66 & 67	none	none	destroyed Manchura 8/45
6 Jul 1943	93	S	202, 203, 204	93 Cav	93 Mtn	ended war in Japan
10 Aug 1943	52	S	69, 107, 150	52 Cav	16 Mtn	ended war on Truk
14 Oct 1943	94	S	256, 257, 258	none	94 Fld	ended war in Malaya
19 Nov 1943	53	S	119, 128, 151	53 Recon	53 Fld	ended war in Burma

** one battalion of artillery or mortars added 1945*

from regular divisions in the triangular reorganization. The "B" divisions were to be new formations, raised in seven of the old divisional districts to replace regular divisions that were overseas. The earlier practice of rotating regular divisions between home station and Manchuria had broken down with the expansion of the Kwantung Army. To form these new units five reservist divisions (101st, 106th, 108th, 109th, and 114th) were brought home from China, disbanded and stripped of cadre personnel and used to form the bulk of the new divisions. Once in place these new "B" divisions took over the homeland districts.

51st Division (ex-14th Divisional District)
52nd Division (ex-9th Divisional District)
53rd Division (ex-16th Divisional District)
54th Division (ex-10th Divisional District)
55th Division (ex-11th Divisional District)
56th Division (ex-12th Divisional District)
57th Division (ex-8th Divisional District)

Of these, four (54–57) were immediately activated as field divisions in their home districts on 10 July 1940, leaving similarly numbered depot divisions home to handle the administrative duties of the district. One more (the 51st) was activated in September, while the remaining two remained as depot divisions until 1943. With this reorganization in place the 53rd Depot Division, for instance, was now the "parent" of the 16th Division, responsible for keeping it up to strength. Similarly, the 57th Depot Division was responsible for the 57th Division and the 8th Division.

Regiments from the dissolved reservist divisions not utilized in forming the "B" divisions, along with other newly-raised reservist infantry regiments, were grouped together in Japan in independent infantry groups.

One further division was raised in 1940, the 48th. This was formed through expansion of the former Formosa Brigade by the addition of a regiment shed by the 6th Division and was officially activated on 30 November. It was unusual for combining a partially-motorized reconnaissance regiment with a pack artillery regiment under the same division.

Surprisingly, given the drubbing Japanese forces had taken at Nomonhan, it was not until July 1940 that the first separate anti-tank units were formed, and then only in the form of five small battalions.

Aside from the 29th Division, already mentioned, no further divisions were formed until early 1942, and indeed no completely new divisions would be formed until mid-1943. Clearly, the Japanese felt the force structure to win wars was in place.

Further, Japan had just about reached its threshold of pain in terms of peacetime conscription. In 1936 some 170,000 men had been conscripted or volunteered, but under the pressures of creating new units and replacing war losses in China this had jumped to 320,000 in 1938 and stayed there through 1940. Compounding this was an apparent policy of sending the best recruits to the regular divisions, leaving the "wartime" divisions with the lower quality personnel, including officers.

Decision Time

With the invasion of the Soviet Union by Germany attention suddenly turned northward again. The Japanese had learned at Nomonhan not to underestimate the Red Army, but if the Germans were successful the eastern portion of the Soviet Union could well fall into Japanese hands like a ripe fruit, if only the IJA were ready. Determined not to miss this golden opportunity the Kwantung Army was enlarged from 400,000 to 700,000 men during the summer of 1941.

For the most part the reinforcements consisted of masses of logistics and communications units, leavened by a few combat units. IGHQ ordered the first increment of reinforcements to Manchuria on 11 July, comprising five anti-tank battalions, four medium artillery regiments, a dozen anti-aircraft battalions, and a bewildering variety of transport, signal and medical units. The order for the second increment was issued on 1 August, and it provided the 51st and 57th Divisions, the 3rd Cavalry Brigade, two independent mountain artillery regiments, three heavy artillery regiments, eight medium artillery regiments, three mortar battalions, and five AA battalions, along with another large group of service support elements. Finally, on 8 August, IGHQ ordered the 7th and 8th Tank Regiments from China to Manchuria.

With war with the Soviets hanging in the balance the order of battle for the Kwantung Army was as follows in early August 1941:

3rd Army	
four divisions	8, 9, 12 & 57
one Tank Group	1
four border garrison units	1, 2, 10 & 11
five medium artillery regiments	4, 8, 9, 17 & 22
two heavy artillery regiments	2 & 3
five heavy artillery battalions	1, 2, 4, 6 & 7
one mountain artillery regiment	4
four field anti-aircraft battalions	48, 50, 51 & 55
three mortar battalions	11, 12 & 13
4th Army	
one division	1
one independent garrison unit	8
four border garrison units	5, 6, 7 & 13
one heavy artillery regiment	1
one heavy artillery battalion	8
one field anti-aircraft battalion	52
5th Army	
three divisions	11, 24 & 25
one Tank Group	2
one independent garrison unit	6
three border garrison units	3, 4 & 12
one cavalry regiment	3
one field artillery regiment	1
four medium artillery regiments	5, 7, 12 & 20
two heavy artillery battalions	5 & 9

two field anti-aircraft battalions	53 & 54
6th Army	
one division	23
one border garrison unit	8
Kwantung Defense Force	
five independent garrison units	1, 2, 3, 5 & 9
Kwantung Army direct control	
four divisions	10, 14, 28 & 51
three medium artillery regiments	2, 10 & 18
ten field anti-aircraft battalions	32–36, 40, 44, 45, 49 & 56
one mortar battalion	10
four anti-tank battalions	3, 4, 6 & 7

This was to be the last pre-war reinforcement for the Manchuria theater and represented the peak for the Kwantung Army. The Kwantung Army now contained half of the army's 22 "A" divisions (with another two in support in Korea), plus two "B" divisions and only a single "wartime" division. On 9 August, however, the decision was made that Japan would start looking south again. From now on the Kwantung Army, pride of the IJA, would serve primarily as replacement and reinforcement pool for the rest of the Army.

The reason for the shift was not hard to find. On 26 July the US had announced a freeze on Japanese assets and on 1 August all existing licenses to export oil were revoked. The Dutch East Indies followed suit and Japan was left with the choice of a humiliating withdrawal from China and Indochina or a complete cut off of oil. On 6 September the Imperial Conference approved preliminary planning for a war against the Western powers should diplomacy fail and on 1 November the final diplomatic offer was drafted, under which Japan would essentially keep her China conquests. The United States rebuffed this offer and the die was cast.

A Pacific War

Already, on 6 November, before the US had responded to the Japanese note, the IJA had begun organizing for a strike to the south. On that date IGHQ directed the disbanding of the current 25th Army and the establishment of the Southern Army HQ, four field army HQs (14, 15, 16 and 25) and the "South Seas Detachment" to control the southern operations. In late November the Army assumed responsibility for occupying Hong Kong, Malaya, Sumatra, Java, British Borneo, the Philippines and Burma; while the Navy would take the rest of the East Indies, New Guinea, the Bismarks and Guam. On 1 December IGHQ Army Section Order No.569 notified major subordinate commands that "Japan has decided to wage war against the United States, Great Britain and the Netherlands."

The new structure vested command in the Southern Army, a coequal command with the Kwantung Army and China Expeditionary Forces. It was directed to initially seize Malaya with the 25th Army and the Philippines with the 14th Army in the first phase, and then the Dutch East Indies with the 16th Army, and Burma with the 15th Army in the second phase. The force structure for this venture was as follows:

HQ, Southern Army	
14th Army [Philippines]	
two divisions	16 & 48
one brigade	65
two tank regiments	4 & 7
two medium artillery regiments	1 & 8
one independent medium artillery battalion	9
one naval SNLF	
later arrivals:	
three divisions [10–15 Mar 42]	*4, 5 & 18*
one independent mountain artillery regiment	*3*
one heavy artillery regiment	*1*
one independent mountain artillery battalion	*20*
two mortar battalions	*2 & 3*
one artillery mortar battalion	*14*
one artillery intelligence regiment	*5*
15th Army [Thailand/Burma]	
two divisions	33 & 55
16th Army [Dutch East Indies]	
three divisions	2, 38 & 48
one infantry regimental group	56
seven naval SNLFs	
25th Army [Malaya]	
three divisions	Guards, 5 & 18
one tank group	3
three medium artillery regiments	3, 18 & 21
two mortar battalions	3 & 5
one artillery battalion	14
one field air defense unit	17 (four battalions)

The allocation of forces represented a faith bordered on arrogance in the Japanese doctrine of offensive spirit. Facing the 25th Army in Malaya would be three divisions, two Indian and one Australian, plus the equivalent of a division in the Singapore fortress. In the Philippines were one US division and ten Filipino divisions, although the latter were only partially formed and essentially untrained. On Java the Dutch had three divisions, with another division equivalent on Sumatra. The only area in which the Japanese had numerical superiority was in Burma, and even there the Burma Division and the 16th Indian Brigade (plus supporting elements) gave them only a slight edge.

Table 1.3: Type and Distribution of Divisions, 7 December 1941

		Division	Infantry Regts		Artillery Regt		Cav/Recon Regt	Trans Regt
			(designation)	(type)	(desig)	(type)		
Japan & Korea	Eastern Army	52nd Division	69th, 107th, 150th	B2	16th Mtn	B	52nd Cav	52nd (B2)
	Central Army	53rd Division	119th, 128th, 151st	B1	53rd Fld	B	53rd Recon	53rd (B1)
		54th Division	111th, 121st, 154th	B1	54th Fld	B	54th Recon	54th (B1)
	Northern Army	7th Division	25th, 26th, 27th, 28th	A1	7th Mtn	A1	7th Recon	7th (A1)
	Korea Army	19th Division	73rd, 74th, 75th, 76th	A1	19th Mtn	A1	19th Recon	19th (A1)
		20th Division	77th, 78th, 79th, 80th	A1	26th Fld	A1	20th Recon	20th (A1)
Manchuria (Kwantung Army)	3rd Army	9th Division	7th, 19th, 35th	A2	9th Mtn	A1	9th Cav	9th (A2)
		12th Division	24th, 46th, 48th	A1	12th Fld	A2	12th Recon	12th (A1)
	4th Army	1st Division	1st, 49th, 57th	A1	1st Fld	A2	1st Recon	1st (A1)
		14th Division	2nd, 15th, 59th	A1	20th Fld	A2	14th Recon	14th (A1)
		57th Division	52nd, 117th, 132nd	B1	57th Fld	B	57th Recon	57th (B1)
	5th Army	11th Division	12th, 43rd, 44th	A2	11th Mtn	A1	11th Cav	11th (A2)
		24th Division	22nd, 32nd, 89th	A1	24th Fld	A2	24th Recon	24th (A1)
	6th Army	23rd Division	64th, 71st, 72nd	A3	17th Fld	A3	23rd Recon (mot)	23rd (A3)
	20th Army	8th Division	5th, 17th, 31st	A1	8th Fld	A2	8th Recon	8th (A1)
		25th Division	14th, 40th, 70th	A2	15th Mtn	A1	25th Cav	25th (A2)
	KA Direct	10th Division	10th, 39th, 63rd	A1	10th Fld	A2	10th Recon	10th (A1)
		28th Division	3rd, 30th, 36th	A1	28th Mtn	A2	28th Cav	28th (A1)
		29th Division	18th, 38th, 50th	S	29th Mtn	S	29th Cav	29th (S)

Table 1.3: Type and Distribution of Divisions, 7 December 1941

			Division	Infantry Regts		Artillery Regt		Cav/Recon Regt	Trans Regt
				(designation)	(type)	(desig)	(type)		
China	North China Area Army	1st Army	36th Division	222nd, 223rd, 224th	S	36th Mtn	S	36th Recon	36th (S)
			37th Division	225th, 226th, 227th	S	37th Mtn	S	none	37th (S)
			41st Division	237th, 238th, 239th	S	41st Mtn	S	none	41st (S)
		12th Army	32nd Division	210th, 211th, 212th	S	32nd Fld	S	none	32nd (S)
		Occupation Army	26th Division	11th, 12th, 13th	S	11th Fld	S	26th Recon	26th (S)
		NCAA Direct	27th Division	1st, 2nd, 3rd China Occ	S	27th Mtn	S	27th Recon	27th (S)
			35th Division	219th, 220th, 221st	S	35th Fld	S	none	35th (S)
			110th Division	110th, 139th, 140th, 163rd	S	110th Fld	S	110th Cav	110th (S)
		11th Army	3rd Division	6th, 18th, 34th, 68th	A1	3rd Fld	A1	3rd Recn	3rd (A1)
			4th Division	8th, 37th, 61st	A1	4th Fld	A1	4th Recon	4th (A1)
			6th Division	13th, 23rd, 45th	A1	6th Fld	A1	6th Recon	6th (A1)
			13th Division	58th, 65th, 104th, 116th	S	13th Mtn	S	13th Cav	13th (S)
			34th Division	216th, 217th, 218th	S	34th Fld	S	34th Recon	34th (S)
			39th Division	231st, 232nd, 233rd	S	39th Fld	S	39th Recon	39th (S)
			40th Division	234th, 235th, 236th	S	40th Mtn	S	40th Cav	40th (S)
		13th Army	15th Division	51st, 60th, 67th	S	21st Fld	S	none	15th (S)
			17th Division	53rd, 54th, 81st	S	23rd Fld	S	none	17th (S)
			22nd Division	84th, 85th, 86th	S	52nd Mtn	S	none	22nd (S)
			116th Division	109th, 120th, 133rd, 138th	S	122nd Fld	S	120th Cav	116th (S)
		23rd Army	38th Division	228th, 229th, 230th	S	38th Mtn	S	none	38th (S)
			51st Division	66th, 102nd, 115th	B1	14th Fld	B	51st Recon	51st (B1)
			104th Division	108th, 137th, 161st	S	104th Fld	S	104th Cav	104th (S)

Table 1.3: Type and Distribution of Divisions, 7 December 1941

		Division	Infantry Regts		Artillery Regt		Cav/Recon Regt	Trans Regt
			(designation)	(type)	(desig)	(type)		
Southern Areas	14th Army	16th Division	9th, 20th, 33rd	A1	22nd Fld	A2	16th Recon	16th (A1)
		48th Division*	1st & 2nd Formosa, 47th	A1	48th Mtn	B	48th Recon	48th (S)
	15th Army	33rd Division	213th, 214th, 215th	S	33rd Mtn	S	none	33rd (S)
		55th Division	112th, 143rd, 144th	B1	55th Mtn	S	55th Cav	55th (B1)
	16th Army	2nd Division	4th, 16th, 29th	A1	2nd Fld	A1	2nd Recon	2nd (A1)
	25th Army	Guards Division*	3rd, 4th, 5th Gds	A1	Gds Fld	A1	Gds Recon	Gds (A1)
		5th Division*	11th, 21st, 41st, 42nd	A1	5th Fld	A1	5th Recon	5th (A1)
		18th Division	55th, 56th, 114th, 124th	S	18th Mtn	S	22nd Cav	18th (S)
		56th Division	113th, 146th, 148th	B1	56th Fld	S	56th Recon	56th (B1)
	Direct	21st Division	62nd, 82nd, 83rd	S	51st Mtn	S	21st Recon	21st (S)

**modified to motor-type divisisons*

Further, the IJA had little practical experience in opposed amphibious landings. The 5th, 11th and 12th Divisions had been designated for amphibious duties in the 1920s and had received some training but that had ceased by the early 1930s as attention turned to the Asian mainland. As a result, doctrine stagnated, calling for landings well away from enemy defenses, followed by a fast overland march to the objective. The IJA did conduct sixteen landing operations in China between August 1937 and March 1941, but almost all followed the pattern of avoiding defended beaches in favor a relatively unopposed landing, often at night, followed by a march to the objective.

They restarted amphibious training in October 1940, when the 5th Division was once again tasked to specialize in that art under the existing doctrine. The Guards and 18th Divisions received similar orders a few months later, followed by the new 48th, 55th and 56th Divisions in 1941. In fact, however, the most potentially significant change was the replacement of horse transport in the Guards, 5th and 48th Divisions with trucks. These motorized divisions would have been useful in the vast spaces of Manchuria, but once the initial offenses in the Pacific had past they would spend the rest of the war sitting in the tropical East Indies.

The Japanese would rely on speed, surprise and highly aggressive infantry tactics to sweep away enemy forces in the Far East. The two initial ventures, against the Philippines and Malaya, were to be executed rapidly to permit redeployment of forces for the rest of the campaign. In Malaya they were stunning successful, their three amphibious-trained "A" divisions landing on 8 December against light opposition well away from Singapore and sweeping all of the Malay Peninsula clear by the end of January. On 15 February Singapore surrendered. They were initially just as successful in the Philippines, but ignored intelligence reports that the Fil-American forces were withdrawing into the rugged jungles of the Bataan peninsula and instead drove into the open city of Manila. The apparent early victory led IGHQ to release the 48th Division, as planned, to the 16th Army in January for the East Indies phase of the campaign.

The second phase of the campaign began in January 1942, with the seizure of Borneo, the Celebes, Timor and other smaller islands in the East Indies by naval and small army units of the 16th Army; and the advance into Burma of the 15th Army. Later 16th Army operations in Java went equally smoothly, the Indonesians who formed the bulk of the Dutch East Indies Army proving less than enthusiastic supporters of colonial rule. Similarly, the 15th Army's advance into Burma was rapid and entirely successful, occupying Rangoon in early March. Five brigade-size independent garrison units were formed in early 1942 to occupy the southern holdings of the empire.

The zone to the east was the Navy's responsibility, but with only a handful of battalion-size landing forces they would need some help. The Army lent them a reinforced infantry regiment denominated the South Seas Detachment, which seized Guam in the first phase, and Rabaul in January, followed by the Navy's occupation of Lae and Salamaua on the northern coast of New Guinea in March.

Thus far the Japanese campaign had proceeded exceptionally smoothly. True, resistance was continuing in the Philippines, but significant reinforcements in the form of the 4th Division and the 62nd Infantry Regiment from China in February and the 41st Infantry Regiment from the motorized 5th Division in Malaya theater in April, combined with starvation and disease, brought about the surrender of Bataan in late April and Corregidor in early May.[7] By May the Japanese had every reason to be confident about the war they had launched. Indeed, the 4th Division, an A formation, was actually brought home and demobilized.

Figuring Out the Next Step

With the conclusion of the early campaign of conquest heated debate flared up about the next logical step. The Navy wanted to seize the Hawaiian islands, Australia, Fiji, New Caledonia, and otherwise complete the conquest of the Pacific. The Army, however, recognized the impracticality of these designs and fought for a more conservative approach that would establish an "outer defense perimeter" in the conquered areas and build up a self-sufficient empire. The Army won and on 29 June 1942 IGHQ ordered the Southern Army commander to simply "stabilize important areas" in his theater.

In spite of their wariness at further expansion, however, complacency ruled within the Army on the matter of defending the current empire. The Army's eyes returned to Manchuria. They still entertained hopes that a German victory in Europe against the Soviet Union would enable them to gobble up the Maritime Province and place the Manchukuo puppet regime on a firm footing. Thus in the late spring of 1942 most of the tank and medium artillery units in the south were returned to the Kwantung Army and the 4th Division returned to the homeland.

The bulk of the new conscripts were also sent to the Kwantung Army, initially enabling them to raise a new division (the 71st) and completely reorganize their command structure. On 4 July 1942 IGHQ ordered the activation of two examples of a new level of command, the "area army." These, together with a new mechanized army HQ and a 2nd Army HQ, gave the Kwantung Army the following organization in mid-1942:

HQ, Kwantung Army	
1st Area Army	
2nd Army	71st Division
3rd Army	9th & 12th Divisions
5th Army	11th & 24th Divisions
20th Army	8th & 25th Divisions
2nd Area Army	
4th Army	1st & 57th Divisions
6th Army	23rd Division
14th Division	
Mechanized Army	1st & 2nd Armored Divisions
29th Division	
2nd Air Army	

The most striking feature of the reorganization, however, was not the added layers of

7 The need for additional artillery was also recognized and the reinforcements included the 1st Heavy Artillery Regiment (240mm howitzers), the 3rd Ind Mountain Artillery Regiment, the 20th Ind Mountain Artillery Battalion, the 14th Artillery Mortar Battalion (300mm), the 3rd Trench Mortar Battalion, and the 5th Artillery Intelligence Regiment.

command, but the creation of two armored divisions, spurred in all likelihood by German successes in Europe. The potential of these forces was never to be realized. The tanks were obsolescent, command and control poor, tactics primitive and numbers few.

Personnel and assets were also diverted to the defense of the homeland for the first time. The Doolittle raid had exposed the weakness of the Japanese air defense system and in November 1942 a massive reorganization of the homeland air defenses was ordered, including the expansion of the anti-aircraft force to 20 regiments.

Meanwhile problems, small at first, began to appear to the south. The 15th Army had successfully seized most of Burma, but there was no logical end point for the operation. The IGHQ order had simply directed the army commander to "occupy the strategic areas in Burma." The 18th and 56th Divisions were added to the 15th Army in April, the latter from the 25th Army.

In the southeast IGHQ had ordered the South Seas Detachment to sail from Rabaul and capture Port Morseby on the southern coast of New Guinea. The naval force carrying the detachment, however, was turned back in the battle of Coral Sea. This was seen as only a temporary setback, however, for IGHQ planned to continue their drive to the southeast in order to cut the US–Australia lines of communication, seizing the Fiji islands, Samoa and New Caledonia. For this purpose the 17th Army was ordered activated on 18 May 1942 to control a variety of small formations, primarily the 35th Infantry Brigade (less the 144th Regiment), the South Seas Detachment (built around the 144th Regiment), the Aoba Detachment (4th Infantry Regiment plus a field artillery battalion), the 41st Infantry Regiment, an independent mountain artillery battalion, a tank company, two "type B" field AA battalions, and an engineer regiment.

A reassessment of the offensive plans for the 17th Army in the wake of the losses at Midway caused IGHQ to relieve the army commander of his responsibility for the Fiji, New Caledonia and Samoa operations on 11 July 1942 and instead directed him to seize Port Morseby by overland operations and consolidate the Japanese hold on New Guinea and the Solomons. An attempt by the South Seas Detachment to march over the Owen Stanley mountains failed disastrously in the face of Australian opposition and poor logistical support, while on 7 August the US landed Marines on Guadalcanal in the Solomons and drove out navy construction units building an airfield there.

As the situation deteriorated reinforcements were rushed in, including two of the prized "A" divisions, the 2nd Division from Java and the 6th from China. On 16 November a new command structure was implemented for the southeast Pacific with the activation of two new headquarters, the 8th Area Army and the 18th Army, to control the larger force.

HQ, 8th Area Army
6th Division
21st Independent Mixed Brigade
17th Army
2nd Division
38th Division
51st Division
35th Infantry Brigade (-144th Regiment)

Ichiki Detachment
18th Army
South Seas Detachment
41st Infantry Regiment

Shortly thereafter the 65th Brigade in the Philippines was also transferred to the 17th Army, as was a three-battalion force from the 5th Division. With these forces the 17th Army was to secure the Solomons and the 18th Army the northern coast of New Guinea.

As the Allies counterattacked the 51st Division was shifted to the 18th Army to replace the South Seas Detachment and 41st Regiment decimated at Buna while elements of other forces were fed piecemeal into an attempt to take Guadalcanal. On 16 August the Ichiki Detachment (a battalion from the 28th Infantry) was landed and destroyed a week later. In early September the Kawaguchi Detachment (the newly-arrived 35th Infantry Brigade) landed, but it too was mauled within the month. In October the crack 2nd Division began its move to Guadalcanal but was rendered ineffective in the next two months. Elements of the 38th Division followed, but fared no better. By the end of November the 17th Army had exhausted itself on Guadalcanal.

On 23 December IGHQ took further steps, ordering another of the "A" divisions, this time the 20th in Korea, to join the 8th Area Army, along with the 41st Division from China. At the same time the 6th Division was moved to the 17th Army. On 31 December 1942 an IGHQ conference agreed that Guadalcanal and Port Morseby were lost causes, for lack of shipping to move reinforcements and secure the logistical supply, and directed that forward troops be withdrawn to the Solomons north of New Georgia and Lae and Salamaua on New Guinea.

The 51st Division was moved to Lae, suffering disastrous losses at sea, while the 20th and 41st Divisions moved to Wewak in January 1943, the former from Korea and the latter from northern China.

There was contentious disagreement about the Solomons, with the Army wanting to hold only the northern portion, Bougainville and Shortland, while the Navy felt it necessary to defend the central part of the chain. As a result, the Army moved its 6th Division to Bougainville, while the Navy created two combined Special Naval Landing Forces (SNLFs) for duty to the south, the 7th for Santa Isabel and the 8th for New Georgia.[8] The remnants of the 2nd Division were evacuated from Guadalcanal and brought to the Philippines where the division was rebuilt.

Although the area of New Guinea and the Solomons was the most fluid, it was clearly not the only one in which the war was being waged. On 21 April 1942 IGHQ had ordered the China Expeditionary Army to destroy Chinese bases in Chekiang and Kiangsi provinces. For this operation the 13th Division (at Shanghai), the 11th Division (at Hangkow) and the 30th Army (15th, 22nd, 70th and 116th Divisions) were chosen. The offensive was launched in mid-May and was to have been followed by an advance to Chungking, but by the autumn of 1942 the events in the Solomons had turned sufficiently grave that IGHQ placed further offenses in China on hold until the situation in the south eastern Pacific had stabilized.

8 Of course, the naval forces were completely insufficient for the task and 8th Area Army eventually acceded to the suggestions from Tokyo and sent two infantry regiments to New Georgia a battalion to Santa Isabel.

The Army's standard tactical transport was the Type 39 Cart, a single-horse, one-axle vehicle with a rated capacity of 187kg. It was widely used in China and could go most places, although not everywhere as seen here. Horses were not taken to hot, humid jungles and were not needed on small atolls, so they were rarely encountered in the Pacific theater.

After the turmoil of 1942, the year 1943 was actually surprisingly quiet for the bulk of the IJA. This proved a boon for the army, for no substantive increase in strength had been accomplished since well before Pearl Harbor. The six divisions (58, 59, 60, 68, 69 and 70) raised in China in February 1942 had been nothing more than consolidation of existing independent mixed brigades. The same had been true for three more occupation divisions (62, 64 and 65) raised in China in May 1943. A few other divisions were activated, primarily from existing units in Manchuria and China, but no real change came until August 1943 when the 52nd and 53rd Divisions were told to activate from their peacetime depot status in Japan. The 53rd Division was sent to Burma and the 52nd was readied for combat in the Pacific. In the southern theater, the East Indies, Philippines and Indochina, eleven IMBs were created for occupation duties, using the same pattern of lightly-supported infantry as had previously been used to occupy China. These represented little accretion to strength, since they were formed largely by reorganizing existing garrison units.

By this time, however, it was becoming clear that exceptional measures would be required for the Pacific theater. In October 1943 the Army Staff issued a new TO&E for a novel formation called the ocean division. This would be built around two static defense regiments and an amphibious regiment, with a few divisional troops. The 36th Division (in China) and the 52nd Division (in Japan) converted immediately to this new organization, and shortly thereafter the former was shipped to Halmahera and New Guinea and the latter to Truk.

Early reports were apparently encouraging for on 18 November IGHQ issued reorganization order A-106 for a massive strengthening of the Pacific theater. Under this order the 3rd, 5th and 13th Divisions in China were to be converted to the ocean type, with the 2nd Guards Division in Sumatra being partially configured to this standard, and four amphibious brigades formed. The three China-based divisions, however, could not be withdrawn for reorganization so instead two divisions in Manchuria (14th and 29th) and one in Japan (46th) were substituted. When reorganization was complete the 14th Division was moved to Palau, the 29th to the Marianas, and the 46th (less one regiment) to the lesser Sunda islands in the East Indies.

The Last Gasp of the Offensive Theme

If 1943 was a relatively peaceful year for the army, 1944 was to prove its antithesis and disastrously so. The Allies had spent most of 1943 building up strength and by 1944 were ready to unleash it in Burma and the south and central Pacific. The Japanese, their complacency still intact even after the losses at Guadalcanal and eastern New Guinea, had done practically nothing in the year's grace period to prepare for the onslaught that was awaiting them.

Indeed, it was not until the very end of 1943 the government lowered the draft age from 20 to 19, as well as making Koreans eligible for the draft and ending student deferments. As a result, during 1944 the Army actually drafted two years worth of young men, providing a one-time boost so that the number of inductees rose from the 320,000 to 360,000 that had held from 1938 to 1943, up to a million in 1944.

The Kwantung Army had managed to remain relatively intact during the first two years of the Pacific War and by the end of 1943 still held nine of the crack "A" divisions (1, 8, 9, 10, 12, 14, 23, 24 and 28). The Kwantung Army's primary mission at the start of 1944 was still the defense of Manchuria from the Soviet threat, but by mid-year it had been reduced to that of a holding area for strategic reserves.

The Kwantung Army's "A" divisions were the heart of this reserve. Although still filled with some of the best personnel of the army, these units were not at full strength, probably due to the need to divert replacement personnel to the more active theaters. Beginning in December 1942 the divisions in the Kwantung Army had slowly reduced to peacetime strength. An example of this was the reduced-strength TO&Es adopted by the 1st Division, for instance, reducing rifle company strength from the 190 of the mobilization plan to 117 and infantry battalion strength from 1,273 to 745 in late 1943.

The strength of the Kwantung Army was further reduced when nine divisions were directed by IGHQ in February 1944 to split off a third of their strength (one battalion per regiment) for the formation of eight "expeditionary units" for service in the Pacific. The lost units were to be reconstituted in the divisions by new personnel, but it is not clear how far this progressed before the divisions were shipped out. The expeditionary units, in fact, contributed little to the war effort. Of the 12 eventually raised and shipped out to the Pacific only three (the 1st on Saipan, the 6th on Guam and the 12th in the Philippines) did anything other than sit out the war on bypassed islands.

After these half-hearted efforts it finally came the time of the divisions themselves. By the end of the year not a single "A" division was left in Manchuria. A total of 12 divisions were ordered taken from the Kwantung Army in 1944, essentially gutting it as an effective fighting formation.[9]

1 Feb 44	27th Division	to China
10 Feb 44	14th Division	to Palau
	29th Division	to Guam
26 Jun 44	9th Division	to Okinawa and Formosa
	28th Division	to the Ryukyus
5 Jul 44	24th Division	to Okinawa
18 Jul 44	8th Division	to Philippines
24 Jul 44	1st Division	to China, then the Philippines
10th Division	to Formosa	
	2nd Armored Division	to Philippines
22 Sep 44	23rd Division	to Philippines
22 Dec 44	12th Division	to Formosa

The lost divisions were replaced by six new divisions (107th, 108th, 111th, 112th, 119th and 120th) raised by the Kwantung Army, although a significant proportion of the personnel for these units came from existing units in Manchuria and the gain to strength was thus not as great as might be assumed.

The causes of the drawdown were a combination of offensive and defensive campaigns planned by or forced upon the IJA for 1944. In China Operation *Ichigo* was launched in April 1944 to seize control of the railway routes and clear US airpower out of eastern China. It began with a drive down the Yellow River against a gap in the Peking–Hankow railway held by the Chinese. Chinese forces collapsed and this first phase was an immediate success. The second phase, a southward drive from Hankow to drive the US 14th Air Force from its bases in east China, from which it was bombing Japan, jumped off a month later. On 23 August a new 6th Area Army was organized to control this campaign, consisting of the 11th Army (3rd, 13th, 34th, 40th, 58th and 116th Divisions), the 23rd Army (22nd and 104th Divisions) and the 34th Army (39th Division). As was common in the China theater the aggressive 11th Army undertook the core of the attack. Although this phase of the operation went more slowly, due to bitter Chinese defense of Hengyang and US bombing attacks, it too was successful. It was an empty victory, however, for the arrival of new B-29 bombers with their much increased range meant that the 14th US Air Force could now use bases in Szechuan, comfortably out of reach of Japanese forces.

In Burma, as well, it was an offensive plan that drove Japanese designs. The drive into Burma by the 15th Army in 1942, although successful, had been launched without any clear ultimate objective. British counterattacks in 1943, although for the most part unsuccessful, had forced the issue to the fore. In mid-1943 IGHQ acceded to 15th Army's plan to seize northeastern India, in particular Kohima and Imphal, with the three objectives of cutting off British entry into Burma, cutting the supply route to China, and setting up Chandra Bose and his renegade nationalist government in exile in Indian territory. A new Burma Area Army had already been established in March 1943 consisting of the 15th Army HQ and four divisions (18th, 33rd, 55th and 56th) and reinforcements were directed into Burma at a steady pace, including the 15th and 31st Divisions in August 1943, the 54th Division (via

9 The dates are those the orders were issued by IGHQ. Actual movement came anywhere from a few weeks to a few months later.

Table 1.4: Divisions activated 1944

Activation	Div	Type	Inf Regts	Cav/ Recon	Art Regt	Fate
1 Feb 1944	42	S	129, 130, 158	42 Recon	12 Fld	ended war in Kuriles
7 Apr 1944	43	S	118, 135, 136	43 Recon	43 Fld	destroyed Saipan 7/44
12 Apr 1944	30	S	41, 74, 77	30 Recon	30 Fld	destroyed Mindanao 3/45
12 Apr 1944	91	S	Bdes 73 & 74	none	two bns	ended war in Kuriles
3 May 1944	50	S	301, 302, 303	none	50 Mtn	ended war on Formosa
16 May 1944	107	S	90, 177, 178	107 Recon	107Fld	destroyed Manchuria 8/45
22 May 1944	109	S	Mxd Bdes 1&2	none	none	destroyed Iwo Jima 3/45
27 May 9144	49	S	106, 153, 168	49 Cav	49 Mtn	ended war in Burma
15 Jun 1944	102	S	Bdes 77 & 78	none	one bn	destroyed Leyte 12/44
15 Jun 1944	103	S	Bdes 79 & 80	none	one bn	destroyed Luzon 6/45
15 Jun 1944	105	S	Bdes 81 & 82	none	one bn	destroyed Luzon 6/45
6 Jul 1944	3 Gds		8, 9, 10 Gds	none	3 G Fld	ended war in Japan
6 Jul 1944	44	S	92, 93, 94	none	44 Fld	ended war in Japan
6 Jul 1944	72	S	134, 152, 156	none	72 Fld	ended war in Japan
6 Jul 1944	77	S	98, 99, 100	none	77 Mtn	ended war in Japan
6 Jul 1944	84	S	199, 200, 201	none	84 Fld	ended war in Japan
6 Jul 1944	86	S	187, 188, 189	none	86 Fld	ended war in Japan
10 Jul 1944	47	S	91, 105, 131	47 Cav	47 Mtn	ended war in China
10 Jul 1944	73	S	196, 197, 198	none	73 Fld	ended war in Japan
10 Jul 1944	112	S	246, 247, 248	none	one bn	destroyed Manchuria 8/45
10 Jul 1944	114	S	Bdes 83 & 84	none	none*	ended war in China
10 Jul 1944	118	S	Bdes 89 & 90	none	none*	ended war in China
12 Jul 1944	66	S	249, 304, 305	none	one bn	ended war on Formosa
12 Jul 1944	111	S	243, 244, 245	none	one bn	destroyed Manchuria 8/45
15 Jul 1944	81	S	171, 172, 173	none	81 Fld	ended war in Japan
15 Jul 1944	115	S	Bdes 85 & 86	none	none*	ended war in China
15 Jul 1944	117	S	Bdes 87 & 88	none	none*	destroyed Manchuria 8/45
12 Sep 1944	108	S	240, 241, 242	none	one bn	destroyed Manchuria 8/45
11 Oct 1944	119	S	253, 254, 255	none	one bn	destroyed Manchuria 8/45
21 Nov 1944	120	S	259, 260, 261	none	one bn	destroyed Manchuria 8/45

**artillery or mortar battalion added 1945*

Java) in September, and the 2nd Division (reconstituted after its mauling at Guadalcanal) in January 1944. In January 1944 the 28th Army was activated to command the 2nd, 54th and 55th Divisions in the Akyab region of southern Burma. The final division to be added to the Burma Area Army was the 53rd, which arrived from Japan in April 1944. Finally, the 33rd Army HQ was activated in April as well to help control the expanded force.

The plan called for the 28th Army (2nd, 54th and 55th Divisions) to advance along the southern coast through the Akyab region of Burma, the 15th Army (15th, 31st and 33rd Divisions) to strike in the center and seize Kohima and Imphal, and the 33rd Army (18th and 56th Divisions) to operate in the Hukawng Valley to the north and prevent Chinese interference with 15th Army's crucial battle. The plan was seriously flawed, however, by an overly optimistic view of logistics in the Burmese jungle and the usual underestimation of the fighting capacity of the opposition.

In ferocious fighting the 31st Division seized Kohima and the 15th and 33rd Divisions advanced on and laid siege to Imphal in April. This, however, is where the offensive ended. The British and Indian forces stood fast. Japanese units, starved of supplies, short of ammunition and decimated by jungle diseases, began to crack. The 31st Division commander retreated his division against orders to his supply dumps and other units quickly followed. The 18th Division in the north, assaulted by massive Chinese forces, was saved from annihilation only by a rapid improvised retreat. For the first time in the war a major Japanese force had not only been defeated, but routed. Field Marshal Slim, seeing what had happened, followed up with lightning thrusts through the jungle, over the jungle by glider, and around the jungle by amphibious maneuvers. The hasty dispatch of the 49th Division from Korea in mid-year did little to restore the situation and by the autumn the 15th and 33rd Armies consisted of little more than bands of sick, starving, unarmed soldiers wandering the jungles.

Reality Sets In – Forced Onto the Defensive

Things were no better in the Pacific. By early 1944 Rabaul, the bastion of the IJA in the southeast theater, had been completely isolated and the 8th Area Army based there was no longer able to control the 18th Army on New Guinea. As a result the 18th Army was removed from that command and assigned to the 2nd Area Army with orders to hold the central northern coast of New Guinea with the 20th Division (Aitape), the 41st Division (Wewak) and the 51st Division (Hollandia). In April 1944 US forces landed at Aitape and Hollandia, cutting the 18th Army off. On 9 May IGHQ redefined the area in the south Pacific that was to be held as Biak island, the western peninsula of New Guinea and Halmahera. Three weeks later, however, the US assaulted and captured Biak. By June the commander of 18th Army had assembled his three divisions near Aitape and launched a desperate attack to break out to the west. It was repulsed with heavy losses and the 18th Army finished the war in the jungles of New Guinea, its strength by then reduced to about 10,000 men.

Things progressed no better to the north. The Mariannas were declared an "absolute strategic zone" requiring defense at all costs in September 1943 and the Philippines were placed on alert in March 1944 but little progress was made in preparing for the defense of either. In both cases this appears to have been due to complacency, by both the Navy, which had overall responsibility for the central Pacific, and the 14th Army charged with the defense of the Philippines.

Few plans had been made for the ground defense of the central Pacific islands. Four naval guard units, battalion size elements, had been dispatched to the Mandates immediately before Pearl Harbor in accordance with plans, but no further action had been taken. Under the Navy's doctrine of viewing these islands as unsinkable aircraft carriers the only ground personnel needed were those necessary to maintain the aircraft, man some anti-aircraft guns and provide administrative support.

The Gilberts, farthest to the east, were captured by the Imperial Japanese Navy (IJN) in early 1942 but not actually garrisoned in any strength until the ill-advised US raid on Makin in August. Two SNLFs were quickly dispatched to the area, primarily to Tarawa. Both were destroyed in the November 1943 US assaults on Makin and Tarawa. The next line of defense was the Marshall Islands, the most important being Mille, Maloelap and Wotje in the east and Jaluit, Kwajalein and Eniwetok in the west. The defense of these islands, as well as the rest of the Pacific, was clearly beyond the capabilities of the naval ground forces. IGHQ finally responded and on 16 November 1943 issued two orders creating army forces for the defense of the central Pacific. The first ordered the activation of six "South Seas Detachments" of battalion to regiment strength for service in the Pacific. The second, Order A-106, previously mentioned, was much more far-reaching. It ordered the activation of four amphibious brigades, to be used for the mobile reinforcing and counterattack roles, and the reorganization of four infantry divisions into ocean divisions. Two such conversions had been ordered a month earlier with a view towards employment in the south Pacific, but the new threat forced their redeployment as well.

The Marshall Islands received the 1st South Seas Detachment, the 1st Amphibious Brigade, and the 107th (amphibious) Regiment of the 52nd (ocean) Division. The bulk of these were destroyed on Kwajalein and Eniwetok in January–February 1944, and the rest left behind by the US advance. The rest of the 52nd Division went to Truk, the 36th Division to Halmahera, the 14th Division to Palau, the 29th Division to Guam, and the 46th and 2nd Guards Divisions to the East Indies. The assault on the Marshalls apparently spurred IGHQ to further efforts, for on 20 February they ordered the Kwantung Army to form eight brigade-size "expeditionary units" for the central Pacific, these arriving in late March and early April. Finally, a seventh ocean division, the 46th, was formed and shipped to Saipan in April. To control these new central Pacific forces the army activated the 31st Army HQ on Saipan, which immediately launched a major rationalization and reorganization drive, disbanding the south seas detachments and expeditionary units in favor of 7 new independent mixed brigades and 6 new independent mixed regiments, scattered about the region.

Notable in the build-up was the Army's apparent sanguine attitude towards air defense. The new 31st Army, responsible for the whole central Pacific, was given only one regiment, one battalion and two batteries of AA guns, a total of eight 6-gun and three 4-gun batteries of 75mm guns, to be spread out over nine widely-separated islands. For low-level air defense there were two 6-gun batteries of 20mm guns. The Navy, having been on the receiving end of US air power to its considerable cost, was more worried and in July issued an emergency mobilization order for the formation of no less than 312 AA batteries, 39 coastal batteries and 48 battery- and battalion-size air defense units for use in the Pacific.

With the Marshalls breached the next blocking position comprised the Carolines (primarily Truk, Yap and Palau) to the south and the Marianas (mainly Saipan and Guam) to the north. The Allied drive continued without letup. In June Saipan and Guam were invaded, seeing the annihilation of the 29th and 46th Divisions. In September the southern portion of the Palau Islands was invaded and seized, destroying about half of the 14th Division.

The way was now open for an invasion of the Philippines. There was, however, considerable dispute within IGHQ over whether the US would attempt to retake the Philippines or seize Formosa. Both were poorly defended and IGHQ scrambled frantically to shift troops to cover them.

Formosa had raised the 48th Division, but with the departure of that formation in late 1941 no further major combat units had been stationed there. In May 1944 the 50th Division was finally activated there from depot troops already present and with the fall of Saipan the 66th Division was also activated. In September an "A" division, the 10th, was transferred from Manchuria to fill out the garrison.

The situation in the Philippines was little better. Only one of the original invading divisions, the 16th, had stayed on there, occupying central Luzon. The rest of the archipelago was in the hands of three brigade-size independent garrison units. In November 1943 a modest increase in capability had been ordered through the reorganization of the garrison units into four independent mixed brigades. In March 1944 the 16th Division was moved to Leyte in apparent response to the Allied drive through the central Pacific.

Following the US invasion of Hollandia the army sped up its preparations in the Philippines. The four independent mixed brigades were expanded to occupation divisions (100th, 102nd, 103rd and 105th) and May saw the first new division to arrive in the Philippines since the original Japanese invasion landed. This was the 30th Division, assigned to defend northern Mindanao. In June 1944 five new IMBs were formed for the Philippines, mostly from new personnel, thus representing additional strength. In August to October the reinforcement of the Philippines accelerated, with the 26th Division arriving from north China, the 8th Division from Manchuria and the 2nd Armored Division also from Manchuria, all for Luzon.

There were now nine divisions in the Philippines (including three "A" divisions) and to simplify command the 14th Army was elevated to Area Army and a new army, the 35th, created and made subordinate. The order of battle in September 1944 was thus:

HQ, 14th Area Army
35th Army
16th Division (Leyte)
30th Division (Mindanao)
100th Division (Mindanao)
102nd Division (Visayas)
103rd Division (Luzon)
105th Division (Luzon)
26th Division (Luzon)
8th Division (Luzon)
2nd Armored Division (Luzon)

Finally, in late September, another "A" division, the 1st from the Kwantung Army, was assigned and arrived late the next month.

Meanwhile, IGHQ had initiated a plan to provide additional fire support to existing garrisons. In June and July 1944 no fewer than 26 machine gun battalions were raised, mostly for the Pacific. Shortly thereafter 21 anti-tank battalions were activated, mostly for active theaters. These proved to be one-time events, however, as no further such units were raised except a few for Manchuria in 1945.

On 20 October two US divisions assaulted Leyte and inflicted heavy casualties on the resident 16th Division. On 22 October Southern Army directed that Leyte be regarded as the crucial battle for the Philippines and in accordance with this 14th Area Army ordered the 26th Division and the 1st Division to that island, the latter arriving from Shanghai via Manila on 2 November and the former from Luzon on 10 November. In addition, the 35th Army commander moved the 102nd Division to Leyte in stages during late October and early November. On 18 December, however, US units landed on Mindoro, next door to Luzon, and further efforts to reinforce Leyte were abandoned.

The campaigns of 1944 had indeed proven disastrous for an unprepared IJA. Two divisions (29th and 46th) had been annihilated in the Central Pacific; the Burma Area Army of nine divisions had been badly mauled in Burma; three divisions were cut off, disease-stricken and surviving on roots and small fish in northern New Guinea; and four divisions had been effectively eliminated on Leyte. A further two divisions were cut off in the Solomons and two more in the central Pacific.

Operation Sho-Go

On top of this fears were starting surface about the safety of the homeland itself. With the fall of Saipan concern heightened and on 21 July 1944 IGHQ issued a broad defensive plan known as the *Sho-Go* (Victory) Operation. Four variants were contemplated, with Sho-Go 1 defining the defense of the Philippines, Sho-Go 2 for Formosa and Ryukyus, Sho-Go 3 for Honshu, Kyushu and Shikoku, and Sho-Go 4 for Hokkaido, the Kuriles and Karafuto. Within Sho-Go 3 there were five sub-variants, "A" for Hachinohe, "B" for Sendai, "C" for Tokyo, "D" for Hamamatsu, and "E" for Kyushu. In the event, Sho-Go 1 was never fully implemented, but the other three remained the basis of planning for the rest of the year.

Sho-Go 3, the defense of the inner homeland, was the core of the plan. Assumptions were based on the forces available in the last half of 1944, with the Eastern Army Command responsible for Sho-Go 3A, 3B and 3C; the Central Army for Sho-Go 3D, and the Western Army for Sho-Go 3E. The Eastern Army held the full quota of field forces in the homeland, with the sole field army (the 36th, activated on 21 July 44) primarily responsible for Tokyo and environs. The homeland order of battle for late 1944 was as follows:

HQ, General Defense Command
Eastern Army Command
36th Army
44th Division (in Utsonomiya (51st) District)
81st Division
93rd Division
4th Armored Division
Hirosaki (57th) Depot Division
Sendai (2nd) Depot Division
Kanazawa (52nd) Depot Division
Tokyo (Guards) Depot Division

Central Army Command
- Nagoya (3rd) Depot Division
- Osaka (4th) Depot Division
- Kyoto (53rd) Depot Division
- 84th Division (in Himeji (54th) District)

Western Army Command
- Zentsuji (55th) Depot Division
- Kurume (56th) Depot Division
- Kumamoto (6th) Depot Division
- Hiroshima (5th) Depot Division

The forces for Sho-Go 4 were under the command of the Northern District Army, directly under IGHQ, until 27 March 1944 when the 5th Area Army HQ was established, with a subordinate 27th Army HQ. The 5th Area Army controlled the 7th Division and the Karafuto (southern Sakhalin) Mixed Brigade directly, while the 27th Army, responsible for the Kuriles, held the 42nd and 91st Divisions and the 3rd and 4th Amphibious Brigades.

1945: From Bad to Worse

If 1944 was a bad year, things were to get no better in early 1945. In the Philippines the 35th Army now consisted of the 30th and 100th Divisions on Mindanao and elements of the 102nd Division on Palawan and Panay. After the Leyte battle the 14th Area Army had only the 105th Division in southern Luzon, the 8th Division (less one regiment destroyed on Leyte) and the 2nd Armored Division in central Luzon, and the 103rd Division in the northern part of the island. To strengthen defenses of this crucial north–south link three more "A" divisions were hurriedly dispatched to the island in December 1944: the 10th, 19th and 23rd Divisions. These were the last substantive reinforcements to reach the Philippines.

By the end of the year the 14th Area Army had positioned itself for the coming battle, with the three newly-arrived "A" divisions deployed to defend Lingayen Gulf, the logical invasion site, and the 2nd Armored Division in the center of the broad valley to the south. Like the Fil-American forces in 1941/42 the 14th Area Army had no intention of trying to hold Manila, but where the Allied forces in 1942 withdrew into the Bataan Peninsula, the Japanese chose to withdraw the bulk of their forces north and east into the mountains of northern Luzon. The exception was the Shimbu Group (built around the 8th Division) which would defend the area to the east of Manila. With the American landing in January 1945 these plans were executed. The 2nd Armored Division was caught in the central plains and destroyed as an effective force, but the 10th, 19th and 23rd Divisions managed to pull back, supplemented by the 103rd and 105th Occupation Divisions. Pressed by US forces further back into the mountains and harassed by Philippine guerrillas, these had been rendered largely ineffective by May.

Things were going no better further south. By the start of 1945 it had become apparent to IGHQ that the battle areas were drawing closer to the homeland and that Southern Army's responsibilities were quickly becoming a secondary concern. On 27 January IGHQ issued new orders to Southern Army essentially placing it on the defensive for the first time. Vital areas were defined as Luzon, Indochina, Thailand, Malaya and Sumatra. All other areas had value only insofar as they constituted advanced defensive lines for the vital areas.

Table 1.5: Divisions activated January–April 1945

Activation	Div	Type	Inf Regts	Cav/ Recon	Art Regt	Fate
16 Jan 1945	121	S	262, 263, 264	none	one bn	destroyed Manchuria 8/45
16 Jan 1945	122	S	265, 266, 267	none	one bn	destroyed Manchuria 8/45
16 Jan 1945	123	S	268, 269, 270	none	one bn	destroyed Manchuria 8/45
16 Jan 1945	124	S	271, 272, 273	none	one bn	destroyed Manchuria 8/45
16 Jan 1945	125	S	274, 275, 276	none	one bn	destroyed Manchuria 8/45
16 Jan 1945	126	S	277, 278, 279	none	one bn	destroyed Manchuria 8/45
16 Jan 1945	127	S	280, 281, 282	none	one bn	destroyed Manchuria 8/45
16 Jan 1945	128	S	283, 284, 285	none	one bn	destroyed Manchuria 8/45
1 Feb 1945	131	S	Bdes 95 & 96	none	one bn	ended war in China
1 Feb 1945	132	S	Bdes 97 & 98	none	one bn	ended war in China
1 Feb 1945	133	S	Bdes 99 & 100	none	one bn	ended war in China
6 Feb 1945	79	S	Bns 289, 290, 291	79 Cav	79 Mtn	destroyed Manchuria 8/45
6 Feb 1945	96	S	292, 293, 294	none	none	ended war on Saisho Is
8 Feb 1945	89	S	Mx Bdes 43 & 69	none	none	ended war in Kuriles
15 Feb 1945	88	S	25, 125, 306	none	one bn	ended war on Sakhalin
28 Feb 1945	140	Coast	401, 402, 403, 404	none	one bn	ended war in Japan
28 Feb 1945	144	Coast	413, 414, 415, 416	none	one bn	ended war in Japan
28 Feb 1945	145	Coast	417, 418, 419, 420	none	one bn	ended war in Japan
28 Feb 1945	146	Coast	421, 422, 423, 424	none	one bn	ended war in Japan
28 Feb 1945	147	Coast	425, 426, 427, 428	none	one bn	ended war in Japan
28 Feb 1945	150	Coast	429, 430, 431, 432	none	one bn	ended war in Korea
28 Feb 1945	152	Coast	437, 438, 439, 440	none	one bn	ended war in Japan
28 Feb 1945	154	Coast	445, 446, 447, 448	none	one bn	ended war in Japan
28 Feb 1945	155	Coast	449, 450, 451, 452	none	one bn	ended war in Japan
28 Feb 1945	156	Coast	453, 454, 455, 456	none	one bn	ended war in Japan
28 Feb 1945	160	Coast	461, 462, 463, 464	none	one bn	ended war in Japan
20 Mar 1945	127	S	280, 281, 282	none	n/a	destroyed Manchuria 8/45
2 Apr 1945	201	Mobile	501, 502, 503	none	201 Fld	ended war in Japan
2 Apr 1945	202	Mobile	504, 505, 506	none	202 Mtn	ended war in Japan
2 Apr 1945	205	Mobile	507, 508, 509	none	205 Fld	ended war in Japan
2 Apr 1945	206	Mobile	510, 511, 512	none	206 Mtn	ended war in Japan
2 Apr 1945	209	Mobile	513, 514, 515	none	209 Mtn	ended war in Japan
2 Apr 1945	212	Mobile	516, 517, 518	none	212 Mtn	ended war in Japan
6 Apr 1945	143	Coast	409, 410, 411, 412	none	one bn	ended war in Japan
8 Apr 1945	153	Coast	441, 442, 443, 444	none	one bn	ended war in Japan
9 Apr 1945	142	Coast	405, 406, 407, 408	none	one bn	ended war in Japan
9 Apr 1945	151	Coast	433, 433, 434, 435	none	one bn	ended war in Japan
10 Apr 1945	157	Coast	457, 458, 459, 460	none	one bn	ended war in Japan
12 Apr 1945	129	S	Bdes 91 & 92	none	129 Fld	ended war in China
12 Apr 1945	130	S	Bdes 93 & 94	none	130 Fld	ended war in China

Activation	Div	Type	Inf Regts	Cav/ Recon	Art Regt	Fate
12 Apr 1945	161	S	Bdes 101 & 102	none	one bn	ended war in China
30 Apr 1945	214	Mobile	519, 520, 521	none	214 Fld	ended war in Japan
30 Apr 1945	216	Mobile	522, 523, 524	none	216 Fld	ended war in Japan

Indochina was quiet but the presence of 60,000 colonial French troops there (including 48,000 Indochinese), whose last identifiable loyalty was the Vichy regime that had collapsed several years before, was cause for concern. These had been restricted to their posts for three years by the occupying Japanese and prevented from training, but had not been disarmed. The 38th Army was responsible for garrisoning Indochina with the 21st Division, but the fear that the Allies might land on the Vietnamese coast after Luzon brought about the transfer in of the 22nd and 37th Divisions from south China in January 1945. On 9 March the Japanese ambassador demanded that the French colonial army be placed under Japanese command, the French refused, and the Japanese attacked ruthlessly, butchering the French soldiers and their families in their small posts.

In Burma the situation continued to deteriorate through early 1945, with Commonwealth forces pushing forward and capturing Rangoon in May. On 7 July the 15th Army was withdrawn from the Burma Area Army and the 39th Army (the garrison HQ for Thailand) redesignated the 18th Area Army in its place. The new order of battle for this force was:

HQ, 18th Area Army
15th Army
4th Division
56th Division
15th Division
22nd Division
37th Division
53rd Division

These forces were to defend Thailand from the British–Indian thrust from the west, but were saved only by Japan's surrender.[10]

In China the IJA had ceded control of much of the countryside to Chinese nationalist and Communist forces. They did manage to get some personnel from the homeland, mostly untrained, and used them to form 12 IMBs for the southern part of the country and 13 brigade-size independent guard units. Poorly-trained and underequipped, these new forces were suitable only for passive occupation duties.[11]

10 In a telling note on the state of IJA command and control the author of the Japanese Monograph on Thailand Operations, a staff officer assigned to 18th Area Army, noted that "53rd and 56th Divs were still in the Burma area at the end of the war and unable to communicate with the 18th Area Army." Japanese Monograph No.177, Chart 1, pp. 29–30.

11 The shortage of high-quality troops in north China was so pronounced that the officers of the 110th Division, a long-serving "special" formation had taken to referring to the unit as a "B" (regular) formation by 1945.

Reinforcements were desperately needed for the approaching climactic defense of the homeland, but few could be found. Of the 62 field (non-occupation) infantry divisions raised by the end of 1943 no fewer than 37 had either been destroyed or were bypassed, including seven (1, 8, 10, 16, 19, 23 and 26, all "A" divisions) in the Philippines, eight (15, 18, 31, 33, 53, 54, 55 and 56) in Burma, five (2nd Guards, 5, 32, 46 and 48) in the East Indies, five (20, 35, 36, 41 and 51) on New Guinea, three (6, 17 and 38) in the Solomons, three (14, 29 and 52) in the Central Pacific and four (2, 21, 22 and 37) in Indochina. In addition, 30 percent of the Army supplies shipped overseas in 1944 had been lost in transit, mainly sunk, and during the war the Army lost 182,000 personnel at sea, almost all of them in 1943–44.

Although plans for the Sho-Go operations had been drawn up in mid-1943, actual preparations had been undertaken with any degree of urgency only in the area covered by Sho-Go 2, defense of Formosa and the Ryukyus. In late 1944 the 10th Area Army was established to command the forces envisioned under Sho-Go 2, this commanding the 32nd Army in the Ryukyus and, from January 1945, the 40th Army for forces in Taiwan. The order of battle of this force was:

HQ, 10th Area Army
32nd Army
9th Division (moved to Formosa and 40th Army in January 1945)
24th Division
28th Division
62nd Division
40th Army
12th Division
50th Division
66th Division
71st Division

A separate command, built around the new 109th Division, was activated for the defense of the Ogasawara (Bonin) islands.

On February 19th the US Marines landed on Iwo Jima and in mid-March the 109th Division was finally destroyed. On 20 March US forces landed on Okinawa and in a three-month campaign the 24th and 62nd Divisions, as well as 32nd Army HQ, were annihilated. The ring had closed around Japan.

Defending the Homeland

On 20 January 1945 the now-defunct Sho-Go plans were replaced by a new "Outline of Army and Navy Operations" that explicitly assumed that the final decisive battle of the war would be fought on the Japanese islands. In accordance with this IGHQ completely reorganized the defense command of the homeland, forming a General Defense Command to control all tactical army elements in Japan (except Hokkaido) and transferred the operational missions of the old district armies to a number of new tactical area armies, while establishing new district army commands to handle the routine logistical and administrative matters. The new layout was:

Northern Honshu
11th Area Army
Northeastern District Army
East-Central Honshu
12th Area Army
Eastern District Army
West-Central Honshu
13th Area Army
Tokai District Army
Western Honshu and Shikoku
15th Area Army
Central District Army
Kyushu
16th Area Army
Western District Army
Mobile Reserve (Kanto plain)
36th Army

On 26 February, however, the plan was modified. The new goal called for the abolition of the General Defense Command and its replacement by two General Army HQs and the filling out of the force structure through a massive three-phase mobilization effort. This program was to concentrate on the creation of two new types of divisions, the coastal defense division and the mobile infantry division.

	Line Divs	Mobile Divs	Coastal Divs	Ind Mixed Bdes	Tank Bdes
First Phase					
Honshu/Shikoku/Kyushu	0	0	13	1	0
Hokkaido	2	0	2	1	0
Korea	0	0	1	1	0
Second Phase					
Honshu/Shikoku/Kyushu	0	8	0	0	6
Third Phase					
Honshu/Shikoku/Kyushu	0	7	9	14	0

In addition, the second phase was to raise the two general army HQs and eight army HQs for command purposes, while the third phase added four more army HQs. Further, the 11th, 25th and 57th Divisions and the 1st Armored Division were ordered home from Manchuria in April.

The first phase mobilization was ordered immediately, on 28 February. On 2 April the mobile divisions of the second phase were ordered activated, followed by the tank brigades four days later. The third mobilization was scheduled for activation in late July and early August, but fear of an American invasion earlier than expected caused the activation order to be issued two months early, on 23 May. In addition, one coastal division and one mobile division were added to the mobilization order to fill out new requirements, along with 15 IMBs, 11 medium artillery regiments and 32 trench mortar battalions. Because of equipment shortages, however, the third-wave divisions were mobilized on new, reduced-scale TO&Es.

Table 1.6: Divisions activated May–August 1945

Activation	Div	Type	Inf Regts	Cav/ Recon	Art Regt	Fate
23 May 1945	224	Mobile	340, 341, 342	none	none	ended war in Japan
23 May 1945	230	Mobile	319, 320, 321	none	none	ended war in Japan
23 May 1945	231	Mobile	346, 347, 348	none	none	ended war in Japan
23 May 1945	234	Mobile	322, 323, 324	none	none	ended war in Japan
23 May 1945	303	Coast	337, 338, 339	none	none	ended war in Japan
23 May 1945	312	Coast	358, 359, 360	none	one bn	ended war in Japan
23 May 1945	320	Coast	361, 362, 363	none	one bn	ended war in Korea
23 May 1945	321	Coast	325, 326, 327	none	one bn	ended war in Japan
23 May 1945	344	Coast	352, 353, 354	none	one bn	ended war in Japan
23 May 1945	354	Coast	331, 332, 333	none	one bn	ended war in Japan
10 Jun 1945	222	Mobile	307, 308, 309	none	none	ended war in Japan
10 Jun 1945	229	Mobile	334, 335, 336	none	none	ended war in Japan
10 Jun 1945	308	Mobile	310, 311, 312	none	none	ended war in Japan
10 Jun 1945	351	Coast	328, 329, 330	none	one bn	ended war in Japan
20 Jun 1945	225	Mobile	343, 344, 345	none	none	ended war in Japan
10 Jul 1945	134	S	365, 366, 367	none	134 Fld	destroyed Manchuria 8/45
10 Jul 1945	135	S	368, 369, 370	none	135 Fld	destroyed Manchuria 8/45
10 Jul 1945	136	S	371, 372, 373	none	one bn	destroyed Manchuria 8/45
10 Jul 1945	137	S	374, 375, 376	none	137 Fld	destroyed Manchuria 8/45
10 Jul 1945	138	S	377, 378, 379	none	138 Fld	destroyed Manchuria 8/45
10 Jul 1945	139	S	380, 381, 382	none	139 Fld	destroyed Manchuria 8/45
10 Jul 1945	148	S	383, 384, 385	none	none	destroyed Manchuria 8/45
10 Jul 1945	149	S	274, 386, 387	none	149 Fld	destroyed Manchuria 8/45
10 Jul 1945	221	Mobile	316, 317, 318	none	none	ended war in Japan
10 Jul 1945	322	Coast	313, 314, 315	none	one bn	ended war in Japan
10 Jul 1945	355	Coast	355, 356, 357	none	one bn	ended war in Japan
23 Jul 1945	316	Coast	349, 350, 351	none	one bn	ended war in Japan
10 Aug 1945	158	S	389, 390, 391	none	n/a	n/a

In addition, the IJN began reorganizing its massive base facilities into combat units, although it is unlikely that these sailors-turned-infantrymen were well trained or equipped.

The fevered last-minute preparations, however, were irrelevant. On 6 August 1945 the USAAF dropped the first atomic bomb on Hiroshima, largely destroying the city. The next day the Soviet Union declared war on Japan and launched its army, now hard and battle-tested against the Germans, into Manchuria. The hapless Kwantung Army was a mere shadow of its former proud self, with not a single "A" division left, short of equipment, and filled largely with partially-trained recruits and older reservists. They barely slowed the Red Army juggernaut as it smashed its way through Manchuria. On 9 August a second atomic bomb was dropped on Nagasaki, convincing all but the hard-core militarists of the futility of continued resistance. On 15 August 1945 the emperor announced his acceptance of the terms of the Potsdam Declaration. The Second World War was finally over.

A Plan For Failure

The Japanese war effort was, of course, doomed from the beginning. It hinged almost completely on German ability to neutralize the powers of the British Empire, the Soviet Union and the United States.

With the benefit of hindsight it can be argued that only one expansionist course of action could have succeeded. Had Tokyo accepted the Western demands in 1940 and withdrawn from China, concentrated its energies on building up the Kwantung Army, and invaded the Soviet Union in concert with Germany in mid-1941, it might well have tipped the scales to complete Axis domination of the Eurasian land mass. Even this, however, is open to question, for the IJA was simply not a modern army in 1941, and was incapable of logistics support over even medium distances.

This is not to say, by any means, that the IJA was an ineffectual army. In keeping faith with the code of *Bushido* it was perhaps the most ferocious close-quarters infantry force in the world. The Samurai spirit, however, valued only direct combat with the enemy. All else was regarded as slightly below the standards expected of a warrior. The result was neglect of all the other branches vital to a modern army.

Of most immediate impact, the artillery branch failed the infantry throughout the war. Except in the few siege situations in which it fought, such as at Bataan, the artillery was unable to reliably provide fire support. Shifting of massed fire, as opposed to spreading out assets along the line, was almost never practiced; light field artillery was often fired in the direct-fire mode in support of infantry; and communications were poor so that much of the efforts of the artillery were spent on unobserved missions fired against map grids.

Reconnaissance was often neglected in favor of aggressive offensive action by low-level infantry units. The two are not interchangeable, especially given the poor communications of the IJA. Infantry regiments, especially in the south Pacific, often thrashed about in the jungle, uncertain as to where they were or where they were supposed to go. As a result it was almost unheard of for two separated columns to reach the jump-off point for an attack at the same time. Innate infantry aggressiveness was often enough to overcome this deficiency in the early years of the war when fighting against a poorly-trained enemy, but against a determined and skillful foe it repeatedly led to disasters first seen at Guadalcanal.

The emphasis on direct, aggressive action also tended to devalue the role of staff planning. Staffs were smaller at all headquarters levels than in their Western counterparts and the individuals placed in staff positions were not only overworked but also suffered from a lack of prestige that made it difficult to bring logistical and other non-combat aspects to the fore. There were few repercussions in the intermittent combat in China or during the brief campaigns of early 1942, but in the prolonged heavy combat of Burma and the south Pacific Japanese logistics would prove woefully inadequate.

Many of the failures can be attributed to Japanese complacency. Already contemptuous of foreigners the early victories against China and the Western Allies reinforced a self-confidence bordering on arrogance. New tanks and anti-tank guns were developed at a leisurely pace, as were crucial anti-aircraft guns and field artillery.

Further, the Japanese seemed to have failed to comprehend the nature of total war. They took little advantage of the chance to build up their army during the period before Pearl Harbor or during the relatively quiet period of late 1942 and 1943. In fact, of the males reporting for induction who were found physically fit in 1941 only 51 percent were actually called to service, and this rose to but 60 percent for 1942 and 1943. Only in 1944 was there a significant increase, to 89 percent and then to 90 percent in 1945.[12] As a result no new divisions were formed, except through consolidation of existing units, between September of 1940 and March of 1943.

Without the atomic bomb the end would have been much bloodier but the outcome the same. The militarists had begun a crusade they simply did not have the tools to complete.

12 This represented about 50% of those examined in 1942/43, rising to 68% in 1944, indicating a probable relaxation of physical standards in the interim as well.

2

Infantry

There was never any doubt that the infantry would be not merely the preeminent arm of the Imperial Japanese Army, but the dominant one, almost to the exclusion of others. That was the arm that would close with the enemy and destroy him in the ferocious close-quarters combat demanded by the martial tradition. The cavalry had no such tradition and the artillery certainly lacked the mythical cachet of hand-to-hand fighting.

Although the empire's involvement in the First World War had been peripheral attention was certainly paid to the lessons learned by the Western powers, if only selectively. To the extent that the IJA thinkers looked specifically to one Western nation, it was the French, whose belief in the moral power of the offensive, at least early in the war, most closely approached Japanese views.

The Post WWI Reorganization

Studies on infantry organization carried out in the immediate aftermath of the First World War led to a massive reorganization of the branch in 1922, accompanied by the adoption of a new family of weapons, mostly modeled on French patterns.

Rifle company strength, which had reached 250 under the 1914 organization, was now to be kept within a 200-man ceiling to facilitate command and control. Although this maximum figure was sometimes exceeded, the more normal company strength worked out to 198 men.

In keeping with Japanese practice the organization tables did not specify any details below company level. That a company commander simply had a standard list of personnel and equipment available to him, however, did not mean he could organize his command any way he wished. The infantry manuals were explicit that he was to divide his company up into a small HQ and three equal-size rifle platoons.

The 1922 organization called for a company HQ built around the company commander, a warrant officer and a master sergeant for personnel duties, a supply sergeant and an armorer sergeant. Four attached medical orderlies and 5–6 runners and orderlies drawn from the line platoons supplemented their efforts.

Each of the three rifle platoons was made up of four rifle squads and two light machine gun squads. The rifle squad generally consisted of an NCO and eleven privates, all armed with rifles. A light MG squad was usually made up of an NCO and seven privates with a 6.5mm Type 11 light MG. This organization gave the company an official armament total of 144 rifles and six light MGs.

That much was not controversial, but heated debate flared up during the infantry organization studies on the issue of the number of rifle companies per infantry battalion. The proponents of four line companies finally carried the day with the argument that a 3-company battalion could not maintain sufficient troop strength when breaking through Soviet-style defenses in great depth. Thus, the composition of the infantry battalion was fixed at a HQ, four rifle companies, a machine gun company and an infantry gun company for a total strength of about a thousand men.

The machine gun company was a small unit, consisting of two platoons, each with two 6.5mm Type 3 (1918) medium machine guns, and an ammunition platoon. The battalion gun company was made up of two platoons: one with two 70mm Type 11 (1922) mortars and the other with two 37mm Type 11 infantry guns, the latter analogous to the French TR16 gun. These two weapons were designed to complement each other, with the 37mm providing accurate, flat-trajectory fire against point targets, while the mortars' plunging fire was for use against targets in defilade and area targets.

Although the battalion weapons worked well for their time, they left a dead space directly in front of the infantry. This was quickly remedied by the Type 10 grenade discharger, which threw hand grenades out to 250 meters. This new weapon filled a need but, for reasons that are still not clear, was not included in the official TO&Es of the time. Their usefulness was such, however, that field commands authorized their procurement and within a few years all rifle companies had formed a seven-man grenadier squad out of existing personnel in each rifle platoon to man two of these weapons, giving the rifle company a revised (unofficial) armament of 138 rifles, 6 light MGs and 6 grenade dischargers.

The battalion HQ carried out command of the battalion and also included a combat trains platoon and a field trains platoon. The former used 40 pack horses to carry ammunition, chemical warfare materiel, medical supplies and entrenching tools. The field trains used twenty of the standard Type 39 one-horse carts to carry rations, baggage, and tentage.

Above that the infantry regiment had a headquarters, a signal platoon, a machine gun company and trains elements as well as the three line battalions. The signal platoon comprised four telephone sections, one for the regiment HQ and one for attachment to each battalion HQ. The machine gun company consisted of three platoons, each with three medium MGs.

Each of the 17 regular divisions of the 1920s was built around two brigades (each consisting of a small brigade HQ and two infantry regiments), a field or mountain artillery regiment, a cavalry regiment, an engineer regiment (actually a battalion in size), a transport regiment, a company-size signal unit and service support units.

New Weapons and New Organization in the 1930s

If the 1922 organization closely followed international patterns, the IJA started to diverge from the norm as a result of maneuvers in the late 1920s, developing and introducing two new, uniquely Japanese, weapons.

An improved grenade discharger, the Type 89 with a specially-designed projectile with a 650-meter range and good fragmentation effect, replaced the Type 10 grenade dischargers in the rifle platoons and solidified the status of this unique type of weapon. The earlier weapons (subsequently referred to as light grenade dischargers) were then distributed to battalion and regiment HQs for primary use as signal flare launchers.

The other change resulted from the Army's decision in the mid-1920s to develop a weapon that combined the HE effectiveness of the 70mm mortar with the accuracy and direct-fire capabilities of the 37mm infantry gun. The result was the 70mm Type 92 (1932) infantry howitzer. This weapon, often referred to as the "battalion gun," was capable of both high-angle and low-angle fire. It proved quite effective, its only drawback being the relatively large gun crew required compared with much lighter mortars. With the introduction of these new weapons the former battalion machine gun company was expanded to a heavy weapons company, now consisting of two machine gun platoons each of two weapons, and an infantry gun platoon of 64 men with two 70mm howitzers.

In 1935 the decision was made to double the number of medium machine guns in the battalion and the company reverted to a pure machine gun unit, the infantry guns being split off into a separate platoon directly under the battalion commander. The new machine gun company had a strength of 149 men and was split into four firing platoons (each of a lieutenant and two 11-man squads) and an ammunition platoon (six 8-man squads). Each gun squad received two pack horses (one for the gun and one for 2,400 rounds of ammunition) and each ammunition squad four pack horses.

The infantry gun platoon consisted of a headquarters (a lieutenant, a master sergeant for supply, an aiming circle sergeant, a liaison sergeant, a bugler, three privates for observation duties, and a groom), two gun sections (each a sergeant and 12 privates with a 70mm howitzer and two draft horses) and an ammunition section (a sergeant, four carters each with a cart and draft horse, and 24 reserve gun crew to help manhandle the gun and ammunition as needed). Although regarded as highly successful, the 70mm howitzers were primarily draft-transport weapons (they could be broken down into three loads for pack transport, but this feature appears to have been little used at this stage) and efforts were begun to develop a conventional mortar for the use of the two divisions (9th and 11th) that used pack transport. This eventually resulted in the Type 97 (1937) 81mm mortar.

Typical Transitional Infantry Battalion, 1935

Battalion Headquarters

4 OFF: 1 CO (MAJ)*, 1 adjutant (CPT)*, 1 medical officer, 1 intendance officer *; 9 NCO: 1 chief clerk, 1 supply SGT*, 1 chief bugler*, 1 mess SGT**, 3 clerks**, 1 medic, 1 intendance SGT**

Four Rifle Companies, Each

Company Headquarters:
- 1 OFF: CO (CPT)*; 1 WO: HQ section leader*; 2 NCO: 1 clerk**, 1 supply SGT**; 6 PVT: 4 runners/buglers**, 2 medics

Three Rifle Platoons, each:
- Platoon Headquarters:
  - 1 OFF: platoon leader (LT)*
- Four Rifle Squads, each:
  - 1 NCO: squad leader**; 12 PVT: riflemen**
- Two Light Machine Gun Squads, each
  - 1 NCO: squad leader**; 7 PVT: 2 machine gunners*, 5 riflemen**

Machine Gun Company
Company Headquarters:
1 OFF: CO (CPT)*; 1 WO: HQ section leader*; 2 NCO: 1 clerk**, 1 supply SGT**; 4 PVT:
2 runners/buglers**, 2 medics
Two Machine Gun Platoons, each:
Platoon Headquarters:
2 OFF: 1 platoon leader (LT)*, 1 assistant platoon leader (LT)*
Four Gun Squads, each:
1 NCO: squad leader*; 10 PVT: 6 gun crew**, 4 gun crew*
one pack horse carrying medium machine gun, one carrying 2,400 rounds ammunition
Ammunition Platoon:
Platoon Headquarters:
1 OFF: platoon leader (LT)*; 1 PVT: runner**
Six Ammunition Squads, each:
1 NCO: squad leader**; 7 PVT: horse handlers/ammunition bearers**
four pack horses, each carrying 2,400 rounds of MG ammunition

Infantry Gun Platoon
Platoon Headquarters:
1 OFF: platoon leader (LT)*; 3 NCO: 1 supply SGT*, 1 liaison SGT*, 1 instrument operator*;
4 PVT: 1 bugler*, 3 observers*
Two Gun Squads, each:
1 NCO: squad leader*; 12 PVT: 10 gunners**, 2 carters**
two draft horses towing 70mm howitzer w/ limber, 1 draft horse towing caisson w/ limber
Ammunition Squad:
1 NCO: squad leader*; 28 PVT: ammunition bearers/carters**
four draft horses, each towing caisson w/ limber
(Note: each platoon carries a total of 200 rounds of 70mm ammunition in caissons & limbers)

Battalion Combat Trains
1 NCO: Transport SGT*
Three Transport Squads, each:
16 PVT: 1 squad leader**, 10 carters, 5 baggage handlers
twelve pack horses
(Note: of the trains total 21 horses each carry 2800 rounds of rifle/LMG ammunition, 8 each carry 100 hand grenades, 2 carry medical equipment, 5 carry pioneer and entrenching tools)

Battalion Field Trains
1 NCO: Transport SGT*
Two Transport Squads, each:
16 PVT: 1 squad leader**, 10 carters, 5 baggage handlers
ten draft horses, each pulling a two-wheeled cart
(Note: the field trains carry one day's rations and forage , as well as officers' baggage, organizational records, field clothing and shoe repair kits. Generally, 7 carts carried rations, 3 carried forage, 4 carried officers' baggage, 3 carried records and 3 were spares).

**armed with pistols*
***armed with rifles*

Note: unofficially, all rifle companies formed a grenade launcher squad in each rifle platoon using existing personnel.

Regimental fire support was dramatically improved with the introduction in 1934 of two new weapons: the 37mm Type 94 (1934) anti-tank gun and, indirectly, the 75mm Type 94 mountain gun. The AT guns were organized into two-platoon, four-gun companies and incorporated into the infantry regiments. As these weapons were phased in, the regimental MG companies were gradually disbanded and the personnel and equipment decentralized to the battalion MG companies, which increased in size to incorporate four 2-gun platoons or three 4-gun platoons.

The new mountain guns were used to re-equip the mountain artillery regiments and the old 75mm Type (Meiji) 41 pack guns thus made redundant were handed over to the infantry. The Kwantung Army began experimenting with the Type 41s in the regimental support role in 1934 and in September of that year GHQ issued the "Provisional Training Regulations for the Regimental Infantry gun." The organization it specified for non-pack-type units, which remained in effect with only minor changes through to the end of the war, called for a regimental infantry gun company consisting of a 23-man HQ, two 36-man firing platoons and a 28-man ammunition platoon. Each Type 41 mountain gun was drawn by two horses in tandem with no limber. A gun section consisted of a sergeant and 16 privates, with one such gun and two one-horse carts each carrying 21 rounds of ammunition. A firing platoon consisted simply of a lieutenant on a riding horse, a groom, and two gun sections. The ammunition platoon was made up of five 3-man teams (each with two carts each with 24 rounds), and twelve reserve gun crewmen. The company HQ included the signal section of a sergeant and six privates and a reconnaissance section of similar composition.

Two other changes occurred in the regimental structure at the same time. One was the expansion of the signal platoon to company strength. The new signal company consisted of a wire platoon (with 16km of wire strung by seven teams) and a radio platoon of six teams. The second change was the inclusion of a platoon of pioneers in the regiment HQ. The duties of the pioneer platoon were defined as assault preparations (using bangalore torpedoes, flamethrowers and mine clearing tools), close-in anti-tank combat, construction and destruction of obstacles and field fortifications, and smoke and anti-gas duties. Heavy construction, however, was beyond them as they were equipped only with light hand tools.

Needless to say, not all of these new weapons became available at the same time. Thus, there were a number of transitional organizations in effect between 1933 and 1938. A typical organization as it appeared in 1935 for a flatland type formation is shown in Table 2.1. Units that relied exclusively on pack horses at the regiment level and below were, of course, much more manpower-intensive. It should be noted that this shows an official organization, and actual infantry of the period would have reorganized the rifle platoons by taking seven men out of the rifle and LMG squads to form a grenadier squad with two Type 89 weapons.

The Redesigned Divisions

A small cavalry regiment of a headquarters, two mounted squadrons and a machine gun platoon carried out the division's scouting function. A squadron was made up of a 13-man HQ and four 35-man mounted platoons. Each platoon was composed of the platoon leader, four rifle squads (each a sergeant and six men with rifles) and a light machine gun squad (of

Table 2.1: Typical Transitional Division, 1935 (War Strength)

	Officers	Enlisted	Pistols	Rifles	Light MGs	Medium MGs	70mm Inf Howitzers	75mm Mountain Guns	75mm Field Guns	Riding Horses	Pack Horses	Draft Horses	Carts	Wagons
Division Headquarters	17	91	39	41	0	0	0	0	0	40	0	12	12	0
Signal Company	4	195	25	71	0	0	0	0	0	7	0	18	18	0
Cavalry Regiment														
Regiment HQ	7	14	15	6	0	0	0	0	0	19	0	0	0	0
Two Mounted Squadrons, each	5	148	31	112	4	0	0	0	0	150	4	0	0	0
Machine Gun Platoon	1	37	4	34	0	2	0	0	0	37	12	0	0	0
Field Trains Group	0	53	0	0	0	0	0	0	0	5	0	32	32	0
Two Infantry Brigades, each														
Brigade HQ	3	6	6	3	0	0	0	0	0	3	0	0	0	0
Two Infantry Regiments, each														
Regiment HQ	4	50	9	33	0	0	0	0	0	5	0	8	8	0
Regimental Gun Company	3	135	26	105	0	0	0	4	0	7	0	31	23	0
Regimental Combat Trains Group	0	33	1	2	0	0	0	0	0	3	0	20	20	0
Three Infantry Battalions, each														
Battalion HQ	4	9	6	5	0	0	0	0	0	4	0	0	0	0
Three Rifle Companies, each	4	213	25	190	6	0	0	0	0	0	0	0	0	0
Machine Gun Company	6	143	49	98	0	8	0	0	0	1	40	0	0	0

Table 2.1: Typical Transitional Division, 1935 (War Strength)

	Officers	Enlisted	Pistols	Rifles	Light MGs	Medium MGs	70mm Inf Howitzers	75mm Mountain Guns	75mm Field Guns	Riding Horses	Pack Horses	Draft Horses	Carts	Wagons
Infantry Gun Platoon	1	63	12	52	0	0	2	0	0	1	0	10	8	0
Trains Group	0	91	2	5	0	0	0	0	0	7	36	20	20	0
Field Artillery Regiment														
Regiment HQ	6	81	18	0	0	0	0	0	0	22	2	6	0	2
Regimental Combat Trains	7	160	16	24	0	0	0	0	0	35	0	112	10	31
Three Field Gun Battalions, each														
Battalion HQ	9	83	20	0	0	0	0	0	0	22	2	6	0	2
Three Field Batteries, each	4	120	17	16	0	0	0	0	4	26	0	76	0	18
Trains Group	3	138	10	20	0	0	0	0	0	25	0	96	30	19
Engineer Regiment														
Regiment HQ	4	5	9	0	0	0	0	0	0	2	0	0	0	0
Two Engineer Companies, each	5	178	22	123	0	0	0	0	0	11	18	10	10	0
Transport Regiment	25	2358	67	125	0	0	0	0	0	187	0	1440	1440	0
Medical Unit	22	909	0	0	0	0	0	0	0	11	0	73	73	0
Four Field Hospitals, each	25	720	0	0	0	0	0	0	0	15	0	106	106	0
Veterinary Depot	3	44	0	0	0	0	0	0	0	23	0	24	12	0
Summary:														

Table 2.1: Typical Transitional Division, 1935 (War Strength)

	Officers	Enlisted	Pistols	Rifles	Light MGs	Medium MGs	70mm Inf Howitzers	75mm Mountain Guns	75mm Field Guns	Riding Horses	Pack Horses	Draft Horses	Carts	Wagons
Cavalry Regiment	18	400	81	264	8	2	0	0	0	361	20	32	32	0
Infantry Regiment	88	3692	543	2900	72	24	6	4	0	54	228	149	135	0
Infantry Battalion	27	1158	169	920	24	8	2	0	0	13	76	30	28	0
Field Artillery Regiment	85	1984	277	228	0	0	0	0	36	432	8	1108	100	258
Engineer Regiment	14	361	53	246	0	0	0	0	0	24	36	20	20	0
For "Pack-Type" Infantry Division, Substitute:														
Infantry Battalion Trains Group	0	116	2	6	0	0	0	0	0	8	76	0	0	0
Infantry Regimental Gun Company	3	168	24	82	0	0	0	4	0	7	48	0	0	0
Infantry Regimental Combat Trains Group	0	68	0	0	0	0	0	0	0	5	42	0	0	0
Artillery Regimental Combat Trains	7	577	23	60	0	0	0	0	0	18	310	0	0	0
Three Mountain Artillery Battalions, each														
Battalion HQ	9	97	20	0	0	0	0	0	0	20	22	0	0	0
Three Mountain Batteries, each	4	176	17	16	0	0	0	4	0	16	81	0	0	0
Trains Group	3	259	10	34	0	0	0	0	0	20	167	0	0	0
Two Engineer Companies, each	5	194	22	124	0	0	0	0	0	12	36	0	0	0
Transport Regiment	41	6633	142	864	0	0	0	0	0	547	4050	0	0	0

Table 2.1: Typical Transitional Division, 1935 (War Strength)

	Officers	Enlisted	Pistols	Rifles	Light MGs	Medium MGs	70mm Inf Howitzers	75mm Mountain Guns	75mm Field Guns	Riding Horses	Pack Horses	Draft Horses	Carts	Wagons
Medical Unit	24	1047	0	0	0	0	0	0	0	11	96	0	0	0
Four Field Hospitals, each	29	908	0	0	0	0	0	0	0	11	96	0	0	0
Veterinary Depot	3	74	0	0	0	0	0	0	0	26	42	0	0	0

Note: warrant officers are included in "enlisted" strength.

Note: table does not show unofficial allocation of six Type 89 Grenade Launchers per rifle company.

a sergeant and five men with a light MG and a pack horse).[1]

The regimental machine gun platoon consisted of two firing sections (each of a sergeant and nine privates with a medium MG and two pack horses) and an ammunition section (a sergeant and six privates with six pack horses). Strangely, the regiment had no ammunition trains, doctrine calling for it to draw ammunition from the nearest infantry unit. Signal facilities were also weak, especially for a reconnaissance unit, the only trained signalers being a sergeant and six privates in the regimental HQ, these presumably operating semaphore flags.

An artillery regiment of three battalions each of twelve 75mm guns provided fire support. In two of the permanent peacetime divisions (9th and 11th) this was a mountain artillery unit using pack horses for transport, in the other fifteen divisions it was a field artillery unit drawn by draft horses.

Each draft battery had four sections each with two six-horse teams, one pulling a limber and field gun and the other a limber and caisson. With two more caisson/limber sets in the battery trains, a battery carried 744 rounds of 75mm ammunition as its basic load, supplemented by the battalion combat trains that brought holdings up to 261 rounds per gun. A battery headquarters included the "battery detail" that moved forward from the gun position to the front line, trailing telephone wire behind it, to command the firing elements from up front. This supported the battery commander with the reconnaissance/observer officer, a driver with a cart, a five-man instrument/observation squad and a wire signal squad of nine men (three of them mounted).

For communications each battalion and regiment HQ included a signal platoon with a 34-man signal (telephone) section, an 11-man radio section for two pack-carried radios, and a rarely-used 6-man air signal unit with ground panels. The regimental trains included five ammunition sections, each with three caissons and limbers for 300 rounds for a further 1,500 rounds.

As the 105mm Type 91 howitzers came into service fourth, general support battalions were added to divisional artillery regiments, identical to the field gun battalions but equipped with the new howitzers.

A proposal was also raised in 1937 to incorporate a machine gun battalion in each division, to consist of a 17-man HQ and three 103-man, eight-gun companies. This, however, was not approved.

Ten regular divisions were temporarily mobilized to war strength in 1937 for service in China and Manchuria. Eight (3, 5, 6, 8, 10, 14, 16 and 20) had draft-type infantry regiments and field artillery regiments, and a strength of slightly over 25,000 each. The other two divisions (9 and 11) had pack-type infantry regiments and mountain artillery regiments to yield a strength of 886 officers and 28,687 enlisted each. As new special divisions were raised, the regular divisions were returned to the homeland and reverted to their peacetime strength.

The Wartime Divisions

Through 1936 the reorganization plans had been applied evenly throughout the Army, which consisted of 17 regular divisions. In 1937, however, the pressures of fighting in China, guarding Manchuria and garrisoning the homeland forced the government to consider the

1 The squadron could also be configured as three platoons armed solely with rifles and a light machine gun platoon.

A Type 92 medium machine gun prepares its position outside Tientsin in the opening days of the war. The soldier closest to the camera is holding the sight.

raising of six additional "wartime" divisions, five "special divisions" (13th, 18th, 101st, 108th and 109th) of reservists and one "new type" (26th) by expansion of an early mixed brigade. The "wartime" divisions were finally ordered in July and actually mobilized in September 1937. Two of these divisions (13th and 109th) used mountain artillery regiments, the others field artillery regiments for fire support. These "wartime" divisions were both weaker in firepower and less mobile than their regular counterparts.

The infantry regiments of the conscript divisions varied considerably from one division to the next, probably as a result of the availability of personnel and varying ratios of pack and cart transport.

	Off	WO	MSG	SGT	PVT	Total	Horses
In 13th Division	83	16	9	311	2980	3399	557
In 18th Division	83	16	9	313	2801	3222	370
In 26th Division	88	27	15	230	2208	2568	179

The rifle companies were identical to the regular forces, using the old organization of three rifle platoons, each of three rifle squads, two machine gun squads (each with a 6.5mm Type 11 light MG) and a grenadier squad (with two Type 89 grenade dischargers). The company totaled 194 men with 168 rifles, six grenade dischargers and six light machine guns. In the regular division the machine gun company had four platoons each with two 7.7mm Type 92 medium machine guns, while in the special division it had only two platoons, equipped with the older and heavier 6.5mm Type 3 weapon. The regular infantry

Table 2.2: Typical Regular (3rd) and Special (18th) Divisions, 1937

Regular Division					Special Division				
	Men	Horses	Men	Horses		Men	Horses	Men	Horses
Division Headquarters	330	165			Division Headquarters	330	165		
Signal Unit	255	47			Signal Unit	246	45		
Cavalry Regiment	452	428			Cavalry Regiment	451	431		
Regiment HQ			121	108	Regiment HQ			120	110
Two Rifle Squadrons, each			144	134	Two Rifle Squadrons, each			144	134
Machine Gun Platoon			43	52	Machine Gun Platoon			43	53
Two Infantry Brigades, each	7,569	1,102			Two Infantry Brigades, each	6,497	756		
Brigade HQ			75	20	Brigade HQ			53	16
Two Infantry Regiments, each			3,747	526	Two Infantry Regiments, each			3,222	370
Regiment HQ			224	102	Regiment HQ			174	74
Three Infantry Battalions, each			1,091	115	Three Infantry Battalions, each			974	88
Battalion HQ			120	78	Battalion HQ			114	73
Four Rifle Companies, each			194	0	Four Rifle Companies, each			194	0
Machine Gun Company			139	28	Machine Gun Company			84	15
Regimental Gun Company			161	49	Regimental Gun Company			126	32
Anti-Tank Company			89	21					
Field Artillery Regiment	2,894	2,269			Field Artillery Regiment	1,922	1,508		
Regiment HQ			124	72	Regiment HQ			83	47
Three Field Gun Battalions, each			634	499	Three Field Gun Battalions, each			571	447
Battalion HQ			160	94	Battalion HQ			129	73

Table 2.2: Typical Regular (3rd) and Special (18th) Divisions, 1937

Regular Division					Special Division				
	Men	Horses	Men	Horses		Men	Horses	Men	Horses
Three Batteries, each			128	108	Three Batteries, each			127	106
Battalion Trains			90	61	Battalion Trains			61	56
Field Howitzer Battalion			634	499					
Regiment Trains			234	201	Regiment Trains			126	105
Engineer Regiment	672	99			Engineer Regiment	672	99		
Regiment HQ			100	61	Regiment HQ			100	61
Two Engineer Companies, each			286	19	Two Engineer Companies, each			286	19
Transport Regiment	3,461	2,612			Transport Regiment	1,898	1,451		
Regiment HQ			27	10	Regiment HQ			26	10
Six Transport Companies, each			562	326	Four Transport Companies, each			453	349
Veterinary Hospital			62	46	Veterinary Hospital			60	45
Medical Regiment	1,101	128			Medical Regiment	1,095	128		
Three Field Hospitals, each	236	75			Three Field Hospitals, each	236	75		
Field Hospital	243	79			Field Hospital	243	79		
Total	25,179	8,177			Total	20,087	5,493		

battalion also included a gun platoon with two 70mm Type 92 battalion guns (or Type 11 mortars in the pack-transported units) that the special infantry battalions lacked.

The regular infantry regiment was now supported by both an infantry gun company (with four Type 41 mountain guns, replaced by Type 11 infantry guns in the pack regiments) and an anti-tank company with two 2-gun platoons of 37mm Type 94 AT guns, although the latter was not present in the pack-transported infantry regiments. The special infantry regiments had only a single infantry gun company of three 2-weapon platoons, one platoon being equipped with 37mm Type 11 guns and the other two with 70mm Type 11 mortars.

The special divisions lacked the general support 105mm field howitzer battalion found in the regular units and their scale of transport was lower at all levels.

In compensation the 18th, 101st, 108th and 109th divisions were authorized an additional increment of a machine gun company of 160 men and 39 horses, and an anti-tank company of 143 men and 30 horses in each infantry regiment to yield an authorized strength for the regiment of 3,525 men.[2]

The regimental machine gun companies differed significantly from the battalion MG company. In both cases the basic unit, the machine gun platoon, consisted of a platoon leader and two gun squads (with the battalion machine gun company having two such platoons and the regimental company four), and in both cases the machine gun squad consisted of a corporal and eight privates. The difference, however, was in transport, for the squads of the regimental company each also had a driver with a horse-drawn cart, while in the battalion-level company with its greater emphasis on close accompaniment of the infantry, each squad received instead two leaders with pack horses. Each kind of company also had an ammunition platoon with either two squads (each nine men and two pack horses) in the battalion company, or four squads (each ten men and two carts) in the regimental company.

The anti-tank company had two platoons, each with two 37mm Type 94 AT guns.

The 13th Division, which had infantry regiments with a TO&E strength of 3,399 men received an identical increment, boosting their authorized strength to 3,762.

The 26th Division, on the other hand, received no such increment to its rather small regiments of 2,568 men each. The small size of these regiments was due to the fact that they had only three rifle companies per battalion instead of the usual four. They did, however, receive a new TO&E in April 1939 boosting their strength to 2,915 along with an increment of 190 men to yield an authorized strength of 3,105. This permitted the addition of the fourth rifle company to each battalion.

In 1938 nine more "wartime" divisions were formed, primarily for duty in China. Two special and two new-type divisions (15th, 17th, 21st and 22nd) were officially activated in April, followed by four special divisions (104th, 106th, 110th and 116th) in June.

2 This is an early example of the IJA's proclivity of assigning both a TO&E strength and an authorized strength to a unit. The TO&E strength was the baseline figure and IGHQ could add or delete men (and horses) from that strength as it saw fit to create an authorized strength. In this case the additional personnel were to form specified units within the regiments. Later on, however, these increments or, more commonly, decrements were distributed as the local commander saw fit. A given unit would thus often have both a TO&E strength and an authorized strength, quite often not the same and sometimes very significantly different. Authorized strengths, along with TO&E strengths, were almost always consistent within a division, so that the infantry regiments in a particular division would have the same authorized strength, even though it was not the same as their equally identical TO&E strengths. This use of two methods of manpower allocation, along with sloppy staff work, often makes it difficult to establish nominal strengths of Japanese units.

One more new-type division, the 27th, was raised in June. Two of these divisions (21st and 22nd) had a mountain artillery regiment and the others had field artillery regiments. The June divisions were organizationally identical to the special divisions raised in 1937 (including non-TO&E increments of machine gun and infantry gun companies), but the 27th Division used that of the earlier 26th Division. The April divisions were raised with reconnaissance regiments rather than cavalry regiments for the scouting function. More significantly, as noted below, the conscript divisions were triangular formations, the first in the Japanese Army.

The main components of these divisions had the following TO&E and authorized strengths:

	April 1938 division		June 1938 division		27th Div	
	Men	**Horses**	**Men**	**Horses**	**Men**	**Horses**
Division Headquarters	63	22	292	80	33	6
Signal Unit	178	30	246	45	178	30
Cavalry/Recon Regiment	unk	unk	451	431	unk	unk
Infantry Regiments	3 x 2,909	3 x 179	4 x 3,222	4 x 370	3 x 2,568	3 x 182
Field Artillery Regiment	1,745	1,259	1,922	1,503	n/a	n/a
or Mountain Artillery Regiment	1,841	1,115	n/a	n/a	1,213	472
Engineer Regiment	401	15	672	99	401	15
Transport Regiment	370	113	1,186	763	507	211

An initial increment of 143 men was added to the authorized strengths of the infantry regiments of the April divisions within a few months, bringing their strength up to 3,052; while 142 men were added to each artillery regiment. No horses, however, were added.

The 27th Division was a unique formation, apparently designed for more sedentary garrison duties in China, for its infantry regiments were designated the 1st, 2nd and 3rd China Occupation (*Shina Chuton*) Infantry Regiments. These regiments were raised initially with the same organization as those in the 26th Division the year prior, and additions of 335 men in late 1938 and another 190 in early 1939 to each regiment's authorized strength paralleled the growth of the 26th's regiments.[3]

Optimizing for Manchuria

The final division raised that year, the 23rd, represented the first addition to the ranks of the regular divisions. This division, raised specifically to garrison the vast plains of western Manchuria, had a unique organization. It was built around three infantry regiments instead of the normal four and featured a partly-motorized transport regiment and a partly (and lightly) armored reconnaissance regiment. The division had the following strength on activation:

	Men	Horses	(men)	(horses)
Division Headquarters	90	22		
Signal Unit	178	30		
Reconnaissance Regiment	304	185		

3 In both the 26th and 27th Divisions the growth in infantry regiment strength served to bring them into conformity with those of the 32nd-41st Divisions raised in 1939 and discussed shortly.

	Men	Horses	(men)	(horses)
Regiment HQ			29	30
Mounted Squadron			175	155
Tankette Squadron			100	0
Infantry Group HQ	7	3		
Three Infantry Regiments, each	3,147	119		
Regiment HQ			33	11
Signal Company			96	19
Three Infantry Battalions, each				
Battalion HQ			9	4
Four Rifle Companies, each			193	0
Machine Gun Company			176	35
Regimental Gun Company			147	32
Field Artillery Regiment	1,745	1,259		
Regiment HQ			47	29
Three Battalions, each				
Battalion HQ			5	5
Two Field Gun Batteries, each			187	135
Two Field Howitzer Batteries, each			187	135
Engineer Regiment	401	15		
Regiment HQ			23	7
Two Engineer Companies, each			189	4
Transport Regiment	507	211		
Regiment HQ			31	15
Two Draft Companies, each			137	98
Motor Transport Company			202	0
Medical Regiment	1,101	128		
Two Field Hospitals, each	236	75		
Veterinary Hospital	47	11		
Total	14,072	2,358		

On its arrival in western Manchuria it was found to be little better than the regular IJA divisions. Aside from the 32 trucks in the single motor transport company the division still relied on horses for transport. The single armored reconnaissance company with twelve tankettes was something of an improvement, but the reconnaissance regiment was far too small for the vast flat spaces the division was to watch, and lacked fast, wheeled vehicles for scouting or powerful radios to relay their findings. The 75mm Type 38 unmodernized field guns were old and both they and the howitzers had short range, a critical defect in the area of intended employment. A few changes were made to the organization in 1938/39, including the addition of a small anti-tank company with four 37mm Type 94 guns to each regiment, replacing one of the draft transport companies with a second motor company and the splitting off of the battalion gun platoon in each battalion, but the critical tactical communications net was ignored. The infantry regimental signal companies still consisted of two non-motorized platoons, one for wire (five squads, each of ten men and two pack horses), and the other for radio (four squads each of five men and one pack horse carrying a radio). The division was still poorly organized and equipped for the Nomonhan Incident that was to engulf it shortly.

Thus, by the end of 1938 there were 18 regular army divisions that had been subjected to the various reorganization plans and fifteen "wartime" divisions that generally used the organization in effect from about 1930.

The Change to Triangular Divisions

These new divisions differed in one very significant respect from the regular divisions. A decision had been made in 1936 that henceforth all divisions were to be triangular (consisting of three infantry regiments) rather than square (two brigades each of two infantry regiments) as had been the practice up to that point. Implementation of this decision, however, was carried out at an agonizingly slow pace. In fact, all but one of the divisions raised in 1937 and four of the special divisions of 1938 were all mobilized as square divisions and it was only with the other five "wartime" divisions and the one regular division of 1938 that the triangular organization made its substantive appearance.

The triangular division differed from the older square division only in the loss of one infantry regiment and one infantry brigade HQ. The remaining infantry brigade HQ was retitled an infantry group HQ and served to exercise tactical command over the three line elements, although this additional layer of command between the regiments and the division commander did not prove particularly useful, especially since the group HQ had no organic signal assets.

Reorganizing at the Platoon/Company Level

In late 1937 a proposal was adopted that the rifle platoon organization be completely overhauled and fire support for the battalion be increased by adding an anti-tank platoon to the infantry gun platoon to form a heavy weapons company. The latter proposal was never implemented, but the former was adopted with relative alacrity.

The new rifle platoon was to consist of a small headquarters, three rifle squads and a grenadier squad. The platoon HQ was usually made up of the platoon leader (a lieutenant or warrant officer), a liaison NCO (enjoying less authority than his Western equivalent, a platoon sergeant), and a runner. Since the day of a squad armed only with rifles had long since passed, the new rifle squad was usually made up of an NCO and thirteen privates with a total of eleven rifles and one light machine gun. The grenadier squad had a similar strength with ten rifles and three grenade dischargers. The reduced size of the rifle platoons allowed additional permanent personnel to be assigned to the rifle company HQ while keeping within, or at least near, the 200-man ceiling. Thus, 10–12 runners and buglers were added to the company HQ strength, reducing the detail drain on the rifle platoons. These changes gave the rifle company a heavy weapons strength of nine light machine guns and (officially, for the first time) nine heavy grenade dischargers.

The official sanction for grenade dischargers as infantry weapons also brought forth a manual on their use. The most notable feature of the manual was the insistence that the grenadier squad be employed as a whole, and not split up to provide one team per rifle squad. This stricture seems to have been adhered to throughout the war and the grenade discharger quickly established itself as one of the most respected weapons in the Japanese inventory.[4]

4 The grenade dischargers proved so popular that units in the field sometimes added these weapons unofficially, enlarging the grenadier squad from three to four 3-man teams. This was noted in the China

The new rifle company organization was applied to the regular divisions during 1938–39 and to the special divisions in 1939–41.

Force Expansion 1939/40

Ten more "wartime" divisions were raised in February–June 1939, designated the 32nd to 41st Divisions, and known to the Japanese as security divisions. Like the earlier "wartime" divisions, they were deployed primarily in China. Raised as triangular divisions these units had the following components and TO&E strengths and simultaneously-authorized increments:

	Off	WO	MSG	SGT	PVT	Total	Horses
Division Headquarters	36	0	25	13	60	134	22
Infantry Headquarters	n/a	n/a	n/a	n/a	n/a	14	3
Signal Company	7	2	3	21	168	201	27
Reconnaissance Regiment	23	5	12	48	302	390	193
Three Infantry Regiments, each	101	33	22	275	2,484	2,915	188
Field Artillery Regiment	91	20	25	183	1,558	1,877	1,337
or Mountain Artillery Regiment	91	20	25	156	1,752	2,044	1,130
Engineer Regiment	21	6	11	50	504	592	20
Transport Regiment	24	6	15	48	414	507	211
Medical Unit	n/a	n/a	n/a	n/a	n/a	356	46
Two Field Hospitals, each	15	0	24	0	199	238	79
Ordnance Duty Unit	2	1	10	6	102	121	0
Veterinary Unit	4	1	7	1	36	49	11
Authorized Increments:							
each infantry regiment	+10	-	-	+14	+166	+190	-
each artillery regiment	+10	-	-	+14	+166	+190	-
transport regiment	+8	-	-	+11	+131	+150	-

Other elements of the divisions received no such increments and, indeed, the increments included no additional horses, vehicles or weapons so their impact must have been rather small.

The infantry was organized into battalions, each of four rifle companies and a machine gun company. The rifle companies initially used the old organization of four rifle and two LMG sections per platoon, but were quickly reorganized into the new pattern as equipment came available. The machine gun company had three platoons, each with two medium MGs, plus a battalion gun platoon with two Type 92 infantry guns. The infantry regiment had three battalions, plus a signal company and a regimental gun company.

The small reconnaissance regiment consisted of an armored car (or tankette) company, a motorized reconnaissance company and a mounted reconnaissance company. The artillery regiment, field or mountain, consisted of a signal unit and three battalions. In

theater as early as July 1939.

a field artillery regiment each of these battalions consisted of two batteries of 75mm field guns and one of 105mm howitzers; in the mountain regiment it was made up of three batteries of 75mm mountain guns.[5]

At the same time five of the earlier reservist divisions were brought back to Japan and demobilized in late 1939: the 114th in August, the 109th in September, the 101st and 108th in November, and finally the 106th in March 1940. Their infantry was held in the parent depot divisions as reduced-strength (2-battalion) regiments until 1941 when they were formed into independent infantry groups.

The final division raised in 1939, the 24th, was the first product of the triangularization decreed earlier. It was formed from two infantry regiments shed from existing square divisions (the 32nd Regiment from the 8th Division and the 22nd Regiment from the 11th Division) and one newly raised infantry regiment, and as such it joined the ranks of the regular divisions.

In June–July 1940 six more regular divisions were added to the Army's order of battle. Two of the divisions were built around infantry regiments split off from square divisions: the 25th Division (one regiment each from the 4th, 10th and 12th Divisions) and the 28th Division (one regiment each from the 1st, 2nd and 7th Divisions).

The other four new divisions, 54th, 55th, 56th and 57th, were organized to a slightly lower standard of equipment. A fifth such division, the 51st, was formed in September. These units were held at peacetime strength at their homeland depots until needed. The 57th Division left for Manchuria at its full war strength in July 1941 and the 51st Division, also at full war strength, left in September.[6]

One more regular division, the 48th, was created by the addition of the excess 47th Regiment from the 6th Division to the two-regiment Formosa Mixed Brigade in November 1940.

The 1941 Mobilization Plan Order

October 1940 saw the publication of the 1941 Mobilization plan. This document provided the authority for the organizing and equipping of the regular divisions (but not the "wartime" divisions, which were raised under "temporary mobilization orders"). Henceforth, there were to be two broad types of regular divisions known as the Type A and Type B. The Type A division had a lower percentage of short-service conscripts and recalled reservists than the Type B, and also had a slightly higher scale of equipment.

Within these broad categories the Army also distinguished between formations using draft (cart) transport (A1 and B1) and those using pack horses (A2 and B2). There was also a specially-configured organization (A3) optimized for the Manchurian plains. The 23rd Division was the sole formation adopting the A3 organization, while the recently-raised 50-series divisions were placed on the B pattern. The rest of the regular divisions were organized as A units.

This should have been their opportunity to remedy the shortcomings made apparent by the disastrous handling of the 23rd Division at Nomonhan. Indeed, several high level

5 Although raised with an eye towards the China theater by the end of the war only two, the 34th and 40th, were still in that theater, the others having been shipped out, mostly to the Pacific.

6 These were the only two B divisions to mobilize at their full 1941 mobilization plan strength. By the time the others were mobilized from their depots for the combat zones of the south in 1941–44 new organizational structures had been imposed.

studies had been produced in late 1939 and early 1940 that highlighted a number of weaknesses in organization and equipment. The most common complaint seems to have been that the infantry was too lightly-armed, particularly with regard to anti-tank weapons. All agreed that the assignment of only four 37mm AT guns per infantry regiment was completely inadequate. Similarly, artillery support for the division was noted as insufficient, with the 75mm guns being both too few and too light. Further, the need to motorize the artillery was recognized. Services of supply also needed to be motorized, to increase the distance that could be accommodated between the supply head and the division. In fact, the new organization order did not increase the anti-tank capabilities of the division at all, and motorization only modestly. The one area that did see some improvement was division artillery, where the standard allocation went from 9 batteries to 12, in the A divisions at least.

The 1941 mobilization plan established the new-model rifle company as the standard that would remain the baseline configuration through the rest of the war. A company was defined as a company commander, three platoon leaders, a warrant officer administrative specialist, and 185 enlisted. Internal composition was not fixed by TO&E, but strict guidelines were set forth by tactical doctrine. A company was divided into a headquarters group and three rifle platoons, with each rifle platoon being divided, in turn, into four 13-man squads. Three of the squads in each platoon were rifle squads with eleven rifles and a light machine gun, while the fourth was a grenadier squad with ten rifles and three Type 89 heavy grenade dischargers. The platoon HQ comprised the platoon leader, liaison NCO and 2–3 runners, yielding a platoon strength of 54–55 men. The remainder of the company was held in the company HQ, where the enlisted personnel were used as runners, load-carriers and guards or detached to support other elements of the regiment.

This organization, of three platoons each of three LMG-armed rifle squads and a 3-weapon grenadier squad, remained the standard to the end of the war. What changed over time and from unit to unit was the overall strength of the company, which in turn dictated the size of the squads. While the full-strength rifle company had 190 men, those of the later 50-series IMBs had only 114, reducing rifle squad strength to 7 men and grenadier squads to 8. Similarly, IGHQ could simply order a reduction or enlargement of each rifle company in a division or brigade by a set number of privates (usually ten) and leave it to the commander to distribute the losses or gains. In all cases except some units in China, the basic organization remained unchanged.

The A-type machine gun companies, except in the 23rd Division, consisted of three machine gun platoons (each of four squads), an anti-tank rifle platoon (two squads) and an ammunition platoon (three squads). A 13-man squad, machine gun or anti-tank rifle, was provided with a pack horse to carry the weapon and a second for ammunition. In addition, each machine gun platoon HQ was provided a pack horse for gun tools and equipment. The ammunition platoon had three squads, each with nine pack horses for MG ammunition.

The company had a total of 41 ammunition pack horses, of which 39 carried 77,760 rounds of machine gun ammunition, and two carried 200 AP and 105 HE rounds for the 20mm AT rifles. The company was provided with a single light stereoscopic rangefinder but, like the rifle companies, no signal equipment.

The battalion gun platoon, divided into two gun sections and an ammunition section, was provided with three pack horses for each of its 70mm infantry howitzers, along with a further 15 to carry 280 rounds of HE ammunition. The platoon had no signal equipment at all, limiting its useful role in most cases to engagement of targets that were visible from

Table 2.3: Inf – Mob Plan Inf Regts Oct40

	Personnel		Pistols	Carbines	Rifles	Light Machine Guns	Medium Machine Guns	Type 10 Grenade Lnchrs	Type 89 Gren Lnchers	20mm AT Rifles	37mm AT Guns	81mm Mortars	70mm Inf How	75mm Regt Guns	Riding Horses	Pack Horses	Draft Horses	Wagons & Carts
Infantry Regiment, Type A1																		
Regiment Headquarters	209		2	12	259	0	0	1	0	0	0	0	0	0	27	0	101	96
Signal Company	142		0	132	0	0	0	0	0	0	0	0	0	0	3	20	0	0
Three Infantry Battalions, each																		
Battalion Headquarters	223		2	12	26	0	0	1	0	0	0	0	0	0	18	78	37	35
Four Rifle Companies, each	190		0	0	153	9	0	0	9	0	0	0	0	0	0	0	0	0
Machine Gun Company	223		28	0	0	0	12	0	0	2	0	0	0	0	1	58	0	0
Infantry Gun Platoon	69		4	0	0	0	0	0	0	0	0	0	2	0	1	21	0	0
Artillery Battalion																		
Battalion Headquarters	49		1	2	11	0	0	1	0	0	0	0	0	0	2	0	16	16
Regimental Gun Company	156		8	0	0	0	0	0	0	0	0	0	0	4	3	28	46	25
Anti-Tank Company	106		8	0	0	0	0	0	0	0	4	0	0	0	3	16	23	11
Infantry Regiment, Type A2																		
Regiment Headquarters	480		2	26	279	0	0	1	0	0	0	0	0	0	42	269	0	0
Signal Company	143		0	130	0	0	0	0	0	0	0	0	0	0	3	20	0	0
Three Infantry Battalions, each																		

Table 2.3: Inf – Mob Plan Inf Regts Oct40

	Personnel		Pistols	Carbines	Rifles	Light Machine Guns	Medium Machine Guns	Type 10 Grenade Lnchrs	Type 89 Gren Lnchers	20mm AT Rifles	37mm AT Guns	81mm Mortars	70mm Inf How	75mm Regt Guns		Riding Horses	Pack Horses	Draft Horses	Wagons & Carts
Battalion Headquarters	289		2	16	26	0	0	1	0	0	0	0	0	0		22	159	0	0
Four Rifle Companies, each	190		0	0	153	9	0	0	9	0	0	0	0	0		0	0	0	0
Machine Gun Company	223		28	0	0	0	12	0	0	2	0	0	0	0		1	58	0	0
Mortar Platoon	69		8	0	0	0	0	0	0	0	0	4	0	0		0	21	0	0
Artillery Battalion																			
Battalion Headquarters	138		1	6	11	0	0	1	0	0	0	0	0	0		10	56	0	0
Regimental Gun Company	103		8	0	0	0	0	0	0	0	0	0	0	4		5	66	0	0
Anti-Tank Company	127		8	0	0	0	0	0	0	0	4	0	0	0		3	36	0	0
Infantry Regiment, Type A3																			
Regiment Headquarters	306		2	12	335	0	0	1	0	0	0	0	0	0		27	4	132	126
Signal Company	142		0	130	0	0	0	0	0	0	0	0	0	0		3	9	0	0
Three Infantry Battalions, each																			
Battalion Headquarters	265		2	16	26	0	0	1	0	0	0	0	0	0		22	97	44	41
Four Rifle Companies, each	260		7	0	179	9	2	0	9	2	0	0	0	0		0	16	0	0
Machine Gun Company	82		8	0	0	0	4	0	0	0	0	0	0	0		1	18	0	0
Infantry Gun Company	137		8	0	0	0	0	0	0	0	0	0	4	0		1	44	0	0

Table 2.3: Inf – Mob Plan Inf Regts Oct40

	Personnel		Pistols	Carbines	Rifles	Light Machine Guns	Medium Machine Guns	Type 10 Grenade Lnchrs	Type 89 Gren Lnchers	20mm AT Rifles	37mm AT Guns	81mm Mortars	70mm Inf How	75mm Regt Guns	Riding Horses	Pack Horses	Draft Horses	Wagons & Carts
Regimental Gun Company	156		8	0	0	0	0	0	0	0	0	0	0	4	5	28	8	6
Anti-Tank Battalion																		
Battalion Headquarters	49		1	2	11	0	0	0	0	0	0	0	0	0	5	0	17	16
Three Anti-Tank Companies, each	107		8	0	0	0	0	0	0	0	4	0	0	0	3	17	0	0
Infantry Regiment, Type B1																		
Regiment Headquarters	197		2	10	289	0	0	1	0	0	0	0	0	0	24	0	92	88
Signal Company	142		0	130	0	0	0	0	0	0	0	0	0	0	3	20	0	0
Three Infantry Battalions, each																		
Battalion Headquarters	177		2	10	26	0	0	1	0	0	0	0	0	0	16	53	26	24
Four Rifle Companies, each	190		0	0	153	9	0	0	9	0	0	0	0	0	0	0	0	0
Machine Gun Company	144		16	0	0	0	8	0	0	0	0	0	0	0	1	36	0	0
Infantry Gun Platoon	69		4	0	0	0	0	0	0	0	0	0	2	0	1	21	0	0
Regimental Gun Company	139		8	1	0	0	0	0	0	0	2	0	0	2	3	23	12	12

Table 2.3: Inf – Mob Plan Inf Regts Oct40

	Personnel		Pistols	Carbines	Rifles	Light Machine Guns	Medium Machine Guns	Type 10 Grenade Lnchrs	Type 89 Gren Lnchers	20mm AT Rifles	37mm AT Guns	81mm Mortars	70mm Inf How	75mm Regt Guns	Riding Horses	Pack Horses	Draft Horses	Wagons & Carts
Infantry Regiment, Type B2																		
Regiment Headquarters	341		2	15	279	0	0	1	0	0	0	0	0	0	27	200	0	0
Signal Company	143		0	130	0	0	0	0	0	0	0	0	0	0	3	20	0	0
Three Infantry Battalions, each																		
Battalion Headquarters	256		2	10	26	0	0	1	0	0	0	0	0	0	16	134	0	0
Four Rifle Companies, each	190		0	0	153	9	0	0	9	0	0	0	0	0	0	0	0	0
Machine Gun Platoon	144		16	0	0	0	8	0	0	0	0	0	0	0	1	36	0	0
Mortar Platoon	69		4	0	0	0	0	0	0	0	0	4	0	0	1	21	0	0
Regimental Gun Company	187	8	0	0	0	0	0	0	0	0	2	0	0	2	7	80	0	0

Note: the rifles shown in regimental headquarters include 200 as undistributed self-defense weapons for the regiment.

the guns or in defilade behind visible obstructions.

The only difference between the A-1 (draft) and A-2 (pack) infantry battalions was to be found in the battalion HQ. In the draft-type infantry the battalion headquarters had 78 pack horses (four with specialized medical equipment packs and the remainder with general cargo packs) and 35 Type 39 carts for carriage. It carried a further 480 rounds of grenade discharger ammunition (to complement the 144 carried in each rifle company), 21,600 rounds of medium MG ammunition, 200 rounds of 70mm ammunition and 80 anti-tank mines. The A-2 version had similar reserves, but carried entirely on pack horses.

General fire support for the regiment was provided by the regimental (mountain) gun company. The company was provided with 28 pack horses that could either tow the guns or carry them. Surprisingly, although the mountain guns could be carried by pack the ammunition for them could not, this instead being carried by 20 caissons, each with 21 rounds of HE. Each gun was carried by six horses, one each for the tube, breech and tray, cradle, wheels and axle, trail, and shield. The company also had five Type 39 carts for general transport.

Similarly, the anti-tank company was provided with four pack horses for each of its four 37mm guns, but the ammunition was carried in ten caissons, each carrying 96 rounds (a company total of 192 HE and 768 AP). For general transport the company had a single stores wagon.

Almost all of the regiment's signal assets were found in the regimental signal company. This unit was divided into a radio platoon and a wire platoon, each of six squads. The former had five headquarters squads (each with a Mk 5 and a Mk 6 radio) and direct support squad (with seven Mk 6 radios). This permitted the assignment of the little Mk 6 radios to attached units and sometimes even to a few of the rifle companies in the regiment, as well as Mk 5s to form two-net radio teams for the battalion HQs and regimental artillery. The wire platoon had five direct-support squads each with two telephones and six one-kilometer reels of light wire, and a switchboard squad for the regimental HQ. These supported the regimental net, except for the regimental gun company, which had its own Type 92 telephones and wire.

The regiment HQ included a sizable ammunition train using Type 39 carts, these carrying a further 25,920 rounds of medium MG ammunition, 960 rounds for the grenade dischargers, 1,120 for the AT rifles, 480 for the 70mm howitzers, 216 for the 37mm AT guns and 168 for the regimental mountain guns.

In the pack-transport (A-2) version, used only by the 9th, 11 and 25th Divisions, the battalion gun platoon was replaced by an organizationally identical mortar platoon with two 81mm Type 97 mortars. Four pack horses were provided for the mortars, two for accessories and fifteen for ammunition, the latter carrying a total of 276 HE rounds. In the battalion HQ the Type 39 carts were replaced by 81 more pack horses, with ammunition carriage remaining the same except that 180 rounds of 81mm mortar ammunition was carried instead of the 70mm howitzer ammunition.

In the regimental mountain gun company 32 pack horses (carrying 336 HE rounds) replaced the caissons, while in the AT company 16 pack horses (with 144 HE and 528 AP rounds) replaced its caissons. The signal company remained unchanged, except for the substitution of pack horses for carts, while the regiment HQ company replaced its 96 carts with 269 pack horses. This provided carriage for 30,240 rounds for the medium MGs, 1,120 for the grenade dischargers, 1,400 for the AT rifles, 720 for the mortars, 672 for the 37mm AT guns and 204 for the regimental mountain guns.

Aside from the higher-quality personnel in the A formations, the differences between the A and B classes of infantry were not really great. The A-type infantry battalion had a larger machine gun company (with an additional 4-gun MG platoon and a small platoon with two marginal 20mm AT rifles) and separate infantry gun and anti-tank gun companies at the regimental level while the B-type division had only a composite gun company.

The A-type divisions were also the beneficiaries of greater largesse when it came to trains transport. This allowed somewhat larger stocks of ammunition to accompany the unit. The allocations of rifle/LMG and grenade discharger ammunition were identical in the two types of divisions up through regimental level, with A1 and B1 units each carrying 16 rounds per grenade discharger at the company level, an additional 13 rounds per launcher in the battalion trains and a further nine rounds per launcher in the regimental trains. In the case of the medium machine guns and 70mm battalion guns the initial allocation was 6,480 and 140 rounds per weapon respectively at the company and platoon levels, but in the B1 infantry no additional ammunition was carried in the battalion trains, while the A1 battalion had a further 1,800 rounds of MG and 100 rounds of 70mm ammunition in its combat trains.

The 75mm regimental guns were approximately equally provided for with 150 rounds per gun in both cases, but the A1 infantry regiment carried 294 rounds for each 37mm gun, while the B1 regiment had only 192 rounds per gun. Although the A1 and B1 regimental trains elements were similar, the economies imposed at the battalion level meant that the B-type infantry went into combat with significantly less of certain types of ammunition (notably 70mm and 37mm) as compared with the more fortunate A-types.

The situation was similar in the pack infantry units (A2 and B2), which differed from the A1 and B1 type units only in the substitution of pack horses for Type 39 carts, but carried an almost identical ammunition load, albeit at the cost of higher manpower overhead.

The final variant of the infantry regiment envisioned under the 1941 mobilization plan was a specialized version confined to the 23rd Division and the separate 90th infantry regiment. This type added a medium MG platoon and a 20mm AT rifle platoon (each with two 11-man squads) to each rifle company. The allocation of 70mm guns was also doubled to four per battalion, but the most significant change was an increase in the size of the regimental anti-tank component from a 4-gun company to a 12-gun battalion. A somewhat anomalous feature, considering the A3 regiment used draft transport, was the fact that the 37mm AT guns were carried on pack horses, five horses per gun. It is certainly no coincidence that this modified organization, with 24 AT rifles and twelve 37mm AT guns per regiment along with large numbers of other heavy weapons, was allocated to the 23rd Division, which had been savaged the prior year by Soviet tank and infantry forces at Nomonhan.

Divisional Artillery

Fire support for the divisions was provided by A or B type artillery regiments in field or mountain configuration. Regular divisions were always homogeneous, either all-A or all-B, but could be mixed as regards the type of transport. This is explained by the fact that the field artillery, with its complex limber/caisson/gun arrangement was regarded as the least suitable field unit for employment in mountains or jungles, less so than the draft-type infantry with its simple, one-horse carts. In general, the Japanese regarded the combat branch organizations as being suitable for rough terrain in the following descending order:

Pack-type (A2 or B2) infantry
Mountain artillery
Draft-type (A1, A3 or B1) infantry
Field artillery

No attempt was made to marry the two extremes, pack-type infantry and field artillery, within a single division since it would have been extremely wasteful of manpower to use pack-type infantry in the flat terrain that the field artillery was limited to. All other combinations of infantry and artillery, however, were fielded. Thus, the divisions fell into three general categories:

Open terrain type: draft-type infantry and field artillery
Moderate terrain type: draft-type infantry and mountain artillery
Rough terrain type: pack-type infantry and mountain artillery

Through the mid-1930s the field artillery had been equipped almost exclusively with the 75mm modified Type 38 (1905) field gun. The 105mm Type 91 (1931) field howitzer, introduced on a scale of one, and then two battalions per regiment starting in the 1930s. Shortly thereafter, production began on a new field gun, the 75mm Type 95 (1935) and by 1940 the A divisions had been re-equipped with the new guns and howitzers, although the B divisions had to make do with the older 75mm guns and fewer 105mm Type 91s. Once again, the 23rd Division received special consideration in the form of a unique field artillery organization, the A3, that featured a long-barrel 75mm field gun, the Type 90, that was designed specifically for motorized units and was quite effective in its secondary role as an anti-tank gun.

At the field battery level there was little to choose between the A and B organizations, each having four field pieces (75mm guns or 105mm howitzers), six caissons, a stores wagon, a light observation wagon for FO use and 744 rounds of 75mm HE or 248 rounds of 105mm HE. The artillery carried no smoke or other special ammunition. There was also little difference in the battalion HQs, with the A organization having two observation wagons and the B unit only one, but otherwise they were similar. The major difference, apart from the use of older weapons by the B artillery, was in the battalion service battery (or ammunition trains), which contained nine caissons (each with 100 rds of 75mm or 40 rds of 105mm) in the A-type artillery and only six caissons in the B-type, reducing the battalion basic load about 10 percent in B-type units.

In seven of the A-type field artillery regiments one of the field gun battalions was replaced by a heavy field artillery battalion equipped with 150mm Type (Taisho) 4 (1915) howitzers. These elderly weapons threw a useful shell, but were rather short-ranged (9,000 meters) and were cumbersome, having to break down into two loads for transport.

The 23rd Division's artillery was again unique in having a newer model heavy howitzer, the 150mm Type 96 (1936), which had an 11,600 meter range and was designed for tractor-drawn transport as a single unit.

As has been mentioned, the old Type (Meiji) 41 (1908) pack gun was replaced as the standard weapon of the mountain artillery in the 1930s by the newer 75mm Type 94 (1934). This weapon could be rapidly disassembled into six pack loads, the heaviest of which weighed 100kg, but its flat trajectory and unitary charge system limited its usefulness in

Table 2.4: Regular Army Divisional Artillery, from October 1940

	Personnel		Pistols	Carbines	Rifles	Light Machine Guns	75mm Mountain Guns	75mm Field Guns	105mm Mountain Howitzers	105mm Field Howitzers	150mm Field Howitzers		Riding Horses	Pack Horses	Draft Horses		Carts	Wagons		Cars	Trucks	4-ton Tractors	6-ton Tractors	Repair Vehicles
Field Artillery Regiment, Type A1																								
Regiment Headquarters	189		14	3	700	2	0	0	0	0	0		46	11	28		22	2		0	0	0	0	0
Two Field Gun Battalions, each																								
Battalion HQ	183		11	6	0	0	0	0	0	0	0		34	0	71		65	2		0	0	0	0	0
Three Batteries, each	136		12	0	0	0	0	4	0	0	0		35	0	38		0	8		0	0	0	0	0
Service Battery	87		6	0	0	0	0	0	0	0	0		21	0	30		0	9		0	0	0	0	0
Two Howitzer Battalions, each																								
as above, but with 105mm howitzers	678		53	6	0	0	0	0	0	12	0		160	0	217		65	35		0	0	0	0	0
Regimental Service Battery	255		10	3	0	1	0	0	0	0	0		50	0	109		30	24		0	0	0	0	0
Field Artillery Regiment, Type A2																								
Regiment Headquarters	183		14	0	700	2	0	0	0	0	0		44	11	6		0	2		0	0	0	0	0

Table 2.4: Regular Army Divisional Artillery, from October 1940

	Personnel		Pistols	Carbines	Rifles	Light Machine Guns	75mm Mountain Guns	75mm Field Guns	105mm Mountain Howitzers	105mm Field Howitzers	150mm Field Howitzers		Riding Horses	Pack Horses	Draft Horses		Carts	Wagons		Cars	Trucks	4-ton Tractors	6-ton Tractors	Repair Vehicles
Field Gun Battalion																								
as above	678		53	6	0	0	0	12	0	0	0		160	0	217		65	35		0	0	0	0	0
Two Howitzer Battalions, each																								
as above	678		53	6	0	0	0	0	0	12	0		160	0	217		65	35		0	0	0	0	0
Heavy Howitzer Battalion																								
Battalion HQ	205		11	8	0	0	0	0	0	0	0		36	0	90		81	2		0	0	0	0	0
Three Batteries, each	192		13	0	0	0	0	0	0	0	4		42	0	60		0	11		0	0	0	0	0
Service Battery	116		8	0	0	0	0	0	0	0	0		26	0	45		0	14		0	0	0	0	0
Regimental Service Battery	164		14	87	0	2	0	0	0	0	0		0	0	0		0	0		2	41	0	0	0
Field Artillery Regiment, Type A3																								
Regiment Headquarters	118		15	26	700	2	0	0	0	0	0		0	0	0		0	0		2	7	0	0	0
Three Field Gun Battalions, each																								

Table 2.4: Regular Army Divisional Artillery, from October 1940

	Personnel		Pistols	Carbines	Rifles	Light Machine Guns	75mm Mountain Guns	75mm Field Guns	105mm Mountain Howitzers	105mm Field Howitzers	150mm Field Howitzers		Riding Horses	Pack Horses	Draft Horses		Carts	Wagons		Cars	Trucks	4-ton Tractors	6-ton Tractors	Repair Vehicles
Battalion HQ	145		17	45	0	0	0	0	0	0	0		0	0	0		0	0		2	17	0	0	0
Three Batteries, each	113		11	30	0	0	0	4	0	0	0		0	0	0		0	0		2	1	12	0	0
Service Battery	88		7	28	0	0	0	0	0	0	0		0	0	0		0	0		1	13	0	0	0
Heavy Howitzer Battalion																								
Battalion HQ	166		17	64	0	0	0	0	0	0	0		0	0	0		0	0		2	24	0	0	0
Three Batteries, each	148		12	36	0	0	0	0	0	0	4		0	0	0		0	0		2	3	0	13	0
Service Battery	140		10	66	0	0	0	0	0	0	0		0	0	0		0	0		2	31	0	0	0
Regimental Service Battery	329		16	117	0	2	0	0	0	0	0		0	0	0		0	0		2	65	0	0	5
Field Artillery Regiment, Type B																								
Regiment Headquarters	123		11	1	500	2	0	0	0	0	0		34	5	16		13	1		0	0	0	0	0
Two Field Gun Battalions, each																								
Battalion HQ	146		9	5	0	0	0	0	0	0	0		25	0	62		57	1		0	0	0	0	0

Table 2.4: Regular Army Divisional Artillery, from October 1940

	Personnel	Pistols	Carbines	Rifles	Light Machine Guns	75mm Mountain Guns	75mm Field Guns	105mm Mountain Howitzers	105mm Field Howitzers	150mm Field Howitzers	Riding Horses	Pack Horses	Draft Horses	Carts	Wagons	Cars	Trucks	4-ton Tractors	6-ton Tractors	Repair Vehicles
Three Batteries, each	135	11	0	0	0	0	4	0	0	0	34	0	38	0	8	0	0	0	0	0
Service Battery	67	5	0	0	0	0	0	0	0	0	14	0	21	0	6	0	0	0	0	0
Howitzer Battalion																				
as above, but with 105mm howitzers	618	47	5	0	0	0	0	0	12	0	141	0	197	57	31	0	0	0	0	0
Regimental Service Battery	158	7	1	0	2	0	0	0	0	0	25	0	43	10	10	0	0	0	0	0
Mountain Artillery Regiment, Type A1																				
Regiment Headquarters	217	18	0	800	2	0	0	0	0	0	44	25	42	41	0	0	0	0	0	0
Three Mountain Gun Battalions, each																				
Battalion HQ	247	12	8	0	0	0	0	0	0	0	39	8	120	112	0	0	0	0	0	0
Three Batteries, each	203	13	0	0	0	4	0	0	0	0	20	93	0	0	0	0	0	0	0	0
Service Battery	198	8	0	0	0	0	0	0	0	0	12	98	0	0	0	0	0	0	0	0
Regimental Service Battery	327	12	3	0	2	0	0	0	0	0	24	0	180	170	0	0	0	0	0	0

Table 2.4: Regular Army Divisional Artillery, from October 1940

	Personnel		Pistols	Carbines	Rifles	Light Machine Guns	75mm Mountain Guns	75mm Field Guns	105mm Mountain Howitzers	105mm Field Howitzers	150mm Field Howitzers		Riding Horses	Pack Horses	Draft Horses		Carts	Wagons		Cars	Trucks	4-ton Tractors	6-ton Tractors	Repair Vehicles
Mountain Artillery Regiment, Type A2																								
Regiment Headquarters	221		15	4	800	2	0	0	0	0	0		44	25	42		41	0		0	0	0	0	0
Two Mountain Gun Battalions, each																								
as above	1054		59	8	0	0	12	0	0	0	0		111	385	120		112	0		0	0	0	0	0
Mountain Howitzer Battalion																								
Battalion HQ	254		12	9	0	0	0	0	0	0	0		40	8	124		116	0		0	0	0	0	0
Three Batteries, each	228		13	0	0	0	0	0	4	0	0		20	114	0		0	0		0	0	0	0	0
Service Battery	198		8	0	0	0	0	0	0	0	0		12	96	0		0	0		0	0	0	0	0
Regimental Service Battery	328		15	0	0	2	0	0	0	0	0		24	0	177		167	0		0	0	0	0	0
Mountain Artillery Regiment, Type B																								
Regiment Headquarters	174		11	4	600	2	0	0	0	0	0		41	17	42		41	0		0	0	0	0	0

Table 2.4: Regular Army Divisional Artillery, from October 1940

	Personnel		Pistols	Carbines	Rifles	Light Machine Guns	75mm Mountain Guns	75mm Field Guns	105mm Mountain Howitzers	105mm Field Howitzers	150mm Field Howitzers		Riding Horses	Pack Horses	Draft Horses		Carts	Wagons		Cars	Trucks	4-ton Tractors	6-ton Tractors	Repair Vehicles
Three Mountain Gun Battalions, each																								
Battalion HQ	232		9	7	0	0	0	0	0	0	0		26	6	115		107	0		0	0	0	0	0
Three Batteries, each	200		12	0	0	0	4	0	0	0	0		14	91	0		0	0		0	0	0	0	0
Service Battery	172		8	0	0	0	0	0	0	0	0		13	98	0		0	0		0	0	0	0	0
Regimental Service Battery	311		12	3	0	1	0	0	0	0	0		26	0	180		170	0		0	0	0	0	0

Note: all rifles are undistributed weapons for regiment self-defense.

mountains. Nevertheless, the Type 94 quickly became the near-universal weapon of the mountain artillery. In fact, under the 1941 mobilization plan only one divisional mountain artillery battalion was equipped with anything other than a 75mm gun, a battalion of the 28th Mountain Artillery Regiment, which was equipped with 105mm Type 99 pack howitzers.[7]

The mountain artillery battery had six pack horses to carry each gun (plus a spare set of horses), 48 horses to carry ten rounds of HE ammunition apiece, 8 horses to carry the FO equipment set, and six for medical stores. The battery HQ included an observer section to operate the forward observation post and a signal section with telephones and semaphore flags to carry firing orders back to the gun position. The battalion service battery (ammunition trains) provided a further 72 pack horses with 720 rounds of HE ammunition. The regimental ammunition trains carried 2,880 more rounds in Type 39 one-horse transport carts.

For communications artillery batteries relied primarily on telephones. The battalion HQ had 30 signalmen provided with telephones, while the regimental HQ had 90 signalmen under a captain. The regiment had a total of only six No.5 and four No.6 radios plus two No.3C air–ground radios for communications with spotter aircraft, although these appears to have been rarely used.

As events transpired, these assets were not nearly enough to provide effective internal nets and liaison with FOs and supported units. In particular, the radios had short ranges and were few in number. A continuing characteristic of Japanese Army units throughout the Pacific War was that as soon as the front lines changed significantly artillery support virtually disappeared. Field artillery made little contribution to the battle effort.

Divisional Reconnaissance

A battalion-size cavalry or reconnaissance regiment carried out the divisional reconnaissance function. The cavalry regiment consisted of a headquarters, three mounted squadrons and a machine gun squadron.

The basic unit, the mounted squadron, was built around two mounted rifle platoons. Each of these platoons consisted of a two-man HQ, three rifle squads (each of an NCO and ten privates) and a grenadier squad (of an NCO and two 3-man teams). Each of the mounted platoons included not only a riding horse for each man, but also a pack horse for each grenade discharger (carrying the launcher and 48 rounds) and a gun horse and ammunition horse for the light MG of each rifle squad.

Fire support for the regiment was provided by its small machine gun squadron. This unit consisted of a headquarters, two machine gun platoons, an anti-tank platoon and an ammunition platoon. A machine gun platoon was made up of two squads, each of an NCO, eight gun crew and two horse handlers, with two pack horses – one carrying the Type 92 machine gun and the other with 1,375 rounds of ammunition. The anti-tank platoon was also made up of two squads, this time consisting of an NCO, ten gun crew and six horse handlers. The AT squad was provided with six pack horses, four of which carried the disassembled Type 94 AT gun and two each carried 48 rounds of ammunition, usually in the ratio of four AP to one HE. The ammunition platoon provided twenty pack horses for

7 The 28th Division finished the war on Miyako Island, at the southern end of the Ryukyus, with its 28th Mountain Artillery Regiment still with two 12-gun battalions of 75mm Type 94 guns and one 12-gun battalion of 105mm Type 99 howitzers.

MG ammunition and six for 37mm ammunition.

Regimental HQ consisted primarily of a command section, a signal platoon and a very large field trains. The signal platoon had a strength of one officer, 6 NCOs and 41 privates and provided four radio teams (each of an NCO and seven men with a pack horse-carried radio), two with No.3A long-range radios and two with No.5 medium-range units, along with a 10-man optical/telephone section. The bulk of the remaining personnel were given over to the field trains, which provided about 230 pack horses, each with its own mounted leader, to carry rations, tools, baggage, etc. The field trains carried no additional ammunition for the regiment.

Table 2.5: Division Reconnaissance Units, from August 1943

	Personnel	Pistols	Rifles & Carbines	Light MGs	Medium MGs	Heavy Gren Lnchrs	37mm AT Guns	Riding Horses	Pack Horses	Light Tanks	Cars	Trucks	Lt Repair Vehs
Cavalry Regiment													
Regiment Headquarters	366	19	113	0	0	0	0	191	210	0	0	0	0
Two Rifle Squadrons, each	137	3	109	6	0	4	0	138	20	0	0	0	0
Machine Gun Squadron	139	3	94	0	4	0	2	134	57	0	0	0	0
Reconnaissance Regiment													
Regiment Headquarters	112	13	95	0	0	0	0	0	0	0	2	17	1
Two Motorised Rifle Squadrons, each	190	3	148	6	2	6	1	0	0	0	3	14	0
Light Tank Squadron	85	15	64	0	0	0	0	0	0	12	2	3	0

The divisional cavalry regiment was somewhat small for the role of reconnaissance for a large division. In fact, it mustered a line strength of only six platoons. The heavy weapons squadron was also too weak to support delaying and security missions. The choice of pack-carried AT guns is particularly unusual for a reconnaissance force, since their time-into-action must have been quite slow.

These problems were compounded by limited communications. In particular, since the only link between the platoons and the squadron HQ was by dispatch rider this limited the distance the platoons could reconnoiter from their parent HQ without compromising their mission of providing timely information.[8]

8 The cavalry regiments did not, indeed, prove very popular. The cavalry regiments of the 9th and 11th divisions were disbanded in July 1941, that of the 12th Division in October 1942, the 18th in April 1943, the 40th and 116th Divisions in May 1943 and the 15th Division in March 1944. In addition, the16th Division's cavalry regiment was converted to a reconnaissance regiment in July 1939, followed by that of the 5th Division in November 1940, and of the 1st and 10th Divisions in October 1942. Nevertheless, and inexplicably, small numbers of new divisions were raised with cavalry regiments. The conversion of cavalry to motorized reconnaissance was not always accompanied by an understanding of the role of scouting. The 16th Reconnaissance Regiment in the Philippines with its parent division in 1944–45 spent its entire time as the guard unit for the 14th Area Army HQ until destroyed in April 1945.

Divisions that operated in flatter terrain had begun receiving partially motorized reconnaissance regiments (actually battalions) in the mid-1930s. A reconnaissance regiment consisted of a headquarters, a mounted cavalry squadron, two motorized rifle squadrons, a tankette squadron and a motor transport squadron. The mounted squadron was similar to that of the cavalry regiment but added a third grenade discharger team to each platoon.

The motorized rifle squadron consisted of two rifle platoons, a machine gun platoon and an anti-tank squad. A rifle platoon was made up of three rifle squads and a grenadier squad. A rifle squad consisted of an NCO, eight riflemen and a four-man light MG team, while a grenadier squad had an NCO and three 3-man teams. The platoon HQ was made up of the platoon leader, a liaison NCO and a runner. The machine gun platoon had two squads, each of an NCO and eight privates with a Type 92 MG. The squadron's AT squad was made up of an NCO and ten gun crew. The motorized rifle squadron had no motor vehicles of its own, relying on the assets of the motor transport squadron for mobility.

The tankette squadron was a small unit of a headquarters, two line platoons and a trains platoon. Each line platoon manned three tankettes or, more rarely, armored cars. The HQ was provided with a tankette for command use and a field car for administrative duties, while the trains platoon operated two trucks for fuel, ammunition and supplies.

The motor transport squadron was divided into two platoons, one to support each motorized rifle squadron. Each rifle squad and grenadier squad was provided with a truck. Each MG squad received a truck to carry the crew, MG and 8,640 rounds of ammunition. Similarly, the anti-tank squad was provided a truck to tow the gun and carry the crew and 120 rounds of ammunition. Further, each company HQ was allocated a field car and the company trains four trucks.

The regiment HQ provided signal, maintenance and trains support for the unit. The signal platoon provided four radio squads, two long range and two medium range, each with an NCO and seven privates (including a driver) with a truck. The 20-man repair platoon was provided with two trucks and a repair vehicle. The trains group provided one truck for ammunition, 1 for baggage, and 2 for rations, along with other trucks for medical and general supplies.

As could be expected, the 23rd Division had its own reconnaissance regiment TO&E. This, a fully motorized formation, differed from the normal reconnaissance regiment in that it substituted a second tankette squadron for the mounted rifle squadron.

The reconnaissance regiments were slightly more capable in the scouting role than the cavalry regiments, if for no other reason than the addition of two platoons of tankettes. Communications for the reconnaissance role mission and heavy weapons for security and delaying roles, however, were still weak.

A small number of divisions intermittently included tankette companies, usually retained following the disbanding of the reconnaissance regiments. Thus, they generally followed the organization of the tankette squadron of the reconnaissance regiment. In February 1943 the 21st Division reported that its tankette company consisted of 56 men with five tankettes (two with 37mm guns), three trucks and a staff car. There was no mention of trailers for the tankettes and presumably the resupply role for which they had originally been designed had been abandoned by that point.

Table 2.6: Divisional Transport Regiments, from October 1940												
	Personnel	Pistols	Carbines	Rifles	Light MGs	Riding Horses	Pack Horses	Draft Horses	Carts	Cars	Trucks	Lt Repair Vehs
Transport Regiment, Type A1												
Regiment Headquarters	82	10	21	300	0	34	0	12	12	2	0	0
Draft Transport Battalion												
Battalion HQ	22	5	7	0	0	14	0	1	1	0	0	0
Three Draft Companies, each	398	3	32	0	1	42	0	252	238	0	0	0
Motor Transport Battalion												
Battalion HQ	80	12	57	0	0	0	0	0	0	3	5	3
Three Motor Transport Companies, each	145	12	113	0	1	0	0	0	0	4	38	0
Transport Regiment, Type A2												
Regiment Headquarters	82	11	14	500	0	34	0	1	1	1	0	0
Draft Transport Battalion												
Battalion HQ	22	5	7	0	0	14	0	1	1	0	0	0
Four Draft Companies, each	398	3	32	0	1	42	0	252	238	0	0	0
Pack Transport Battalion												
Battalion HQ	23	5	7	0	0	14	2	0	0	0	0	0
Four Pack Companies, each	431	3	38	0	1	48	303	0	0	0	0	0
Motor Transport Company	150	12	130	0	1	0	0	0	0	4	39	1
Transport Regiment, Type A3												
Regiment Headquarters	101	5	75	0	0	0	0	0	0	3	5	4
Six Motor Transport Companies, each	145	12	111	0	1	0	0	0	0	4	38	0
Transport Regiment, Type B1												
identical to Type A1												
Transport Regiment Type B2												
Regiment Headquarters	82	11	14	400	0	34	0	12	12	1	0	0
Draft Transport Battalion												
Battalion HQ	22	5	7	0	0	14	0	1	1	0	0	0
Four Draft Companies, each	398	3	32	0	1	42	0	252	238	0	0	0
Pack Transport Battalion												
Battalion HQ	23	5	7	0	0	14	2	0	0	0	0	0
Three Pack Companies, each	431	3	38	0	1	48	303	0	0	0	0	0
Note: rifles are undistributed reserve for unit self-defense.												

Service Support for the Division

For logistics support each division was provided with a transport regiment. This was a large unit that, except in the unique 23rd Division, consisted of a draft transport battalion and a pack or motorized battalion. The reliance on one-horse carts and pack horses guaranteed mobility in rough terrain, but imposed a very high manpower overhead.

Each regiment had three or four draft companies each with 238 of the Type 39 single-axle carts. These carts had a rated capacity of 187kg, giving the draft company a total capacity of 49 tons. Flatland-type divisions also had three motor transport companies (each with 38 two-ton trucks), while the mountain/jungle-type divisions had three or four pack transport companies, each with 303 pack horses.

Within each regiment, three companies were given over to the transport of ammunition: one exclusively for infantry ammunition, one for artillery ammunition, and one for a combination of the two. As its basic load, for instance, an "A1-type" transport regiment supporting a division with two battalions each of 75mm field guns and 105mm howitzers carried a basic load of 218,000 rounds of medium MG ammunition; 7,040 rounds of grenade discharger; 912 rounds of 70mm; 1920 rounds of 37mm; 2,100 rounds of AT rifle; 840 rounds of regimental gun; 2,792 rounds of 75mm and 1,126 rounds of 105mm ammunition.

Table 2.7: Division Base Elements, from October 1940

	Personnel	Pistols	Carbines	Rifles	Light MGs	Medium MGs	Light Gren Lnchrs	Riding Horses	Pack Horses	Draft Horses	Carts	Wagons	Cars	Trucks	Lt Repair Vehs
Division Headquarters	300	6	104	56	2	2	1	62	0	0	0	0	8	27	0
Infantry Group Headquarters	86	0	34	32	1	0	1	12	5	8	7	0	0	0	0
Artillery Group Headquarters	164	22	2	20	2	0	0	49	0	29	18	3	0	0	0
Signal Unit	239	0	207	0	0	0	0	6	0	36	35	0	0	0	0
Engineer Regiment															
Regiment HQ	94	1	2	33	0	0	0	9	3	14	13	0	0	0	0
Three Engineer Companies, each	254	0	0	236	1	0	0	5	17	0	0	0	0	0	0
Materiel Platoon	42	0	8	29	0	0	0	1	0	18	17	0	0	4	0
Anti-Gas Unit (A)	187	0	1	0	0	0	0	0	0	0	0	0	1	18	0
or															
Anti-Gas Unit (B)	227	1	1	0	0	0	0	10	58	30	30	0	0	0	0
Ordnance Duty Unit	112	0	25	20	0	0	0	0	0	0	0	0	1	8	2
Medical Unit															
Unit Headquarters	267	7	12	0	0	0	0	19	0	62	60	0	0	0	0

Table 2.7: Division Base Elements, from October 1940

	Personnel	Pistols	Carbines	Rifles	Light MGs	Medium MGs	Light Gren Lnchrs	Riding Horses	Pack Horses	Draft Horses	Carts	Wagons	Cars	Trucks	Lt Repair Vehs
Three Stretcher Companies, each	188	21	0	0	0	0	0	1	0	0	0	0	0	0	0
Ambulance Company	278	3	21	0	0	0	0	4	0	36	26	0	0	0	0
Three Field Hospitals, each	278	1	11	20	0	0	0	12	0	67	64	0	0	0	0
Fourth Field Hospital	284	1	11	20	0	0	0	12	0	71	68	0	0	0	0
Veterinary Hospital (A)	119	1	14	10	0	0	0	9	0	33	31	0	0	0	0
or															
Veterinary Hospital (B)	113	1	12	10	0	0	0	9	0	29	27	0	0	0	0
Water Purification Unit	239	4	76	20	0	0	0	0	0	0	0	0	1	28	0

Also supporting each division was an engineer regiment of three companies and a materiel platoon. Each company, in turn, was divided into four engineer platoons (each of four squads) and a materiel platoon. The engineer platoons had very little in the way of equipment, only small hand tools. Items such as two-man saws, shovels, sledge hammers, etc, were handled by the company materiel platoon, which had thirteen pack horses to carry equipment and four for chemical and general supplies. The regiment materiel platoon had the only power tools in the unit, along with a single generator truck and a compressor truck. Coordination within the regiment was by means of four radio teams, each with a Mk 5 radio on a pack horse, held by regiment HQ.

Officially the engineers were equipped almost exclusively with rifles, only one light machine gun being allocated per company. In practice units in the field picked up additional weaponry, with each platoon having two or three light MGs and each company a couple of grenade dischargers. For instance the in 3rd Engineer Regiment operating in China in 1943 the regimental material platoon carried five medium MGs for distribution if needed.

The division signal unit was made up of two wire platoons, a radio platoon, and an equipment platoon. Each of the wire platoons had four squads, each with a cart, 10km of Type 92 insulated wire, four telephones and a switchboard. The radio platoon was made of ten squads, eight of which manned Mk 3A medium-wave radios with hand-operated generators and the other two special ground–air radios for rarely-used liaison with aircraft.

The divisional ordnance duty unit was responsible for mechanical maintenance within the division and was provided with two specialized vehicle repair trucks and one radio repair truck, as well as trucks for spare parts and tools.

The anti-gas unit was made up of three platoons (each of three squads) plus a trains group. The large medical unit was designed exclusively to provide limited first aid and transport patients to the divisional rear. Each stretcher-bearer company provided 36 stretcher teams, while the ambulance company operated 26 ambulance carts. The division also had four large field hospitals, each capable of caring for 500 patients. One of the field hospitals in each division was equipped with an X-ray machine and was slightly larger.

The 1941 Division

The following table shows a summary of the 8th Division as mandated by the 1941 mobilization plan, typical of flatland-type divisions for 1940/41.[9] The total of 24,000 men is quite high by international standards for a division, and it should be remembered that this was actually the most economical of the divisions in terms of manpower. A mountain-type division would use pack-type infantry regiments (at 5,006 men apiece), a 3,616-man mountain artillery regiment, a 1,011-man cavalry regiment and a 3,123-man transport regiment to give a division total of 27,644 men.

For all its great size, the Japanese division of 1940/41 was not impressively armed. Armament at the platoon level, with three light machine guns and three 50mm grenade dischargers, was certainly adequate, but above that level the commanders had little reserve of firepower with which to influence the battle. The rifle company commander had no heavy weapons. The battalion commander had twelve medium MGs, but these were usually distributed to support the line elements. His two 70mm howitzers were useful weapons, but were far too few in number to do other than engage point-type targets and were ill-served by signals capabilities.

The regiment commander had four anti-tank guns and four elderly 75mm mountain guns. Again, these were handy weapons, but the firepower of four 75mm guns mostly in the direct-fire mode really was not sufficient to support a regiment. The divisional field artillery regiment was theoretically a potent force, at least in those divisions that had 105mm howitzers, but the lack of communications rendered it ineffectual in most circumstances.

Anti-tank firepower was another weakness of the division. Sufficient for operations in China and early offenses against Western colonial holdings, the lack of effective AT weapons took more and more of a toll as the war progressed. Certainly, the Kwantung Army would have been very hard pressed even in 1940 against Soviet tank forces with its allocation of only twelve 37mm guns per division. Air defense for the field forces was another weakness. Although AA mounts were provided for the medium MGs of the machine gun companies, the divisions did not include a single dedicated anti-aircraft gun.

While these organizational patterns applied specifically only to the regular divisions, the "wartime" divisions adopted TO&Es that mimicked the regular "B" divisions closely. As a result they were often informally referred to as "B" divisions, but in fact were a separate group.

The neat organizational patterns prescribed by the mobilization plan, however, did not last long. It was rare, for instance, to find a division in Manchuria or Japan with its full medical unit and more than one field hospital. Units in China usually had two field hospitals, but rarely more.

Nominal Organization Flat-Land Type (8th) Infantry Division, from October 1940

Division Headquarters (312) [2 LMG, 2 MG]

Signal Unit (232)

Infantry Group Headquarters (86) [1 LMG]

9 The strengths given are based on the mobilization plan. Actual strengths of this type of division on activation were slightly different, such as 4,487 men in each infantry regiment, 3,254 men in the field artillery regiment, and 898 men in the engineer regiment.

The heavy weapons of the motor-type divisions were carried in trucks. Seen here is a battalion gun in the Manila victory parade in 1942.

- Artillery Group Headquarters (164) [2 LMG]
- Reconnaissance Regiment (774)
  - Regiment HQ (140)
  - Mounted Rifle Squadron (141) [6 LMG, 6 GL]
  - Two Motorised Rifle Squadrons, each (168) [6 LMG, 2 MG, 6 GL, 1 37mm AT]
  - Tankette Squadron (52) [7 tankettes]
  - Motor Transport Squadron (105)
- Three Infantry Regiments, each (4,487)
  - Regiment HQ (209)
  - Signal Company (142)
  - Three Infantry Battalions, each (1,275)
    - Battalion HQ (223)
    - Four Rifle Companies, each (190) [9 LMG, 9 GL]
    - Machine Gun Company (223) [12 MG, 2 ATR]
    - Infantry Gun Platoon (69) [2 70mm Inf How]
  - Artillery Battalion (311)
    - Battalion HQ (49)
    - Regimental Gun Company (156) [4 75mm Regt guns]
    - Anti-Tank Company (106) [4 37mm AT]
- Field Artillery Regiment (3,254)
  - Regiment HQ (159) [2 LMG]
  - Field Gun Battalion (678)

        - Battalion HQ (183)
        - Three Batteries, each (136) [4 75mm guns]
        - Service Battery (87)
    - Two Field Howitzer Battalions, each (678)
        - Battalion HQ (183)
        - Three Batteries, each (136) [4 105mm how]
        - Service Battery (87)
    - Heavy Field Howitzer Battalion (897)
        - Battalion HQ (205)
        - Three Batteries, each (192) [4 150mm how]
        - Service Battery (116)
    - Regimental Service Battery (164)
- Engineer Regiment (898)
    - Regiment HQ (94)
    - Three Engineer Companies, each (254) [1 LMG]
    - Materiel Platoon (42)
- Transport Regiment (1,801)
    - Regiment HQ (70)
    - Draft Transport Battalion (1,216)
        - Battalion HQ (22)
        - Three Draft Companies, each (398) [1 LMG]
    - Motor Transport Battalion (515)
        - Battalion HQ (80)
        - Three Motor Companies, each (1451) [1 LMG]
- Anti-Gas Unit (180)
- Ordnance Duty Unit (108)
- Medical Unit (1,068)
- Three Field Hospitals, Each (278)
- Field Hospital (284)
- Veterinary Hospital (117)
- Water Purification & Supply Section (226)

Total: 23,891 men

The China Experiments with Divisional Armor

On 13 July 1940 a short-lived experiment was tried when the 15th, 17th and 22nd divisional Reconnaissance Regiments were disbanded and converted into the 15th, 17th and 22nd Divisional Tank Units. These divisions comprised the 13th Army in southern China at the time and this represented the first deployment of tank units organic to an infantry division for the Japanese.

The divisional tank unit was a small battalion consisting of a headquarters, a light tank company, two medium tank companies and a trains company. A light tank company had nine light tanks in three platoons plus one tank in company HQ. The medium companies each had three platoons (each five medium tanks) and a company HQ (two medium and two light tanks). The regiment HQ was provided with one light and two medium tanks for command, while the trains company held three light and four medium tanks as reserve equipment.

These organizations did not last long. Under pressure to form additional independent armored units IGHQ ordered the three divisional tank units disbanded for conversion to tank regiments on 11 June 1941. At the same time, and to the same end, they ordered the dissolution of the 21st, 32nd, 33rd, 35th and 36th Reconnaissance Regiments, stripping those divisions of their scouting elements. All these divisions were in China at the time, where reconnaissance was apparently not considered essential.

The Motor-Type Divisions

Three A-Type divisions were selected for special training in amphibious warfare in October 1940, the 5th, 48th and the Guards Divisions. In these divisions the horses and carts were removed (except in the medical units) and replaced by trucks, which allowed the divisional transport regiment to be reconfigured as three motor transport companies. Along with the horses, the associated personnel were also removed from the strength and replaced by a much smaller number of drivers and mechanics. In addition, the number of rifle companies per battalion was reduced from four to three. The field artillery regiments were reconfigured as three 462-man battalions, two of them each with eight 75mm field guns and four 105mm howitzers, and one with twelve 75mm mountain guns. Thus, unit strengths were reduced considerably in November 1940:

	Off	EM	Cars	Trucks	Horses
Division Headquarters	57	239	7	25	0
Signal Unit	7	180	0	17	0
Reconnaissance Regiment	20	398	29	16	0
Infantry Group HQ	4	83	7	15	0
3 Infantry Regiments, each	114	2,466	4	87	0
Artillery Regiment	63	1,423	8	141	0
Engineer Regiment	29	732	1	47	0
Transportation Regiment	31	463	4	120	0
Medical Unit	139	684	0	0	88
Other	91	40	1	14	0

The 5th and Guards divisions were used in the invasion of Malaya in 1941/42, where they demonstrated their agility in the drive down the peninsula. Thereafter the 5th Division was moved to the Dutch East Indies, where it finished the war, while the Guards Division (redesignated 2nd Guards Division) ended the war in Sumatra. The 48th Division was used in the invasion of the Philippines, and also ended the war in the East Indies. Thus, despite the investment of scarce motor vehicles, none of these divisions saw combat after mid-1942.[10]

10 Other divisions selected for amphibious training in 1940 (18th) and 1941 (55th and 56th) were not motorized.

The 1941 Reworking of Wartime Divisions for China

Revisions to the organization of the special divisions, some minor and some not-so, began in early 1941. The 15th Division was reorganized with the following strengths in mid-March 1941:

	Men	Horses
Division headquarters	452	64
Signal Unit	233	69
Infantry Group Headquarters	92	26
Three Infantry Regiments, each	3,597	628
Field Artillery Regiment	1,975	1,506
Engineer Regiment	913	119
Transport Regiment	1,166	630
Ordnance Duty Unit	122	0
Medical unit	356	46
Three Field Hospitals, each (added 3/43)	301	83
Veterinary Hospital	114	43

Each of the infantry regiments consisted of a headquarters, signal company (one each radio and wire platoons), infantry gun company (four 75mm) and three battalions, each, in turn, made up of four rifle companies and a machine gun company.

Shortly thereafter, most of the remaining "wartime" divisions in China were subjected to minor revisions in infantry organization. On 22 May 1941 IGHQ authorized a new TO&E strength for the infantry regiments of the 17th, 21st, 32nd, 35th, 36th, 37th and 41st Divisions. These had a TO&E and authorized strength of 117 officers, 33 warrant officers, 27 master sergeants, 289 sergeants and 2,724 privates for a total of 3,190 men with 309 horses. This represented an addition of 85 men to the former authorized strengths and, more importantly, a 64 percent increase in the number of horses, making the unit much more mobile.

Table 2.8: Typical (17th Division) "Special" Infantry Regiment, from July 1941

	Personnel	Pistols	Carbines	Rifles	Light MGs	Medium MGs	Light Gren Lnchrs	Heavy Gren Lnchrs	70mm Inf Howitzers	75mm Type 41 Mountain Guns	Riding Horses	Pack Horses	Draft Horses	Carts
Regiment Headquarters	24	0	0	*80	0	0	1	0	0	0	13	0	0	0
Signal Platoon	94	0	89	0	0	0	0	0	0	0	11	16	0	0

Table 2.8: Typical (17th Division) "Special" Infantry Regiment, from July 1941

	Personnel	Pistols	Carbines	Rifles	Light MGs	Medium MGs	Light Gren Lnchrs	Heavy Gren Lnchrs	70mm Inf Howitzers	75mm Type 41 Mountain Guns	Riding Horses	Pack Horses	Draft Horses	Carts
Mounted Platoon	34	1	0	33	2	0	2	0	0	0	35	0	0	0
Infantry Gun Company	139	6	0	67	0	0	0	0	0	3	19	18	6	6
Three Infantry Battalions, each														
Battalion Headquarters	10	0	0	0	0	0	1	0	0	0	4	0	0	0
Four Rifle Companies, each	173	12	0	165	6	0	0	9	0	0	0	0	0	0
Machine Gun Company	172	20	0	83	0	8	0	0	2	0	21	33	0	0
Undistributed Troops**														
Combat Trains Personnel	213	0	74	0	0	0	0	0	0	0	26	0	48	44
Medical Personnel	68	0	0	0	0	0	0	0	0	0	0	0	0	0

** rifles are undistributed weapons for unit self-defense*

*** trains and medical personnel are distributed within the regiment as needed, mostly to regimental and battalion HQs*

In conformity with the new directives the divisions issued their own detailed organization orders. That for the 17th Division's infantry regiments (shown in Table 2.8) is representative. Closely mimicking the "B-type" divisions, the organization is interesting on a number of points. The most striking is the complete absence of anti-tank weapons anywhere in the regiment. This was made possible by the Chinese lack of tanks, and in any event did not really represent much of a departure from the "B-type" divisions, which had only two guns per regiment themselves. Equally interesting was the retention of the old-style rifle company wherein each platoon consisted of three rifle squads, two LMG squads and a grenadier squad.

Ammunition carriage for the new China-type "wartime" division infantry differed somewhat in emphasis from its regular counterparts. Larger stocks of grenade discharger ammunition were carried (20 rounds per gun with the weapons and a further 33 per gun in regiment HQ), while the 107 rounds per gun for the 70mm battalion howitzers was lower than the 140–240 rpg for the A and B type regiments. Ammunition loads for other weapons were about the same as the regular infantry.

Infantry communications were improved, as the regimental signal company was now provided six Type 94 Mk5 radios for use at battalion and regiment HQs, and twelve Type

94 Mk6 radios.

The other component of the divisions to be reorganized was the transport regiment, which was strengthened somewhat to 652 men with 212 horses and 104 wagons. By this time, however, the bulk of the "wartime" divisions seem to have lost their cavalry or reconnaissance regiments, a process begun in late 1940. Nevertheless, the additional mobility conferred on the special division as a whole served, intentionally or coincidentally, to free the regular divisions in China for employment elsewhere.

A typical reorganized "wartime" division (in this case the 37th) had the following authorized strengths:

	Off	WO	MSG	SGT	PVT	Total	Horses
Division Headquarters	n/a	n/a	n/a	n/a	n/a	139	57
Infantry Group Headquarters	n/a	n/a	n/a	n/a	n/a	67	3
Signal Unit	7	2	3	21	168	201	27
Three Infantry Regiments, each	119	33	27	289	2,724	3,190	309
Mountain Artillery Regiment	66	18	25	130	1,449	1,688	942
Engineer Regiment	21	6	11	50	504	592	20
Transport Regiment	29	8	15	60	540	652	212
Medical Unit	n/a	n/a	n/a	n/a	n/a	356	46
Field Hospital	15	0	24	0	199	238	79
Field Hospital	16	0	25	0	204	245	83
Ordnance Duty Unit	n/a	n/a	n/a	n/a	n/a	120	0
Veterinary Hospital	4	1	7	1	36	49	11
Water Purification & Supply Unit	11	1	25	2	157	196	0
Total	14,113	2,380					

Preparing for the Move South

By November of 1940 the divisions envisioned in the 1941 mobilization plan had all been formed, although not mobilized to war strength. The only exception was the 29th Division, which was not raised until 1 April 1941 using excess infantry regiments shed from the 3rd, 14th and 16th Divisions in the triangularization process. As a result, no all-new divisions were formed between November 1940 and February 1942.

Preparations began in August 1941 for a major change in strategic direction, from the northward orientation of the Kwantung Army to a southerly thrust adopted as a contingency by IGHQ. The huge, bulky divisions envisioned as necessary against the fortifications of the Soviets in the Maritime Province were obscenely wasteful of scarce shipping. The divisions earmarked for employment in the south were put on an immediate diet and new TO&Es were issued, reducing strength by almost half. The first divisions to feel the ax were the 2nd, 16th and 55th Divisions all mobilized on 16 September and the 56th Division mobilized on 15 November, all on "special temporary organization orders." The most visible change for all these divisions was a reduction in the number of rifle companies per infantry battalion from four to three, saving almost 600 men per regiment. In addition, ten men (and nine

A rifle section, still equipped with the old Type 11 light machine gun, in action in Mindanao in 1942.

rifles) were shaved from each of the remaining rifle companies.

Further reductions were more severe in the two regular "A" divisions, 2nd and 16th, than in the two "B" divisions, the latter being already at lower strength levels. In the "A" divisions the machine gun companies were reorganized into four smaller platoons (each of a lieutenant and two 11-man gun squads), an 11-man 20mm ATR squad, and a 25-man ammunition section. With a headquarters of 14 men this gave the company a total strength of 142 with eight medium MGs and one anti-tank rifle. Each machine gun squad now was made up of a sergeant, four gun crew, four ammunition bearers, and two handlers each with a pack horse (one for the gun and one for ammunition). The ammunition section had two pack horses and ten Type 39 one-horse carts. The company's ammunition basic load was reduced but with fewer guns to feed they still had 6,480 rounds per gun, the same as the original A-1 type.

At the same time the battalion and regiment headquarters were reduced in size by half mainly through elimination of pack horses and handlers, the regimental artillery battalion HQ disbanded (leaving the infantry gun and AT companies directly subordinate to the regiment commander), the AT company reduced to draft-only, and other economies imposed. Most notably, the 96 Type 39 carts in the regimental HQ, and their carters and associated personnel, were replaced by eleven "Type B" (4x2) cargo trucks. The regimental ammunition reserves, carried by the regimental HQ, were little changed (on a per-weapon basis) for rifles, machine guns and grenade dischargers, but for the battalion guns it fell from 120 rounds per gun to 70, for regimental guns from 42 per gun to 30, and for the anti-tank guns from 54 per gun (42 AP and 12 HE) to 27 (21 AP and 6 HE).

The streamlining of the two non-motorized "A" divisions continued at the divisional level. Notably, the heavy howitzer battalion was deleted from the artillery regiment, the remainder being configured as one tractor-drawn 75mm gun battalion and two horse-drawn 105mm howitzer battalions. The mounted squadron was dropped from the reconnaissance regiment, and the transport regiment cut in half. A comparison of the 16th Division as envisioned by the 1941 mobilization plan (see pages 92–94) and as it (and the identical 2nd Division) actually mobilized for war clearly shows the results of these cuts:

	Off	WO	MSG	SGT	PVT	Horses	Mtr Veh
Division Headquarters	48	2	32	27	194	28	0
Signal Unit	5	1	4	18	159	9	15
Reconnaissance Regiment	23	4	11	61	340	0	*51
Infantry Headquarters	7	0	4	11	65	0	7
Three Infantry Regiments, each	94	15	31	281	2,460	409	11
Field Artillery Regiment	80	12	34	142	1,498	896	81
Engineer Regiment	23	3	14	69	766	38	29
Transport Regiment	23	0	25	35	666	296	87
Medical Unit	34	0	28	27	402	0	36
Two Field Hospitals, each	23	0	36	1	182	0	21
One Field Hospital	24	0	37	1	185	0	22
Ordnance Duty Unit	2	1	15	4	101	0	13
Veterinary Hospital	5	1	6	1	40	10	4
Total**	579	69	339	1,240	11,978	2,504	399

** Includes 16 tankettes*

*** Total does not include water purification and supply unit of 196 men with 25 horses often included for specific operations*

The 16th Division retained this organization for the invasion of the Philippines and, indeed, through the rest of the war. The 2nd Division, on the other hand, reorganized again on 31 January 1942 shortly before embarking on the invasion of Java.

In this case the infantry regiment was further reduced in strength by four sergeants and 158 privates, presumably reflecting a loss of 18 men in each rifle company. The rifle companies still had the standard organization, with three platoons each of three rifle squads (each with a light MG) and a grenadier squad (three grenade dischargers). The machine gun company had four platoons each with two medium MGs, plus a single anti-tank squad with an AT rifle. The battalion gun platoon had two 70mm howitzers, the anti-tank company four 37mm AT guns, and the infantry gun company four 75mm mountain guns. The signal company provided eight Mk 5 and twelve Mk 6 radios, along with 26 telephones. Apparently this organization worked, because no official changes were made to the 2nd Division organization through the rest of the war.[11]

11 Although not an organizational change it is worth noting that in September 1942 each infantry regiment in the 2nd Division did receive an additional allotment of 10 Type 97 sniper rifles, 16 Type 89 grenade dischargers, 3 Type 94 90mm light trench mortars, 13 AT rifles, and no fewer than 180 Type 100 rifle

The September 1941 infantry regiment organization of 2,881 men and three rifle companies per battalion, was also applied to some of the divisions headed to Burma. These included the 56th Division in late 1941, the 53rd Division in November 1943 (with an additional 36-man increment) and the 49th Division in May 1944.

The other B division headed south, the 55th, was also slimmed down, this time on a unique organization table. The TO&E adopted by the division for its infantry regiments on 16 September was:

	Off	WO/NCO	PVT	Total
Regiment HQ	16	19	73	108
Signal Company	3	17	113	133
Regimental Gun Company	3	10	114	127
Three Battalions, each				
Battalion Headquarters	7	13	87	107
Three Rifle Companies, each	4	17	157	178
Machine Gun Company	5	13	126	144
Infantry Gun Platoon	1	6	63	70
Total	97	295	2,541	2,933

The rifle companies used the now-standard organization with three rifle platoons, each of three rifle squads (each with a light machine gun) and a grenadier squad (with three Type 89 grenade dischargers). The machine gun companies were each built around four 24-man platoons. The regimental gun company was divided into a 43-man headquarters, a 36-man mountain gun platoon with two Type 41 regimental guns, a 28-man anti-tank platoon with two Type 94 AT guns and a 20-man ammunition platoon. Apparently this was found satisfactory, for the 55th Division retained this TO&E through the rest of the war in Burma.

The 144th Infantry Regiment, which was temporarily split off from the division to form the South Seas Detachment for the capture of Rabaul and Lae, also used this organization but with its authorized transport reduced from the normal 11 trucks (in the regiment HQ) and 435 pack horses, to 11 trucks and 295 pack horses. In fact, however, the regiment deployed for the move south not only with its authorized 11 trucks but with 359 horses, comprising 37 in each MG company, 23 in each battalion gun platoon, and 33 in each battalion HQ, along with 35 for the regimental gun company, 22 for the signal company and 17 in the regiment HQ.[12]

The "A" and "B" divisions of the Kwantung Army in Manchuria, on the other hand, retained their cumbersome original configurations based on the 1941 mobilization plan.

Reorganization for Burma

Both sides quickly discovered that Burma was a miserable place to fight a war. The combination of tropical jungles, rugged mountains and an array of north–south rivers

grenade launchers for its operations on Guadalcanal.

12 After being mauled in the south Pacific the regiment's survivors were pulled out and the unit reformed on the original (September 1941) TO&E and moved to Burma to join the rest of the division.

and streams made mobility difficult, resupply extremely challenging (especially for the Japanese), and disease rampant.

For the Japanese part of the problem was lack of suitable equipment. The famous "jeep" had proven invaluable for the Allies, and all-wheel-drive trucks useful, but the Japanese had none of the former and few of the latter. Instead, they had to rely mostly on inefficient pack horses, and the even less efficient man portage. In some cases the equipment failures were quite mundane. The mountain artillery batteries, for instance, used a light field wire with thin insulation for communications between the forward observation post and the gun position. In the wet jungles of Burma it was found that the maximum range for the telephones was only about a thousand meters, which would have limited the employment of the batteries considerably.

As has been noted the first division to enter Burma, the 55th in late December, used a reduced-strength variant of its normal "B1" infantry regiment organization, with only three rifle companies per battalion, but the 33rd Division, which followed right behind in January 1942 retained the standard organization of the 30-series "wartime" divisions. It kept its original official organization until mid-December 1944 when the division, already well understrength, was told to shave 807 men and 98 horses from each infantry regiment TO&E, and 520 men and 412 horses from its mountain artillery regiment, an academic exercise at that point.

Two more divisions entered in April 1942, the 18th and 56th. The 18th Division was dispatched with its original September 1937 organization, including the 3,525-man infantry regiments, except it seems likely that the divisional mountain artillery regiment had already been modified. For service in Burma the 18th Mountain Artillery Regiment adopted a unique organization of two battalions of 75mm Type 94 mountain guns and one field artillery battalion with 75mm improved Type (Meiji) 38 guns. All of the batteries were 3-gun units, and in the field artillery each of the three sections of a battery generally consisted of 17 men, a field gun with caisson, an ammunition wagon and 13 horses. The battery HQ included two 11-man sections, one for the observation post and one of signalmen to run wire and operate semaphore flags. By 1943 the field artillery battalion had been reorganized and re-equipped into two batteries each with four 150mm howitzers. In any event difficulties in transportation meant that only half of each type of weapon could be deployed for the operations in the Hukawng Valley of 1943, the others being left in the rear.

In addition the divisional cavalry was decentralized. The 22nd Cavalry Regiment was deactivated in mid-April 1943 and the assets used to provide a mounted platoon to each infantry regiment of the 18th and 31st Divisions. There they were generally used to guard the regimental HQs rather than for reconnaissance in the Burmese jungles. Such a platoon had about 60 men with two light MGs and two grenade dischargers, as well as rifles.

In February 1943 a new organization was mandated for the infantry regiments that reduced the number of rifle companies per battalion from four to three and eliminated the regimental MG company. The result was a reduction in regimental strength to 2,947, although the number of horses authorized actually increased slightly from 439 to 481 until December 1944 when it dropped again to 394, although by mid-1944 the official tables were largely irrelevant.

The 56th Division, on the other hand, retained its original TO&Es throughout the war. No further divisions were sent to Burma for over a year until August 1943 saw the arrival of the 31st Division, followed by the 54th Division from Java in December, neither of which changed their official organization in Burma.

The 15th Division arrived in early 1944, still on its March 1941 organization tables with one exception. In this case one of the two gun platoons in each regimental gun company was re-equipped with 81mm mortars. This new mortar platoon had a lavish scale of manpower, consisting of a 40-man firing section with two mortars, a 40-man ammunition section with 40 pack horses, and a 3-man HQ. Their 21st Field Artillery Regiment partly re-equipped with six Type 41 and three Type 31 mountain guns for the Imphal operation, leaving behind their sixteen 75mm Type 95 field guns and eight 105mm howitzers.

The final division to suffer through Burma campaigning was the 53rd, which retained its November 1943 organization after entering in April 1944.

The Security Divisions of 1942

On 2 February 1942 six new divisions were formally raised: the 58th, 59th, 60th, 68th, 69th and 70th. Although "wartime" divisions in the sense that they were not incorporated into the Army's permanent order of battle, they represented a new sub-type of the "security" division.

These divisions were not raised *de novo*, but were simply expansions of existing independent mixed brigades. The five independent infantry battalions already present were enlarged and three more added from reservists in Japan, the artillery unit was dropped, service support slightly expanded and the resultant force was retitled a division. These divisions were made up of the following elements:

58th Division (ex-18th IMB)
- 51st Inf Bde: 92nd, 93rd, 94th, 95th IIBns
- 52nd Inf Bde: 96th, 106th, 107th, 108th IIBns

59th Division (ex-10th IMB)
- 53rd Inf Bde: 41st, 42nd, 43rd, 44th IIBns
- 54th Inf Bde: 45th, 109th, 110th, 111th IIBns

60th Division (ex-11th IMB)
- 55th Inf Bde: 46th, 47th, 48th, 49th IIBns
- 56th Inf Bde: 50th, 112th, 113th, 114th IIBns

68th Division (ex-14th IMB)
- 57th Inf Bde: 61st, 62nd, 63rd, 64th IIBns
- 58th Inf Bde: 65th, 115th, 116th, 117th IIBns

69th Division (ex-16th IMB)
- 59th Inf Bde: 82nd, 83rd, 84th, 85th IIBns
- 60th Inf Bde: 86th, 118th, 119th, 120th IIBns

70th Division (ex-20th IMB)
- 61st Inf Bde: 102nd, 103rd, 104th, 105th IIBns
- 62nd Inf Bde: 121st, 122nd, 123rd, 124th IIBns

These divisions consisted of little beyond the independent infantry battalions. This series of divisions had the following TO&E strengths:[13]

13 The strengths shown are for the 60th Division. There were minor variations among the divisions in HQ and transport unit strengths.

	Off	WO	MSG	SGT	PVT	Total	Horses
Division Headquarters	27	0	18	12	0	57	17
Signal Unit	15	4	5	36	264	324	44
Two Infantry Brigades, each							
Brigade Headquarters	7	1	3	19	84	114	18
Four Ind Infantry Battalions, each	45	14	16	119	1,054	1,247	82
Engineer Unit	7	2	5	14	147	175	6
Transport Unit	22	6	11	45	324	408	208
Medical Unit	23	0	33	9	184	249	45
Veterinary Unit	4	1	7	1	31	44	11
Total	472	127	213	1,097	9,550	11,459	1,023

The independent infantry battalions consisted of five standard 190-man rifle companies (each with nine light MGs and nine grenade dischargers), a 132-man machine gun company (with eight medium MGs), a 130-man infantry gun company and a 35-man headquarters. The infantry gun company was made up of three 2-gun firing platoons, with two platoons being equipped with 70mm battalion guns and one with 75mm Type 41 regimental guns.[14]

The MG and infantry gun companies were manned at relatively low levels for static garrison duties, their operational shortages for maneuver being made up through the expedient of drawing 30 men from each rifle company for duty as needed. Interestingly, the TO&Es, while providing three officers and five NCOs of the medical branch for each infantry battalion, did not provide any privates as medics. This was remedied for units in combat theaters (and all the divisions were) by simultaneously authorizing an additional 27 medical-branch privates, bringing total authorized battalion strength up to 1,274. Similarly, three medical privates were authorized for the engineer unit, six for the signal unit, three for each brigade HQ, along with 36 medical and veterinary privates for the transport unit. At the same time an additional 30 horses were authorized for each infantry battalion to improve mobility. On 10 December 1943 the infantry battalions received an additional 50 men each (this time infantry privates), bringing strength to 1,324.

Such a security division was supported by a division HQ (for which privates for general duties had to be drawn from the infantry battalions), an engineer unit of three platoons, a signal unit with a wire company and a radio company, a transport unit with two draft companies and a motorized company with 55 trucks, a medical unit and a veterinary hospital. Few changes were made to the division structure until early 1945.

It was anticipated that the division's elements would be spread out over large areas, so the signal unit was actually somewhat larger than normal, consisting of a wire company of three platoons, a radio company with 19 sections, each of about eight men with a Mk 3 radio, and a supply detail of 35 men. The wire company had two wire platoons each of four 13-man sections plus a pigeon platoon of 40 men, the latter unique to the occupation division.

14 The 61st Ind Inf. Bn (and presumably others) reorganized as the war went on. By 1944 its actual organization was a 100-man HQ, five 135-man rifle companies, a 150-man MG company and a 145-man infantry gun company. The rifle companies each had 6 Type 14 pistols, 129 Type 38 rifles, 9 Type 11 light machine guns and 12 Type 89 grenade dischargers. The MG company had 20 Type 14 pistols, 20 Type 38 rifles and 8 Type 92 medium MGs. The gun company had 15 Type 14 pistols, an unknown number of Type 38 rifles, 4 Type 92 battalion guns and 2 Type 41 regimental guns. Other battalions in the 68th Division, however, reported only two 70mm and two 75mm guns in each battalion.

Those wire platoons held 60 Type 92 phones, 120km of Type 92 insulated wire, and two 20-line switchboards. The pigeon platoon was provided with 600 to 1,000 homing pigeons.

The most serious shortcoming of the security divisions was their complete lack of artillery. In this they represented a retrograde step from the earlier IMBs they had been drawn from. As long as they were employed solely for garrison duties this apparently did not prove a critical weakness, but as forces were withdrawn from China for service elsewhere the occupation divisions were pressed into operational roles for which they were manifestly unsuited. Nevertheless, it was not until February 1945 that fire support elements were added to four of these divisions, with the other two (58th and 68th) receiving theirs in July.[15]

In the 58th, 59th and 68th Divisions these were a 598-man mountain artillery battalions each with twelve 75mm mountain guns, while in the other three divisions they were 577-man mortar battalions each with three companies with six 81mm mortars.

At the same time the division engineer unit, a company-size unit, was expanded to a battalion of 901 men with 119 horses and 4 trucks, made up of three engineer companies and a materiel platoon.

This modest increase in capabilities, however, was far too little to enable the divisions to perform normal field operational roles and their awkward binary organization continued to make such deployment largely unprofitable except against the weakest of opposition.

Improvised Divisions

One further division was raised in 1942, the 71st Division, activated in Manchuria in April. This division, characterized by limited mobility and small command elements, was built around one regiment shed from the reorganized 110th Division and two regiments drawn from personnel of the Hunchun garrison. A cavalry regiment and a 2,350-man mountain artillery regiment, along with other division services, supported the three 3,266-man infantry regiments. In July the authorized strength of the infantry regiments was reduced to 3,001. Each battalion now consisted of a 14-man headquarters, four 173-man rifle companies (each nine light MGs and nine grenade dischargers) and a 190-man machine gun company with three machine gun platoons and an AT rifle platoon. Battalion transport comprised only four horses in the headquarters and 35 in the machine gun company. The mountain artillery regiment consisted of a headquarters and three battalions, each consisting of a small 32-man HQ and three 239-man batteries each with 140 horses and four 75mm mountain guns. The ax hit hard in mid-1944. In April the authorized strength of the infantry regiments was reduced by 836 men each, followed in July by a further 570 men (reducing the number of rifle companies per battalion from four to three), resulting in an infantry regiment authorized strength of a mere 1,575. As a result, division assigned strength fell from 22,700 in April 1942 to 9,145 in July 1944, before climbing slightly back to 11,200 in January 1945 as infantry regiments saw their authorized strength inch back to 2,000 men as part of the redeployment to Formosa.

No further divisions were formed until almost a year later, when two more "wartime" divisions were raised, the 31st Division in Thailand for service in Burma, and the 61st Division in China. Two of the 31st Division's regiments (58th and 138th) had been shed from divisions converting to triangular configuration, while the third regiment (124th) represented rebuilt remnants transferred from Guadalcanal via Rabaul and Saigon in May

15 This weakness had been recognized earlier, and the 68th Division, at least, had formed a smaller mountain artillery unit of 224 men with three 2-gun platoons of 75mm mountain guns in April 1944.

1943. The two older regiments had been formed in 1938 with a strength of 3,222 men and 370 horses and equipped with older weapons, including 37mm Type (Taisho) 10 infantry guns (along with mortars) in the regimental gun companies. All three were reorganized for the division to a new strength of 2,947 men and 407 horses and divided into a headquarters, signal company, infantry gun company (two platoons each 2 regimental guns, plus an ammunition platoon), anti-tank company (two platoons each two 37mm guns), a signal company (radio and wire platoons) and three infantry battalions. The battalion had three rifle companies, a machine gun company and an infantry gun platoon. The rifle companies were of the standard configuration, three platoons each of four sections, while the MG company had four 2-gun platoons.

The mountain artillery regiment and the engineer regiment were organized with personnel drawn from units in central and northern China. The 31st Division followed the organizational pattern of a "B" division with a mountain artillery regiment and no reconnaissance/cavalry element. These gave the division an official strength of 611 officers and 14,365 enlisted.

The 61st Division's infantry was organized along the lines of the July 1941 organization of "wartime" infantry regiments in China, with each infantry regiment having 3,359 men in three infantry battalions (each of four rifle companies and an MG company), a three-gun regimental gun company and a signal company. Each rifle company was provided with six light MGs and nine grenade dischargers, while the machine gun company had eight medium MGs and two 70mm battalion guns. In December 1943 a further 10 infantry privates were authorized for each rifle company, bringing regiment strength up to 3,479 but without effecting the organizational structure of the units. The division had no reconnaissance/cavalry element and, initially, had no artillery component. On 1 February 1945 a mortar battalion with three companies of 81mm weapons was provided for divisional fire support, usually supplemented by an attached field artillery battalion (two batteries of 75mm guns and one of 105mm howitzers). The engineer unit was initially a company of 178 men with six horses, but in February 1945 was enlarged to the theater-standard 901-man regiment with 119 horses and 4 trucks. The transport regiment initially had a strength of 508 men with one pack transport company (100 horses) and two motor companies (total 90 trucks). A water transport company was added in February 1945 for use on Chinese rivers, increasing strength to 615 men. The engineer regiment and signal unit were of standard organization and the division included one field hospital and a veterinary detachment, but no medical unit. These elements gave the 61st Division a TO&E strength of 13,444 officers and men.

The Security Divisions of 1943

Three more security divisions (62nd, 64th and 65th) were raised in May 1943, again through expansion of existing IMBs. These followed the organizational pattern of the earlier security divisions but the units all included organic medical personnel, yielding an infantry battalion strength of 1,233 and a brigade HQ strength of 117. Like the earlier occupation divisions, in December 1943 the infantry battalions received the additional allotment of 50 infantry privates each and additional increments were applied to other parts of the divisions.

In January 1944, in apparent anticipation of a move overseas, the 62nd Division was stripped of all non-essential personnel. Rifle company strength was drastically reduced to about 100 men (although the number of light MGs and grenade dischargers remained the same) and one 70mm howitzer platoon was dropped from each infantry gun company. This resulted in a reduction in battalion strength to 865 men and in brigade strength to 3,584. These were supplemented by a 174-man engineer unit, a 283-man signal unit, a 201-man transport unit, a 308-man field hospital and a 21-man veterinary unit, and commanded by a 160-man division HQ for a total division strength of 8,315. In August 1944 the division was stripped of one of its transport companies and some engineering personnel and shipped to Okinawa. On arrival on that island about 3,000 Okinawan conscripts were added to the division, mostly to the service elements, allowing Japanese soldiers to be redistributed to the infantry. This increased infantry battalion strength to about 1,100 men. The division fought ferociously against the Americans, but without artillery support was eventually destroyed.

The components of the three security divisions were as follows:

62nd Division (ex-4th IMB)
- 63rd Inf Bde: 11th, 12th, 13th, 14th IIBns
- 64th Inf Bde: 15th, 21st, 22nd, 23rd IIBns

64th Division (ex-12th IMB)
- 69th Inf Bde: 51st, 52nd, 53rd, 131st IIBns
- 70th Inf Bde: 54th, 55th, 132nd, 133rd IIBns

65th Division (ex-13th IMB)
- 71st Inf Bde: 56th, 57th, 58th, 59th IIBns
- 72nd Inf Bde: 60th, 134th, 135th, 136th IIBns

As was the case with the earlier security divisions, as the war progressed additions were made to the divisions to bring them closer to normal field standards. In 1944 the engineer units were brought up to "regiment" (= battalion) size and the transport units increased to one pack company and two motor transport companies each. In July 1945 a mountain artillery battalion (twelve 75mm guns) was added to both of the divisions still in China (64th and 65th). As a result, the authorized strengths of the two divisions in China steadily increased on the dates shown.

	64th Div	65th Div
1 May 43	11,676	11,604
10 Dec 43	13,637	12,312
1 Feb 43	13,637	12,445
10 Jul 45	15,212	13,028

One triangular division was raised during this period, the 46th Division, a "wartime" division organized along "B-type" lines. This was mobilized in Japan for service on Soemba island in the Dutch East Indies. The infantry regiments of this division were initially raised with authorized strengths of only 775–1100 men, but on division mobilization in May 1943 they were standardized at 2,059 men in three battalions (each three rifle companies and a machine gun company), a regimental gun company and a signal company. In October the division was reorganized as an ocean division.

This was followed in June 1943 by two more security divisions created by expansion of IMBs:

63rd Division (ex-15th IMB):
- 66th Inf Bde: 77th, 78th, 79th, 137th IIBns
- 67th Inf Bde: 24th, 25th, 80th, 81st IIBns

100th Division (ex-30th IMB):
- 75th Inf Bde: 163rd, 164th, 165th, 166th IIBns
- 76th Inf Bde: 167th, 168th, 352nd, 353rd IIBns

The 63rd Division followed the standard organization of the security divisions, consisting of the following elements:

	Off	WO	MSG	SGT	PVT	Total	Horses
Division Headquarters (from 5/43)	29	1	29	15	80	154	17
Division Headquarters (from 12/43)	32	1	30	21	172	256	17
Signal Unit	15	4	5	36	270	330	44
Two Infantry Brigades, each							
Brigade HQ	7	1	3	19	87	117	18
Four Infantry Battalions, each	44	13	14	117	1,045	1,233	108
Mortar Artillery Unit (from 2/45)	21	3	9	45	499	577	271
Engineer Unit (from 5/43)	6	2	5	14	150	177	6
Engineer Unit (from 2/45)	24	3	14	77	783	901	119
Transport Unit	20	3	11	45	423	502	108
Field Hospital	25	0	33	9	290	357	45
Veterinary Hospital	4	1	7	1	32	45	11
Total (from 5/43)	465	117	222	1,094	9,779	11,677	1,131
Total (from 2/45)	507	121	241	1,208	11,003	13,080	1,515

The transport unit consisted of two motor transport companies (each 20 trucks) and a draft company. In February 1945 a mortar artillery unit was added and the engineer unit expanded to a battalion. Ammunition and rations trains were added to each battalion. In June 1945 the division was transferred to the Kwantung Army.[16]

The 100th Division was numbered outside the normal security division sequence because it was raised not in China but in the Philippines. It used the six-battalion 30th IMB as a basis and thus required only two additional battalions to fill out its force structure. Unlike the China-based security divisions, the 100th Division retained the earlier IMB's artillery unit as its own, in this case a field artillery battalion of three batteries. This gave it the following TO&E strength:

	Off	WO	MSG	SGT	PVT	Total	Horses
Division Headquarters	35	0	15	21	164	235	23
Signal Unit	15	4	6	38	283	346	44
Two Infantry Brigades, each							
Brigade Headquarters	6	0	2	9	39	56	3
Signal Unit	3	1	3	15	110	132	16
Four Infantry Battalions, each	30	11	10	103	843	997	20

16 The mortar unit appears to have been left behind in China. In Manchuria the division was supported only by a single battery of 75mm field guns.

Labor Unit	2	1	2	12	142	159	0
Artillery Unit	20	6	7	40	437	510	188
Engineer Unit	61	20	22	176	1,743	2,022	53
Transport Unit	23	4	23	28	791	869	632
Field Hospital	26	0	42	9	288	365	38
Veterinary Hospital	4	1	7	31	6	47	4
Total	446	127	216	1,239	11,038	13,066	1,180

Although IGHQ never sanctioned any changes in the authorized strengths of the 100th Division (or the later 102nd, 103rd or 105th Divisions) some local strengthening appears to have taken place. The original battalion organization tables called for a 95-man HQ, four 176-man rifle companies, a 134-man gun unit, and a 64-man pioneer platoon. The gun unit was to have three four-gun platoons, two with heavy MGs and one with 81mm mortars. After fighting these divisions in the Philippines US Army intelligence gave their independent infantry battalions a strength of 1,080 men divided into four 180-man rifle companies, a 150-man heavy weapons company, an 80-man labor unit and a 130-man HQ. The rifle companies followed the standard organizational pattern, built around three 50-man rifle platoons each with three light MGs and three grenade dischargers. The remaining 30 men were assigned to the company HQ as runners, medics, etc. The heavy weapons company was divided into three platoons each with four squads, usually two platoons of MGs and one of mortars.

Each brigade included not only its four infantry battalions but also a HQ, a signal unit and a 159-man labor unit. The division artillery unit had three batteries, each with four 75mm Type 38 field guns. The transport unit was made up of two companies, while the signal unit had one wire and two radio companies. The engineer unit was exceptionally large, made up of a regimental HQ and ten engineer companies each of about 140 men, although it was provided only with hand tools, dynamite and some mines. For armament each company had rifles and two light MGs. These combined to give the 100th Division a TO&E strength of slightly over 13,000.

Defending the South 1943

On 14 May 1943 the Imperial Guards units were reorganized to provide a contingent for operations in the south. The Guards Brigade (1st and 2nd Guards Infantry Regiments), which had been split off from the Guards Division to campaign in China, had returned to Japan in mid-1941. In the meantime, the 5th Guards Infantry Regiment had been formed in February 1940 to create a triangular division of the Guards Division, which participated in the invasion of Malaya and then moved to Sumatra. Thus, during 1940–43 the Imperial Guards forces consisted of the Guards Division (3rd, 4th and 5th Guards Infantry Regiments, along with the Guards Field Artillery Regiment) in Sumatra and the Guards Brigade (1st and 2nd Guards Infantry and Guards Cavalry Regiments) in Japan. In May 1943 two infantry regiments, a field artillery regiment and division base elements were activated to yield two divisions: the 1st Guards Division (1st, 2nd, 6th and 7th Guards Infantry, 1st Guards Field Artillery and Guards Cavalry Regiments) in Japan and the 2nd Guards Division (3rd, 4th and 5th Guards Infantry, 2nd Guards Field Artillery and Guards Reconnaissance Regiments) on Sumatra.

The new 1st Guards Division (ex-Guards Brigade) was initially formed at low strength.

Infantry regiments were only authorized 1,716 men apiece in two battalions (each three rifle and one MG companies), a regimental gun company, a machine cannon company and a signal company. The 1,031-man field artillery regiment also had only two 3-battery battalions.

The 2nd Guards Division (ex-Guards Division) also went through some organizational turmoil, but was shortly converted to an ocean division discussed later.

In July 1943 a new triangular division was raised in Japan, the 93rd. This "wartime" division was organized along the lines of an "A" type division. The infantry regiments, at 3,946 men, were slightly smaller than the A1 standard, but with no loss of firepower.[17] The division also included a mountain artillery regiment (3,791 men) with two battalions of 75mm Type 94 mountain guns and one battalion of 105mm Type 99 mountain howitzers. Also part of the division was a 672-man cavalry regiment, and the division was strengthened a year later by the addition of a motorized anti-tank battalion, giving the division a total strength of 18,628. The division remained in Japan through to the end of the war.

The 94th Division was raised in October in Malaya. The infantry of this unit followed the organization of the 2nd and 16th Divisions on their mobilization in 1941 but with somewhat fewer horses. The division included a three-battalion field artillery regiment and other normal divisional troops, but no reconnaissance or cavalry elements giving the following strength:

	Off	WO	MSG	SGT	PVT	Total	Horses
Division Headquarters	48	2	32	27	194	303	0
Signal Unit	5	1	4	20	208	238	38
Three Infantry Regiments, each	94	15	31	281	2,460	2,881	309
Field Artillery Regiment	82	12	33	132	1,377	1,636	1,274
Engineer Regiment	25	3	13	74	782	897	94
Transport Regiment	16	0	25	35	666	742	296
Medical Unit	37	0	30	34	598	699	100
Two Field Hospitals, each	23	0	35	1	218	277	79
Ordnance Duty Unit	2	1	9	4	65	81	0
Veterinary Hospital	11	1	25	2	157	196	0
Total	554	65	334	1,173	11,863	13,989	2,887

In August and November of 1943 the 52nd and 53rd Divisions, respectively, were mobilized from their depot status in Japan.[18]

The former was an ocean type division discussed later. The 53rd Division was sent to Burma with the following organization:

	Off	WO	MSG	SGT	PVT	Total	Horses
Division Headquarters	40	2	25	32	194	293	0
Signal Unit	6	1	4	20	208	239	44

17 They presumably, however, did not include the anti-tank rifles found in the A divisions as production of these weapons had fallen precipitously and ended in 1943. Except for the amphibious regiments and brigades it seems unlikely that any new units formed after mid-1943 included these marginal weapons.

18 The infantry regiments of the 52nd Division had actually been formed in December 1941, but division HQ and many of the supporting services were not activated until November 1943. Some cadre elements of the 53rd Division may have been set up in September 1941.

	Off	WO	MSG	SGT	PVT	Total	Horses
Reconnaissance Regiment	20	3	11	53	315	402	0
Three Infantry Regiments, each	95	15	31	283	2,493	2,917	215
Field Artillery Regiment	82	12	33	132	1,377	1,636	1,294
Engineer Regiment	25	3	13	74	791	906	119
Transport Regiment	23	0	25	35	666	749	296
Medical Unit	37	0	30	34	598	699	100
Two Field Hospitals, each	23	0	36	1	182	242	27
One Field Hospital	24	0	37	1	185	247	28
Ordnance Duty Unit	2	1	9	4	65	81	0
Veterinary Hospital	4	1	6	1	40	52	10
Water Supply & Purification Unit	11	1	25	2	157	196	0
Total	605	69	383	1,239	12,439	14,735	2,590

The infantry organization was identical to the 2,881-man regiments pioneered by the 16th Division in 1941 and subsequently used by the 56th Division in Burma, but with an undistributed addition of one company-grade officer, two sergeants and 33 privates.

At the same time (late August) the 54th Division was transferred from Java to Burma and brought its own organization. This division had been mobilized in February 1943 at a lower strength than that envisioned by the 1941 mobilization plan. It was now built around three infantry regiments each of 2,886 men (with an authorized additional increment of 24 men, bringing strength up to 2,910), a 1,636-man field artillery regiment, a 439-man reconnaissance regiment and an 898-man engineer regiment. The infantry regiments were similar to those of the 55th and 56th Divisions above, but featured separate regimental infantry gun and anti-tank companies each with four weapons. In that respect, they resembled "A" infantry more than "B" infantry. To stay within the manpower ceiling the rifle companies were probably reduced in personnel strength.

The 54th Field Artillery Regiment was also reorganized, into two battalions of 75mm field guns and one battalion of 105mm howitzers, each with three batteries.

The 54th Division Reconnaissance Regiment consisted of a HQ, a signal platoon, two motorized companies and two tankette companies. This reflected a considerable difference from the organization originally detailed for these 50-series divisions. The motorized companies lost 24 men, but the basic structure remained the same. Each line platoon fielded two 10-man rifle sections and a 6-man grenadier section. The company was provided with one sidecar motorcycle, six trucks for personnel, two for ammunition, and one as a reserve vehicle.

A tankette company consisted of two 9-man platoons each with three tankettes. The 19-man company HQ had two motorcycles, two trucks and two tankettes. The regimental signal platoon consisted of five radio sections, each with eight men, a truck and a Mark 5 radio. These elements gave the reconnaissance regiment a total strength of 439 men. Notable in this organization is the abandonment of the separate motor transport company to handle the regiment's motor vehicles, an anomalous characteristic of the 1941-pattern reconnaissance regiments.

Conditions in the south Pacific, the Solomons, New Guinea and New Britain, were regarded as even worse than those in Burma. In particular, horses were not expected to survive the foetid jungle conditions, and were ruthlessly stripped from deploying units.

Table 2.9: 54th Reconnaissance Regiment

	Officers	NCOs	Other Ranks	Pistols	Carbines	Rifles	Light MGs	Medium MGs	Grenade Dischargers	37mm AT Guns	Tankette (MG)	Tankette (37mm)	Motorcycles	Cars	Trucks	Bicycles
Regiment HQ	10	12	34	7	17	17	0	0	0	0	0	0	1	1	7	5
Signal Platoon	1	4	44	1	46	0	0	0	0	0	0	0	1	0	5	3
Two Motorized Companies, each				23	105	0	6	2	6	1	0	0	1	0	9	4
Company HQ	2	4	10													
3 Platoons, each	1	2	25													
MG Platoon	1	2	14													
AT Section	0	1	11													
Two Tankette Companies, each				10	25	0	0	0	0	0	5	3	2	0	2	1
Company HQ	2	4	13													
2 Platoons, each	1	2	6													

Lacking sufficient motor vehicles (or, indeed, roads in the area) the divisions were almost immobile upon landing unless they abandoned their heavy equipment and much of their ammunition.

An example of an "A" division reorganized for the south Pacific was the 6th Division, which was deployed from China to the Solomons in December 1942. The infantry regiments, formerly A1-type, were reconfigured to strip a third of their horses from their constituent units and replace the AT rifle platoon with a mortar platoon to each battalion. The artillery regiment was also reduced in size, reconfiguring to three battalions of mountain guns. In part this was accomplished through new TO&Es issued for the division in February 1942 and in larger part through a massive redistribution of personnel within the division once it arrived in the south Pacific. A comparison of the authorized strength of the division and the actual organization as shown in divisional orders captured by the US was:

	Actual		Authorized	
	Off	**EM**	**Off**	**EM**
Division HQ	59	479	41	407
Signal Unit	n/a	n/a		
Cavalry Regiment	21	762	20	432
Infantry Group HQ	11	150	n/a	n/a
Three Infantry Regiments, each	127	3851	108	3735
Regiment HQ	(12)	(164)		
Signal Company	(4)	(140)		
Three Infantry Battalions, each	(34)	(1106)		
Battalion HQ	(7)	(139)		
4 Rifle Companies, each [9 LMG, 9 GL]	(5)	(176)		
Machine Gun Company [12 MG]	(5)	(169)		
Infantry Gun Platoon [2 70mm inf howitzers]	(1)	(55)		
Mortar Platoon [2 90mm mortars]	(1)	(39)		
Infantry Gun Company [4 75mm mtn guns]	(4)	(118)		
Anti-Tank Company [6 37mm AT]	(5)	(111)		
Field Artillery Regiment	84	2439	81	3,105
Regiment HQ	(13)	(163)		
Three Field Battalions, each	(23)	(706)		
Battalion HQ	(10)	(86)		
Three Batteries, each [4 75mm mtn guns]	(4)	(175)		
Ammunition Trains	(1)	(95)		
Ammunition Trains	(2)	(158)		
Engineer Regiment	26	785	n/a	n/a
Regiment HQ	(4)	(59)		
Three Engineer Companies, each	(7)	(226)		
Materiel Platoon	(1)	(48)		

	Actual		Authorized	
	Off	EM	Off	EM
Transport Regiment	37	1367	55	2234
Ordnance Duty Unit	6	74	3	118
Medical Unit	99	1615		
Medical HQ	(4)	(15)		
Medical Unit	(29)	(731)	34	879
Three Field Hospitals, total	(52)	(705)	46	669
Veterinary Unit	(3)	(21)	n/a	n/a
Water Purification Unit	(11)	(143)	7	112
Total	669	19,472	521	20,286

The 6th Division, however, appears to have been the only major formation to have incorporated both 70mm howitzers and mortars in the infantry battalion organization. Given the general ineffectiveness of Japanese field artillery, this may have been an even trade-off.

The 20th Division, shipped from Korea to Palau in January 1943, and then to New Guinea in June, was similarly modified, albeit without the additional mortars. All horses were left behind, so all heavy equipment had to be manhandled through the jungle. The two men in each 11-man machine gun section that formerly led the pack horses, for instance, were redesignated as ammunition bearers. Similarly, in the infantry gun platoons each of the 16-man sections had to manhandle not only the guns but also 100 rounds of ammunition.

The 51st Division, moving to Rabaul in January 1943, was also much modified. The infantry regiments retained their B-1 structure but saw all their horses (in this case, along with associated handlers) stripped away. This meant the loss of two horses in each medium MG squad, reducing it to ten men, four of whom had to carry the gun and four more ammunition. The 75mm regimental guns were carried by 30 men each, although they could get by with fewer for short moves. The regimental ammunition and baggage trains in the headquarters company had to man-porter their cargoes. Infantry regiments of the "wartime" divisions, of course, fared no better. Those of the 41st Division were also stripped of their horses and had to move their infantry and AT guns manually. When told to prepare to move overland the MG companies left six of their twelve Type 92s back at Wewak.

Another significant reduction in capability in an attempt to retain some mobility involved the re-equipping of the field artillery regiments of divisions. In the case of the 20th Division the batteries saw their four Type 38 field guns replaced by three Type 94 mountain guns, although the regiment retained its "field artillery" designation. This same process was applied to the 6th Field Artillery of the 6th Division for Bougainville, the 14th Field Artillery Regiment of the 51st Division on New Guinea, and the 23rd Field Artillery Regiment of the 17th Division on New Britain, all of which then had to manhandle their mountain guns, and ammunition, through the jungles.[19]

19 This had been applied earlier with the 38th Division's field artillery in China, which was reorganized into three 9-gun battalions of mountain artillery in May, 1941. It retained this organization, minus the horses, when shipped to Rabaul in October 1942.

Table 2.10: Infantry Regiment (Island Defense), from 1943

	Officers	Warrant Officers	NCOs	Other Ranks	Pistols	Rifles	Light MGs	Medium MGs	Light Gren Lnchrs	Heavy Gren Lnchrs	37mm AT Guns	70mm Inf Howitzers	75mm Field & Mountain Guns	Riding Horses	Draft Horses	Carts	Bicycles	Trucks
Regiment Headquarters	12	1	27	59	0	23	0	0	1	0	0	0	0	36	0	0	2	0
Signal Company	3	1	19	113	0	122	0	0	0	0	0	0	0	0	5	5	1	0
Supply Company	4	2	14	144	2	111	0	0	0	0	0	0	0	0	20	18	1	6
Engineer Company	5	2	21	211	0	215	0	0	0	0	0	0	0	0	10	10	1	0
Medical Unit	10	2	16	160	8	67	0	0	0	0	0	0	0	0	1	1	1	0
Three Infantry Battalions, each																		
Battalion Headquarters	5	0	12	48	1	27	0	0	1	0	0	0	0	1	15	15	2	0
Three Rifle Companies, each	5	2	20	127	2	126	12	2	0	12	0	0	0	0	0	0	1	0
Infantry Gun Company	3	1	8	71	4	16	0	0	0	0	2	2	0	0	8	6	1	0
Artillery Battalion																		
Battalion Headquarters	9	0	12	66	8	16	0	0	1	0	0	0	0	1	19	16	2	0
Three Batteries, each	4	2	9	120	10	40	0	0	1	0	0	0	4	0	48	6	1	0

A New Organization for the Pacific

The 52nd Division began mobilizing from its depot status in Japan as a regular division on 10 August but was converted two months later into a new and totally different kind of formation, the ocean division.[20]

By mid-1943 it had become clear that the fighting in the Pacific placed unique demands on the army units deployed there. In particular, the trucks and carts that had worked in China and Manchuria quickly proved useless in the trackless jungle interiors of New Guinea and similar islands, and unnecessary on the small coral atolls of the central Pacific. Pack horses, although somewhat more practical, did not hold up well in the face of the tropical climate and diseases. Water transport, using small boats to avoid prowling United States Navy (USN) units, was quickly recognized as the sole practical means of moving troops and supplies about. The Army, chronically short of trucks and horses to start with, responded by forming unique ocean divisions.

In such a division the superfluous reconnaissance unit was dropped, the artillery decentralized to regiment level, most horses and trucks eliminated, and a sea transport unit added. The division retained its triangular layout, but with a difference. Two of the infantry regiments were configured as quasi-static island defense regiments, while the third was reorganized into an amphibious regiment for use as a mobile reserve.

The island defense-type infantry regiment was characterized by a much leaner manpower authorization than the normal infantry regiment. The 154-man rifle company was composed of a headquarters, three rifle platoons and a machine gun platoon. Each rifle platoon consisted of four rifle squads (each with a light MG) and a grenadier squad (four grenade dischargers). For this role the rifle squads consisted of only one NCO and six privates, while the grenadier squad had an NCO and eight privates. The platoon HQ was made up of the platoon leader and a liaison sergeant or corporal. The machine gun platoon was similarly austere, consisting of two 9-man gun squads and a 2-man headquarters.

The battalion gun company consisted of two platoons each of two sections. The first platoon manned 70mm infantry howitzers, while the second was provided with 37mm anti-tank guns. Horses and carts were provided to this company for administrative duties and moves, but tactical movement would have proven impossible.

The command section of the battalion HQ consisted of only three officers: the commander (a lieutenant colonel) and a captain and a lieutenant for staff and adjutant duties. They were supplemented by two technical officers: one from the intendance branch and the other a medical officer. The largest component of the battalion HQ was the transport section consisting of a sergeant and 22 privates who operated 15 one-horse carts. The HQ also included a small signal section that manned four Type 92 telephones and one set of semaphore flags.

Signal facilities at the regimental level were provided by a dedicated company that was equipped with a 10-circuit switchboard, twelve telephones, twelve Mk 6 and eight Mk 5 radios for attachment to subordinate units, and two Mk 3A radios for the command post.

Fire support for the regiment was provided by an artillery battalion equipped with 75mm guns, either field or mountain depending on the type of artillery regiment in the division before the reorganization. The medical company was made up of a stretcher-bearer group of two 63-man platoons, and a 45-man aid station group. The engineer company was

20 Known to Allied intelligence as the "island defense" type division.

Table 2.11: Infantry Regiment (Amphibious), from October 1943

	Officers	Warrant Officers	Enlisted	Type 14 Pistols	Type 99 Rifles	Type 99 Light MGs	Type 92 Medium MGs	50mm Type 10 Gren Lnchrs	50mm Type 89 Gren Lnchrs	20mm Type 97 AT Rifles	Flamethrowers	50mm Type 98 Mortars	81mm Type 99 Mortars	81mm Type 97 Mortars	20mm Type 98 AA Guns	37mm Type 94 AT Guns	75mm Type 41 Mountain Guns	Light Tanks	Motorcycles	Field Cars	Trucks
Regiment Headquarters	13	2	102	2	61	0	0	0	0	0	0	0	0	0	0	0	0	0	1	1	11
Signal Company	3	1	135	0	n/a	0	0	0	0	0	0	0	0	0	0	0	0	0	0	0	0
Engineer Company	5	2	236	0	127	0	0	0	0	0	0	0	0	0	0	0	0	0	0	0	0
Medical Company	10	2	178	8	0	0	0	0	0	0	0	0	0	0	0	0	0	0	0	0	0
Tank Company	3	1	62	12	44	0	0	0	0	0	0	0	0	0	0	0	0	9	0	0	2
Machine Cannon Company	4	1	71	11	0	0	0	0	0	0	0	0	0	0	6	0	0	0	0	0	0
Three Amphibious Battalions, each																					
Battalion Headquarters	7	1	96	2	57	0	0	1	0	0	0	0	0	0	0	0	0	0	0	0	0
Three Rifle Companies, each	6	2	189	10	118	12	2	0	12	1	0	0	0	0	0	0	0	0	0	0	0
Mortar Company	5	2	148	20	34	0	0	0	0	0	0	0	0	12	0	0	0	0	0	0	0
Artillery Company	3	1	116	12	19	0	0	0	0	0	0	0	0	0	0	2	3	0	0	0	0
Infantry Pioneer Platoon	1	0	65	0	63	0	0	0	0	0	2	1	2	1	0	0	0	0	0	0	0

equipped for the most part simply with shovels, picks and saws, but did include a single air compressor, a generator, a blacksmith set, and surveying equipment.

The island defense-type infantry regiment possessed more firepower than most Japanese infantry regiments with a substantially smaller manpower base. The cost that had to be paid, however, was that the rifle squads were so small they had little margin for accepting casualties, and transport facilities were limited to those needed for normal administrative duties, leaving the unit largely immobile in combat.

The amphibious-type infantry regiment was organized in such a way as to minimize the amount of shipping space required. There were no horses and few vehicles in the regiment, nor any field artillery. With only a few exceptions, weapons were limited to those that could be carried by the men themselves. The lack of artillery was to be compensated for by high levels of light weapons support, particularly mortars, and, when possible, by gunfire support from offshore naval vessels.

The rifle company of the amphibious regiment was similar to that of the island defense regiment but added a 12-man anti-tank rifle section to the MG platoon and a 21-man mortar platoon to the company. This latter platoon, unique to the amphibious rifle company, had two 10-man squads each manning a 81mm Type 97 mortar.

An impressive amount of close-in fire support for the battalion was provided by the mortar company, which included three firing platoons, each with four 9-man squads with Type 97 mortars. The firing platoon HQ consisted simply of the platoon leader and had no FDC function. The mortar company also included a 16-man ammunition trains for man-carrying of ammunition and a command section with a 5-man observation squad and a 9-man signal squad. Somewhat longer-ranged fire was provided by the artillery company, which was built around a 52-man mountain gun platoon with three 75mm Type 41 regimental guns and a 41-man anti-tank platoon with two 37mm Type 94 AT guns. Squad size was fairly large in this unit, 17 men in the mountain gun platoon and 20 men in the anti-tank platoon, due to the lack of organic transport.

The battalion's pioneer platoon was an assault force divided into a 7-man HQ, three 13-man sections and two 10-man sections. It was provided with substantial firepower in the form of two Type 100 flamethrowers, a 50mm Type 98 spigot mortar, two 81mm Type 99 short-barrel assault mortars, and one 81mm Type 97 infantry mortar. The amphibious battalion HQ was built around a 32-man command section and a 52-man transport platoon, the latter in turn divided into two 26-man sections.

The regiment was supported by a machine cannon company, a light tank company, and an engineer company. The machine cannon company consisted of three 20-man platoons, each of which was made up of two squads with 20mm Type 98 automatic cannon. The light tank company had three Type 95 light tanks in each of its two line platoons, plus three more in the company HQ platoon. It may have been planned to equip the tank companies with amphibious tanks, but this does not seem to have taken place. An 18-man maintenance section with two trucks supported the company.

Where the battalion pioneer platoons were primarily assault units, the regiment's engineer company was more of a construction unit. It was divided into four 52-man engineer platoons (each of four squads) and a materiel section. The latter maintained a wide variety of equipment for distribution as needed, including four small rubber boats, two 4-man manual piledrivers, a Type 92 light road grader, two Type 95 caterpillar tractors, two 1-kw generators and a small air compressor.

The regimental signal company was divided into a 44-man wire platoon and a 74-man radio platoon. The former was built around six 6-man and one 7-man squads; while the latter featured eight 5-man portable radio squads, a 25-man radio section for the CP, and an 8-man section. The 1st through 8th radio squads and the 1st through 6th wire squads were distributed to subordinate elements to maintain the regiment net, while the remaining signal assets served the regimental HQ.

The final element of the amphibious regiment was the medical company, which consisted of a 17-man headquarters, a 128-man stretcher-bearer group and a 45-man aid station group. The stretcher-bearer group was in turn divided into two 64-man platoons, each with three 21-man sections.

The amphibious infantry regiment was designed for portability on small coastal craft and was well suited for this role. Heavy weapons firepower for the infantry was impressive, but once ashore the amphibious regiment was completely immobile. Except for the eleven trucks held by the regiment HQ, there was no way to move the six anti-tank guns, nine 75mm mountain guns and six 20mm AA guns of the regiment other than for the gun crews to push, pull and carry their weapons. Perhaps even more serious, there was no way to move the required ammunition inland from the landing place other than on backpack loads of porters.

Combat support for the ocean infantry division was limited to a company-size tank unit. This was made up of four 21-man platoons, each with four light tanks.[21] The company also included a 15-man headquarters with a field car and a 31-man maintenance platoon with a reserve light tank, five cargo trucks and a repair truck.

The division's sea transport regiment largely replaced the land transport assets of the normal infantry division and carried out both logistical roles and the transportation of the amphibious regiment as required. It was provided with a total of 150 Daihatsus, along with 10 special large landing barges for this purpose. Three speed boats were provided for command and control and 10 speed boats, each with two cannon, were assigned to the escort company for protection of the coastal barges.

The sea transport unit was not considered a combat unit and was not designed to take part in opposed landings. The transport companies were responsible not only for the operation of the Daihatsu and Shohatsu coastal craft, but also for stevedoring functions, which accounts for their large size. Surprisingly the personnel were drawn not from the Army's shipping forces, but from regular engineer units given hasty training in boat-handling. The landing craft were neither armed nor armored.

Division (Ocean type), from October 1943	
DIVISION HEADQUARTERS (266)	
SIGNAL UNIT (235)	
TWO INFANTRY REGIMENTS (ISLAND DEFENSE), each (3,148)	
	Regiment HQ (99) [1 GL]
	Signal Company (136)
	Three Infantry Battalions, each (610)

21 The fourth platoon set of tanks was nominally held as an equipment reserve, but in fact they were used as line assets in almost every heavy battle.

Division (Ocean type), from October 1943		
		Battalion HQ (65) [1 GL]
		Three Rifle Companies, each (154) [12 LMG, 2 MG, 12 GL]
		Infantry Gun Company (83) [2 37mm AT, 2 70mm Inf How]
	Artillery Battalion (492)	
		Battalion HQ (87) [1 GL]
		Three Batteries, each (135) [1 GL, 4 75mm guns]
	Engineer Company (239)	
	Supply Company (164)	
	Medical Unit (188)	
INFANTRY REGIMENT (AMPHIBIOUS) (3,939)		
	Regiment HQ (117)	
	Signal Company (139)	
	Three Infantry Battalions, each (1,036)	
		Battalion HQ (104) [1 GL]
		Three Rifle Companies, each (197) [12 LMG, 2 MG, 12 GL, 1 ATR]
		Mortar Company (155) [12 81mm mort]
		Artillery Company (120) [2 37mm AT, 3 75mm Mtn guns]
		Infantry Pioneer Platoon (66) [1 50mm mort, 3 81mm mort, 2 FT]
	Machine Cannon Co (76) [6 20mm AA]	
	Tank Company (66) [9 light tanks]	
	Engineer Company (243)	
	Medical Unit (190)	
TANK UNIT (130) [17 light tanks]		
TRANSPORT UNIT (131)		
SEA TRANSPORT UNIT (1,542)		
	Unit HQ (54)	
	Four Transport Companies, each (315)	
	Escort Company (164)	
	Materiel Depot (64)	
ORDNANCE UNIT (107)		
INTENDANCE UNIT (133)		
FIELD HOSPITAL (755)		

The division signal unit consisted of two wire platoons (each of four sections), a radio platoon (of ten sections) and a materiel platoon.

The first divisions to be reorganized on this new TO&E were the 36th Division (then in China) and the 52nd Division (then in Japan), with the 222nd and 107th Infantry Regiments, respectively, converting to the amphibious configuration. These divisions were reorganized in October 1943 and the following month the 36th Division was shipped to

Halmahera and New Guinea, and the 52nd Division to Truk.

The 36th Division was split up on its arrival in the SW Pacific, with the amphibious 222nd Infantry sent to Biak island (and apparently taken from division control), while the two island defense infantry regiments were sent to Wakde-Sarmi. When the Allies landed at Hollandia, an attempt was made to use one of the island defense regiments (224th) in a mobile role, with one of the infantry battalions acting as porters to carry supplies for the remainder of the regiment. On 4 May 1944 this force left Sarmi for the overland march to Hollandia, but on the 17th the landing at Sarmi forced its recall, by which time it had covered about half the distance. Although it fought hard, the division was essentially neutralized at Wakde-Sarmi, with the remnants in the coastal jungle providing no real threat to Allied plans. The 222nd (amphibious) Regiment was destroyed on Biak after tenacious fighting in which their mortars proved highly effective.

A month after these initial conversions, on 16 November 1943, the War Office published their Organization Order A-106 which provided for a massive reorganization of forces to reinforce the Pacific theater. Among the changes mandated by this order was the conversion of the 13th Division to ocean-type in February 1944, followed by the 3rd and 5th Divisions in March. In addition, the 5th Guards Infantry Regiment was to be converted to an amphibious regiment and a sea transport unit added to the 2nd Guards Division as soon as possible, although the rest of the division was not to be reorganized.

Conditions in China, however, did not permit the withdrawal of the three divisions from that theater and as a result the 14th and 29th Divisions (in Manchuria) and the 46th Division (in Japan) were reconfigured in their stead. The 15th and 18th Infantry Regiments became the amphibious units for the first two divisions, respectively. The 46th Division was unique in that it did not include an amphibious regiment, all three infantry regiments being of the island defense type. As a result, the 46th Division also did not include a sea transport unit.

Additional weapons, above and beyond those specified in the organization tables, were provided to these three divisions when they deployed to the Pacific. Most notably, each division received 200 Type 100 rifle grenade launchers and 100 Type 2 anti-tank rifle grenade launchers. Other additions included 40 Type 99 light and 10 Type 92 heavy machine guns, two Type 92 battalion guns and three 37mm AT guns.

The 14th Division was officially assigned to the 2nd Army in February 1944 and shipped out for the SW Pacific in March, but was diverted to the Palau islands *en route* and finished the war there. The defense of the southern island of Peleliu was assigned to the 2nd Infantry Regiment (island defense type), reinforced by 3rd Battalion of the 15th Infantry. A week after the initial US landings on that island the 15th Infantry's 2nd Battalion was called upon to perform the mission it had been designed for and was moved by small boats at night down the island chain to reinforce the Peleliu garrison. Boat losses were substantial and although the bulk of the battalion's personnel reached their destination most of their heavy weapons were lost *en route*. In the end, they were far too few in number to have any effect.

The 29th Division was assigned to the Combined Fleet on 10 February 1944, then to the 31st Army on 25 February. It was transferred from Manchuria via Saipan to Guam, with the 50th Regiment detached to Tinian. The division sea transport unit apparently did not accompany the division to the Marianas and on 12 July 1944 (nine days before the US invasion of Guam) the sea transport unit was officially detached from the division and reassigned to the Shipping Commander in Tokyo. Having suffered heavy losses at sea the division was finally destroyed on Guam and Tinian in July 1944.

The 46th Division (less its 145th Infantry Regiment) was transferred to Soemba Island in the East Indies in October 1943, where the reorganization took place on arrival, then to Flores Island in January 1945, then to Singapore where it ended the war. The wayward 145th Infantry Regiment was shipped to Iwo Jima to bolster defenses there and was destroyed along with the rest of the island garrison.

The 2nd Guards Division was partially reorganized as an ocean division while on garrison duty in Sumatra, with the 5th Guards Regiment becoming the amphibious unit and a sea transport unit being added. The rest of the division retained its original organization except that the 2nd Battalion of the artillery regiment had to be demobilized to provide a cadre of artillery personnel for the 5th Regiment.

The final ocean infantry division was the 43rd, which was raised as such from scratch in April 1944. It was quickly shipped out to Saipan and destroyed there. Like the 46th Division, the 43rd was composed entirely of island defense-type regiments and lacked the sea transport regiment.[22]

The ocean divisions were an attempt to adapt field formations to radically new conditions. They were successful insofar as they facilitated coastal transport of the divisions but suffered a number of serious shortcomings. Probably the most serious was the almost complete absence of land transport, rendering them largely immobile once ashore. This was mitigated somewhat by the fact that even the best-equipped forces were only slightly more mobile through the jungles of New Guinea, but it nevertheless remained a problem. A second deficiency was the lack of fire support. The provision of only two battalions of 75mm guns for a full division, especially when compounded by the usual Japanese ineptitude at infantry–artillery coordination, was entirely insufficient. Finally, in their zeal to reduce personnel overhead by reducing division strength from a normal 20,000 to about 12,500, the Army may have gone too far in reducing the size of the line elements, leaving little margin for accepting casualties.

Reorganization in China, 1943

As the war in China dragged on and the Chinese forces showed no signs of strategic surrender it became clear that the organization of the "wartime" divisions was no longer adequate. This was particularly true as the regular army divisions were pulled out of that theater for service in Burma or the Pacific, leaving the role of mobile reserve force to the "wartime" divisions. Indeed, even by December 1941 there were only two "A" divisions in China, the 3rd and the 6th, and the latter left China a year later. The role of elite shock troops to back up the ill-equipped garrison divisions thus fell by default to the "wartime" divisions.

Two new sets of TO&Es were published on 1 May 1943 in an attempt to give the field divisions in China greater punch at the lower tactical levels. One set was applied to the 3rd, 13th, 17th, 27th, 32nd, 36th and 116th Divisions and created large divisions of about 18,000 men that would provide the backbone of mobile campaigning in that theater, although the 36th Division was soon completely reorganized and shipped to the Pacific,

22 Being raised from depot formations the division suffered some heavy weapons shortages. There were only three field guns per battery (total three 75mm Type 38 improved and six 75mm Type 95 per battalion), and only one AT gun per infantry battalion.

followed by the 17th Division to Rabaul in September.[23] A second group of TO&Es was applied to the 22nd, 26th, 34th, 35th, 37th, 39th, 40th, 104 and 110th Divisions to create slightly smaller divisions of about 14,000 men.

Army Order A-36 of 1 May 1943 created the larger field divisions, which emphasized increasing the firepower of the infantry components. The ancient platoon organization of three rifle squads, two light MG squads and a grenadier squad was finally abandoned in favor of the conventional structure of three rifle squads (each a corporal and twelve privates with eleven rifles and a light machine gun) and a grenadier squad (a corporal and eleven privates with nine rifles and three grenade dischargers). The battalion machine gun company was little changed, consisting of four firing platoons (each of two gun squads) and an ammunition platoon. Each machine gun squad was provided with two pack horses, one for the gun and the other for ammunition. The ammunition platoon held a further 20 pack horses, 16 for ammunition, 2 for ammunition and tools, and 2 as reserve. An infantry gun platoon was added to the battalion structure to give it, for the first time, longer-ranged HE fire support. The platoon had six pack horses to carry the two guns, 14 for ammunition and one reserve.

At the regiment level the infantry gun company was enlarged to the standard four-gun configuration, now consisting of two firing platoons (each two Type 41 mountain guns and four ammunition wagons) and an ammunition platoon (with ten ammunition wagons and four supply wagons).

Added to the regiment was an anti-tank company with three firing platoons and an ammunition platoon. Each platoon consisted of two gun squads, each with a horse-drawn 37mm AT gun and an ammunition wagon. The ammunition platoon, which was divided into two sections, had six ammunition wagons and one supply wagon.

The regimental signal company was expanded to consist of a wire platoon and a radio platoon. The wire platoon was made up of six sections, while the radio platoon had five squads each with a Mk 5 radio and one section with twelve Mk 6 radios. The expansion presumably increased wire-laying capability, for the radio allocation was the same as that specified for the special regiment under the 1941 reorganization.

Actual authorized strengths, however, were below the TO&E strengths throughout the division. The infantry regiment TO&Es, for instance, actually called for a strength of 4,098 men but the same document that announced them also specified a reduction of 8 NCOs and 498 privates to yield an authorized strength of only 3,592 with the structure as shown in Table 2.12. At the same time the number of horses was reduced by over half. The reductions were cancelled and units brought up to full TO&E strength starting with the 17th Division in September 1943, with the other divisions following in February 1944. The only divisions that did not have their reduced authorized strength cancelled were the 36th, which departed for New Guinea in November 1943 at the smaller authorized strength and remained on that organization to its end, and the 36th, which went to the East Indies at an even further reduced strength of 2,569 men.

These larger China theater field divisions had the following composition under the initial reduced authorized strengths and the full TO&E strengths later implemented.[24]

23 In mid-1944 the 17th Division shed its 81st Infantry Regiment, on detached duty on Bougainville, to form the 38th IMB, and received the 2nd and 6th Mixed Regiments in compensation.

24 Although nominally remaining a field artillery regiment, the 122nd FA Regiment of the 116th Division re-equipped with 75mm mountain guns in May 1944, presumably to increase mobility.

Table 2.12: Infantry Regiment Type A-36, from June 1943

	Officers	Warrant Officers	NCOs	Privates	Pistols	Rifles	Light MGs	Medium MGs	Light Gren Lnchrs	Heavy Gren Lnchrs	37mm AT Guns	70mm Inf Howitzers	75mm Regtal Guns	Riding Horses	Pack Horses	Draft Horses	Carts
Regiment Headquarters	13	0	20	39	2	39	0	0	1	0	0	0	0	24	0	0	0
HQ Transport Group (att)	0	0	2	123	0	0	0	0	0	0	0	0	0	0	0	94	88
Signal Company	3	1	15	123	0	130	0	0	0	0	0	0	0	3	21	1	1
Mounted Platoon	1	0	4	30	0	28	2	0	0	0	0	0	0	35	0	0	0
Three Infantry Battalions, each																	
Battalion HQ	8	0	12	32	2	38	0	0	1	0	0	0	0	4	0	0	0
HQ Transport Group (att)	0	0	2	123	0	0	0	0	0	0	0	0	0	12	59	35	24
Four Rifle Companies, each	4	2	18	166	0	153	9	0	0	9	0	0	0	0	0	0	0
Machine Gun Company	5	1	12	126	16	0	0	8	0	0	0	0	0	1	36	0	0
Infantry Gun Platoon	1	0	6	62	4	0	0	0	0	0	0	2	0	1	21	0	0
Infantry Gun Company	3	1	10	128	?	?	0	0	0	0	0	0	4	3	1	37	?
Anti-Tank Company	4	1	9	118	?	?	0	0	0	0	6	0	0	4	1	21	13

	Authorized		TO&E	
	Men	Horses	Men	Horses
Division HQ	345	64	329	64
Signal Unit	233	40	233	40
Three Infantry Regiments, each	3,592	322	4,098	752
Field Artillery Regiment	1,682	1,286	1,959	1,518
Engineer Regiment	872	94	901	119
Transport Regiment	970	431	1,170	631
Ordnance Duty Unit	95	0	95	0
Medical Unit	367	46	400	73
Two Field Hospitals, each	277	79	301	83
One Field Hospital	284	83	309	87
Veterinary Hospital	119	45	119	45
Total	16,297		18,411	

The infantry regiments of the smaller field divisions in China (22nd, 26th, 34th, 35th, 37th, 39th, 40th, 104th and 110th) were generally similar in structure to the larger divisions, but with economies of manpower and transport imposed. In most cases they lacked the anti-tank company in the infantry regiment and the separate infantry gun platoons in the infantry battalions (the latter being replaced by rearming one of the four MG platoons with 70mm or old 37mm infantry guns). More significant were the differences at the divisional level. The artillery regiment was downgraded to an "artillery unit" of variable size, usually consisting of a headquarters and six batteries of field or mountain guns with no intermediate battalion HQs.[25] The engineer regiment was reduced to two companies in most of the divisions and to a single company-equivalent in three more (34th, 35th and 40th). All of these divisions were reorganized into their full TO&E strengths with only few changes made to yield a slightly different authorized strength:

	Authorized		TO&E	
	Men	Horses	Men	Horses
Division Headquarters	221	32	119	23
Signal Unit	219	32	219	32
Three Infantry Regiments, each	3,427	308	3,427	308
Artillery Unit	882	671	882	671
Engineer Regiment	401	14	401	14
or Engineer Unit	(177)	(6)	(177)	(6)
Transport Unit	502	108	502	108
Medical Unit	351	45	351	45
Ordnance Duty Unit	95	0	95	0

25 In the case of the 39th Division, for instance, one battery was taken from each of three battalions in October 1943 and sent to French Indochina, while the battalion HQs were sent back to Japan, reducing strength from 1,688 men and 1,221 horses to 932 men and 671 horses. In many cases the unit continued to use its regiment designation. In the 34th, 40th and 110th Divisions the artillery component was removed completely, although these were reinstated in mid-1945 in the form of 583-man mountain artillery battalions with a dozen 75mm mountain guns.

	Authorized		TO&E	
	Men	Horses	Men	Horses
Veterinary Unit	n/a	n/a	n/a	n/a

The infantry regiments of all these divisions, except the 35th preparing for the New Guinea, received authorization for an additional 120 privates (ten per rifle company) on 10 December 1943, raising authorized regimental strength to 3,547.[26] A further increment of 3 sergeants and 234 infantry privates, along with an officer and two sergeants of the ordnance corps, was authorized on 15 February for each infantry regiment, except those of the 26th Division, bringing authorized strength to 3,787.[27]

The 1944 Mobilization Plan Order

In August 1943 the new Army mobilization plan for 1944 was issued, the first since the 1941 mobilization of three years earlier. It made no significant changes to the organization of the bulk of the divisions activated under the 1941 mobilization plan. The only change made across the board to all divisions was a minor modification to the engineer regiment, which was increased slightly in strength and mobility, with seven one-horse carts added to each engineer company.

Such divisional cavalry regiments as remained were generally left intact, but divisional reconnaissance regiments were reorganized. Under the new organization the former motor transport squadron was abolished and its assets assigned directly to the rifle squadrons. At the same time the mounted rifle squadron was abolished and the armored car/light tank squadron strengthened.

What was intended as the most significant aspect of the new mobilization plan, however, proved in the end to be a mere aberration in the development of IJA divisions. Faced with a declining pool of available manpower and the need to conserve shipping, IGHQ finally produced a new standard divisional organization in the 1944 mobilization plan called the Type C division. The new Type C infantry regiment was similar to the older Type B regiments but had only three rifle companies per battalion and added a mounted platoon for reconnaissance duties. The draft-transport model (Type C1) was specified for the 2nd, 3rd and 6th Divisions (in Burma, China and the Solomons, respectively) and the pack transport model (Type C2) for the 57th Division (in Manchuria).

The Type C regiments were to be the new standard, but there is no evidence that any divisions other than the four specifically mentioned in the mobilization plan were ever directed to convert to the Type C. In fact, the 3rd Division, at least, never actually converted and retained its Type A infantry organization through to the end of the war, although it supplemented it with regimental mounted platoons. Similarly, the 2nd Division retained its Type A infantry organization. In this case, the infantry battalions had only three companies apiece, but this was not due to the influence of the new TO&E, but simply

26 The 35th Division was unusual in winding up at Sorong at the western end of New Guinea with its infantry regiments at full strength of three battalions (each 4 rifle and 1 MG company), infantry gun company, signal company and mounted platoon. They were the only division to bring their horses with them to New Guinea. On the other hand, the artillery regiment had been removed from the division before its departure.

27 The 110th Field Artillery Regiment was disbanded in June 1943 to form the 2nd Independent Field Artillery Regiment, leaving only a composite battery to support the division. It was gradually brought partially back to strength, ending the war as a five-battery unit with twenty 105mm howitzers.

Table 2.13: Division Reconnaissance Units, from August 1943

	Personnel	Pistols	Rifles & Carbines	Light MGs	Medium MGs	Heavy Gren Lnchrs	37mm AT Guns	Riding Horses	Pack Horses	Light Tanks	Cars	Trucks	Lt Repair Vehs
Cavalry Regiment													
Regiment Headquarters	366	19	113	0	0	0	0	191	210	0	0	0	0
Two Rifle Squadrons, each	137	3	109	6	0	4	0	138	20	0	0	0	0
Machine Gun Squadron	139	3	94	0	4	0	2	134	57	0	0	0	0
Reconnaissance Regiment													
Regiment Headquarters	112	13	95	0	0	0	0	0	0	0	2	17	1
Two Motorised Rifle Squadrons, each	190	3	148	6	2	6	1	0	0	0	3	14	0
Light Tank Squadron	85	15	64	0	0	0	0	0	0	12	2	3	0

reflected the fact that the 2nd Division had mobilized that way in 1941 for the move south.

The 57th Division may have adopted the C2 organization, although this is not clear. From an infantry regiment TO&E strength of 4,487 in early 1942 its regiments had suffered the following reductions in authorized strength:

	Off	WO	MSG	SGT	PVT	Horses
Original TO&E strength	115	18	35	382	3,937	927
May 1942	-1	-4	-40			
July 1943	-1	-2	-250			
July 1944	-12	-57	-486			
December 1944	+3	+6	+76			

These yielded an authorized infantry regiment strength of 3,788, very close to the C2 TO&E strength of 3,790. The July 1944 reduction had eliminated one rifle company in each battalion, so it seems likely that the division in fact adopted, if not the C2 organization itself, then a close variant.

The new Type C divisions were also to have new reconnaissance unit organizations, either the new standard reconnaissance regiment or a C-type cavalry regiment with only two rifle squadrons. Otherwise, the C-type divisions were similar to the Type B divisions raised earlier.

Force Expansion 1944

Meanwhile, defense preparations continued elsewhere. The 42nd Division was raised in Sendai in early 1944 and immediately shipped north to garrison the central Kuriles. The division was formed with the following composition:

Table 2.14: Infantry Regiments, C-Type, from 1944

	Personnel	Pistols	Rifles	Light Machine Guns	Medium Machine Guns	50mm T.89 Gren Lnchrs	81mm T.97 Mortars	37mm T.94 AT Guns	70mm T.92 Infantry Howitzers	75mm T.41 Mountain Guns	Riding Horses	Pack Horses	Draft Horses	Carts
Infantry Regiment, Type C1														
Regiment Headquarters	186	2	41	0	0	1	0	0	0	0	24	0	94	88
Signal Company	138	0	130	0	0	0	0	0	0	0	3	21	1	1
Regimental Gun Company	134	8	1	0	0	0	0	2	0	2	6	1	36	22
Mounted Platoon	38	0	28	3	0	3	0	0	0	0	38	3	0	0
Three Infantry Battalions, each														
Battalion Headquarters	177	2	38	0	0	1	0	0	0	0	16	59	35	24
Three Rifle Companies, each	190	0	153	9	0	9	0	0	0	0	0	0	0	0
Machine Gun Company	144	16	0	0	8	0	0	0	0	0	1	36	0	0
Infantry Gun Platoon	69	4	0	0	0	0	0	0	2	0	1	21	0	0
Infantry Regiment, Type C2														
Regimental Headquarters	331	2	46	0	0	1	0	0	0	0	29	203	0	0
Signal Company	139	0	130	0	0	0	0	0	0	0	3	23	0	0
Regimental Gun Company	183	8	2	0	0	0	0	2	0	2	3	83	0	0
Mounted Platoon	38	0	28	3	0	3	0	0	0	0	38	3	0	0
Three Infantry Battalions, each														
Battalion Headquarters	250	2	38	0	0	1	0	0	0	0	16	155	0	0
Three Rifle Companies, each	190	0	153	9	0	9	0	0	0	0	0	0	0	0
Machine Gun Company	144	16	0	0	8	0	0	0	0	0	1	36	0	0
Mortar Platoon	69	8	0	0	0	0	4	0	0	0	38	3	0	0

Note: each regimental HQ had an additional 250 rifles as undistributed self-defense weapons for the regiment.

	Off	WO	MSG	SGT	PVT	Total	Horses
Division Headquarters	46	0	48	29	353	476	15
Signal Unit	14	3	5	38	400	460	3
128th Infantry Regiment	153	39	41	425	3,347	4,005	93
129th Infantry Regiment	209	39	41	434	3,396	4,119	72
130th Infantry Regiment	153	39	41	425	3,286	3,944	72
Independent Tank Company	6	2	4	28	84	124	0
Engineer Regiment	27	3	12	80	839	961	125
Transport Regiment	46	0	41	61	1,692	1,840	1,246
Medical Unit	37	0	30	34	598	699	100
1st Field Hospital	23	0	36	1	182	242	21
4th Field Hospital	24	0	37	1	185	247	22
Ordnance Duty Unit	3	0	20	6	171	200	0
Veterinary Hospital	7	1	8	5	97	119	45
Three Inf Assault Training Cos, each	4	1	0	11	116	132	0

The division was reorganized on 16 July 1945, at which time the infantry regiments were standardized at a strength of 2,881 each and provided with 409 horses apiece to improve mobility. At the same time a field artillery regiment of 1,800 men with 1,400 horses was added to give the division its fire support.

A few months later the forces in the northern Kuriles were strengthened and reorganized to constitute the 91st Division. Responsible for Shumushu and Paramushiro Islands, the division was essentially binary in structure, consisting of the following elements:

	Off	WO	MSG	SGT	PVT	Total	Horses
Division Headquarters	46	0	48	29	353	476	15
Signal Unit	13	3	4	44	264	328	0
Two Infantry Brigades, each							
Brigade HQ	13	0	18	0	58	89	6
Brigade Signal Unit	10	2	4	28	227	271	4
Six Infantry Battalions, each	32	10	9	88	719	858	96
Pioneer Unit	2	1	1	11	142	157	0
Tank Regiment	32	13	13	137	366	561	0
Tank Company	7	2	6	25	54	94	0
1st Artillery Unit	64	26	13	126	1,022	1,251	110
2nd Artillery Unit	39	12	10	87	797	945	40
Anti-Tank Unit	18	3	8	40	350	419	0
Air Defense Unit	87	28	19	215	2,162	2,511	0
Engineer Unit	21	6	7	64	669	767	18
Transportation Unit	11	2	11	20	240	284	0

Hospital	35	0	61	0	221	317	0
Ordnance Duty Unit	3	0	20	6	171	201	0
Other Medical Units	6	0	17	6	140	169	0
Total	816	341	391	1,933	16,291	19,772	1,355

The two primary elements, of course, were the 73rd Infantry Brigade (282nd–287th Ind Inf Bns) and the 74th Infantry Brigade (288th–293rd Ind Inf Bns), backed up by the 11th Tank Regiment and the 2nd Independent Tank Company as a mobile reserve, and two disparate artillery units for artillery support. The only organizational change to the division was a reduction in the divisional air defense unit from a 2,511-man regiment-size unit to an 840-man battalion-size unit in mid-July 1944.

These were followed by two more triangular divisions, the 43rd Division (ocean type for Saipan) and the 30th Division (raised in Korea, then shipped to the Philippines). The latter division was built around a 401-man reconnaissance regiment, three 2,857-man infantry regiments (each of three infantry battalions, an infantry gun company, an AT company and a signal company) and a 1,636-man three-battalion field artillery regiment.

May of 1944 saw the activation of four additional divisions. The 50th Division with a strength of 16,200 was formed on Formosa, the 107th Division (16,424 men) was formed in Manchuria, the 109th Division was activated in Bonins and the 49th Division was activated under the Korea Army and shipped to Burma.

The 50th Division was built around three 2,620-man infantry regiments, each of three battalions, a regimental gun company, an anti-tank company and a signal company. Having only 86 horses on its TO&E (plus 21 more added to authorized strength) these could not have been very mobile units. Fire support was supposed to be provided by a 1,244-man motorized mountain artillery regiment with three battalions totaling 85 horses and 192 motor vehicles. In fact, however, authorized strength for this unit was only 661 with 85 horses and 37 motor vehicles until late September when it was increased to 1,483. Also supporting the division were a 420-man engineer regiment (with 9 horses) and a 419-man transport regiment (with 29 motor vehicles).

The 107th Division was one of the few divisions formed to reinforce the Kwantung Army, which was slowly being stripped of its personnel and equipment to face the Anglo-American threat. This, however, represented only a marginal increase in effective force, as the division was formed through reorganization of the Arshaan Garrison Unit. Each of the division's infantry regiments had a strength of 3,409 men and 734 horses, while the field artillery regiment had 2,809 men. The reconnaissance regiment had 603 men, the engineer regiment 981 men and the transport regiment 1,014 men and 472 horses.

The division appears to have reached full strength by February 1945, but thereafter continual drafts to form other units drew down division strength and the creation of a divisional raiding battalion in June 1944 pulled significant numbers of personnel from the three infantry regiments. Authorized strengths for the various components of the division were changed several times in a futile attempt to keep up with a dizzying succession of personnel transfers into and out of the division, resulting in the following strengths, authorized and actual, on the outbreak of the war with the Soviets:

Table 2.15: 109th Division

	Officers	Warrant Officers	NCOs	Privates		Pistols	Carbines	Rifles	Light MGs	Medium MGs	Type 89 Gren Lnchrs	37mm Type 94 AT Guns	70mm Type 92 Inf Howitzers	75mm Type 38 Field Guns	75mm Type 88 AA Guns	Riding Horses	Pack Horses	Draft Horses	Carts	Wagons	Motorcycles	Field Cars	Trucks
Division Headquarters	29	0	34	93		0	0	89	0	0	0	0	0	0	0	8	0	0	0	0	2	2	3
1st Mixed Brigade																							
Brigade Headquarters	20	5	44	212		0	0	152	0	0	0	0	0	0	0	0	0	0	0	0	2	1	10
Field Artillery Battery	4	2	13	110		11	0	0	0	0	0	0	0	4	0	0	0	0	0	1	0	0	0
Engineer Company	5	2	23	189		0	0	205	0	0	0	0	0	0	0	0	0	0	0	0	0	0	0
Six Ind Infantry Battalions, each																							
Battalion HQ	8	0	15	52		0	0	14	0	0	1	0	0	0	0	0	0	0	0	0	0	0	0
Three Rifle Companies, each	4	2	16	93		0	0	77	9	0	9	0	0	0	0	0	0	0	0	0	0	0	0
Machine Gun Company	4	2	18	99		24	0	0	0	12	0	0	0	0	0	0	0	0	0	0	0	0	0
Infantry Gun Company	4	1	10	71		8	0	0	0	0	0	2	2	0	0	0	0	0	0	0	0	0	0
Pioneer Platoon	1	0	7	59		0	73	0	0	0	0	0	0	0	0	0	0	0	0	0	0	0	0
2nd Mixed Brigade																							
Brigade Headquarters	20	5	44	212		0	0	152	0	0	0	0	0	0	0	0	0	0	0	0	2	1	10
Field Artillery Unit																							

Table 2.15: 109th Division

	Officers	Warrant Officers	NCOs	Privates		Pistols	Carbines	Rifles	Light MGs	Medium MGs	Type 89 Gren Lnchrs	37mm Type 94 AT Guns	70mm Type 92 Inf Howitzers	75mm Type 38 Field Guns	75mm Type 88 AA Guns	Riding Horses	Pack Horses	Draft Horses	Carts	Wagons	Motorcycles	Field Cars	Trucks
Unit HQ	6	0	11	45		7	0	0	0	0	0	0	0	0	0	16	0	0	0	1	0	0	0
Three Batteries, each	4	2	10	110		11	0	0	0	0	0	0	0	4	0	0	0	0	0	1	0	0	0
Engineer Company	5	2	23	189		0	0	205	0	0	0	0	0	0	0	0	0	0	0	0	0	0	0
Six Ind Infantry Battalions																							
as above																							
1st Mixed Regiment																							
Regiment Headquarters	12	0	17	76		0	0	35	0	0	1	0	0	0	0	0	0	0	0	0	0	0	5
Signal Unit	1	1	9	62		0	0	60	0	0	0	0	0	0	0	0	0	0	0	0	0	0	0
Artillery Company	4	2	8	110		2	0	0	0	0	0	0	0	4	0	0	0	0	0	1	0	0	0
Two Infantry Battalions, each																							
Battalion HQ	6	0	11	16		0	0	14	0	0	1	0	0	0	0	0	0	0	0	0	0	0	0
Three Rifle Companies, each	4	2	16	93		0	0	77	9	0	9	0	0	0	0	0	0	0	0	0	0	0	0
Machine Gun Company	4	2	18	99		24	0	0	0	12	0	0	0	0	0	0	0	0	0	0	0	0	0

Table 2.15: 109th Division

	Officers	Warrant Officers	NCOs	Privates		Pistols	Carbines	Rifles	Light MGs	Medium MGs	Type 89 Gren Lnchrs	37mm Type 94 AT Guns	70mm Type 92 Inf Howitzers	75mm Type 38 Field Guns	75mm Type 88 AA Guns	Riding Horses	Pack Horses	Draft Horses	Carts	Wagons	Motorcycles	Field Cars	Trucks
Infantry Gun Company	4	1	10	71		8	0	0	0	0	0	2	2	0	0	0	0	0	0	0	0	0	0
Pioneer Platoon	1	0	7	59		0	73	0	0	0	0	0	0	0	0	0	0	0	0	0	0	0	0
9th Heavy Artillery Regiment																							
Regiment Headquarters	10	0	22	38		0	0	0	0	0	0	0	0	0	0	3	0	0	0	0	0	0	0
One Heavy Battery	4	1	13	122		0	0	30	0	0	0	0	0	0	0	0	0	0	0	0	0	0	0
Three Heavy Batteries, each	4	1	10	93		0	0	30	0	0	0	0	0	0	0	0	0	0	0	0	0	0	0
Divisional Anti-Aircraft Unit																							
Unit Headquarters	7	0	10	84		6	0	20	0	0	0	0	0	0	0	0	0	0	0	0	2	0	18
Three AA Companies, each	4	1	11	128		13	0	20	0	0	0	0	0	0	6	0	0	0	0	0	0	0	0
Searchlight Company	4	1	12	150		13	13	0	0	0	0	0	0	0	0	0	0	0	0	0	0	0	0
Divisional Signal Unit	2	0	15	122		0	108	0	0	0	0	0	0	0	0	0	0	0	0	0	0	0	0
Divisional Security Unit	2	1	9	41		9	0	0	0	0	0	0	0	0	0	0	0	0	0	0	0	0	0
Field Hospital	12	2	16	90		0	0	0	0	0	0	0	0	0	0	0	0	0	0	0	0	0	0

	Authorized	Actual
Division HQ	581	338
Signal Unit	239	200
Reconnaissance Regiment	603	550
90th Infantry Regiment	3,654	2,260
177th Infantry Regiment	3,599	2,599
178th Infantry Regiment	3,478	2,249
Raiding Battalion	1,150	1,010
Field Artillery Regiment	2,809	2,080
Engineer Regiment	907	707
Transport Regiment	824	817
Other units	469	350

Each infantry regiment consisted of a headquarters, three infantry battalions (each of three rifle companies, a machine gun company and a battalion gun platoon), a signal company, a gun company and a mounted platoon. The field artillery regiment was made up of three battalions each of three batteries. The divisional reconnaissance, raiding and engineer units each had three companies, while the transport regiment had four.

The 109th Division was the second division to bear this name, the first 109th Division having been inactivated in 1939 after service in China. A unique organization, this second incarnation represented a reorganization and consolidation of command of the troops in the Bonin Islands. The troops on Chichi Jima were reorganized into the 1st Mixed Brigade (not to be confused with an independent mixed brigade not subordinate to a division HQ), those on Iwo Jima became the 2nd Mixed Brigade, and those on Haha Jima became the 1st Mixed Regiment (similarly, not to be confused with an independent mixed regiment).

The cadre for the artillery units was drawn from the former Chichi Jima Heavy Artillery Regiment. The 75mm field guns and 120mm howitzers of the regiment were transferred, with their crews, to the new field artillery units, while the rest of the regiment was converted to the 9th Heavy Artillery Regiment in the coast defense role, retaining the regiment's four 24cm Type 45 howitzers, two 15cm Type 45 guns, and four 105mm Type 38 guns. All of these heavy weapons were emplaced on Chichi Jima.

The data shown in Table 2.15 represents the organizational orders issued by the 31st Army. The unit strengths as ordered by IGHQ, however, were slightly different. Tokyo authorized the following strengths for the components of the division:

	Off	WO	MSG	SGT	PVT	Total
Division Headquarters	n/a	n/a	n/a	n/a	n/a	n/a
1st Mixed Brigade						
Brigade HQ	13	0	25	0	55	93
Six Infantry Battalions, each	25	9	6	79	460	579
Artillery Unit	4	2	3	10	110	129
Engineer Unit	6	2	3	21	190	222
2nd Mixed Brigade						
Brigade HQ	13	0	25	0	55	93
Six Infantry Battalions, each	25	9	6	79	460	579
Artillery Unit	19	6	5	37	348	415

Engineer Unit	6	2	3	21	189	221
9th Heavy Artillery Regiment	26	4	10	55	439	534
Division Anti-Aircraft Unit	35	6	13	87	920	1061
Division Signal Unit	3	0	23	0	110	136
Field Hospital	12	0	18	0	90	110

The only organizational difference between the two seems to have been in the infantry battalions, where the IGHQ version did not include the 39-man HQ trains detachment found in those in the two brigades, nor the pioneer platoon in all battalions as authorized by 31st Army. Other differences represented simply a fine-tuning of individual personnel allocations. Weapons and equipment allocations, except personal weapons, do not appear to have been effected. Due to subsequent modifications to the organization by local commanders it is difficult to determine which organization was actually in place at the beginning.

Since the 109th Division commander was simultaneously the Bonins Islands commander his command (although apparently not the division itself) was swelled considerably as reinforcements arrived in anticipation of the US invasion. On Iwo Jima alone this included the 145th (island defense type) Infantry Regiment, the 26th Tank Regiment, two medium mortar battalions, two machine gun battalions, five anti-tank battalions, four special machine cannon companies and a rocket artillery battalion, bringing the Iwo Jima garrison up to division strength.

The reinforcements apparently also included individual soldiers, which enabled the 109th Division to significantly strengthen its organic units. A series of TO&Es, apparently from January 1945, shows a number of anomalous divergences from the planned Central-Pacific type infantry battalions. One organization table shows a fourth company of the 309th Independent Infantry Battalion with an officer cadre almost identical to that of the former 65th (B-type) Fortress Infantry Unit. This new company had a strength of 202 men, for the most part divided into three rifle platoons, each consisting of a platoon leader, four 11-man rifle squads (each with a light MG) and a 13-man grenadier squad (with four grenade dischargers). Whether all the companies in the 309th Battalion were this large, compared to their official TO&E strength of only 115 men, is uncertain, as is whether this reorganization was applied to other independent infantry battalions.

Another TO&E, dated 4 January 1945, shows an infantry gun company of 75 men, consisting of an 18-man HQ, a 25-man AT platoon and a 26-man battalion gun platoon. Each of the two platoons was made up of the platoon leader and three gun squads. This again differed from the official TO&E in having three, instead of two, guns per platoon.

Since the division HQ was located on Iwo Jima no command problems arose there, however in the case of the other two islands, where only brigade and regimental commanders were present, the command arrangements must have been somewhat more complicated as reinforcing units arrived.

In June 1944 three more garrison divisions were formed in the Philippines. Like the earlier 100th Division, these represented expansions of existing six-battalion independent mixed brigades:

102nd Division (ex-31st IMB)
 77th Infantry Brigade (170th, 171st, 172nd, 354th IIBns)

Table 2.16: Divisions & Brigades of 35th Army, 23 October 1944

	Authorized	Actual	
		Japanese	Captured
16th Division			
Type 14 Pistols	482	352	309
Type 38 Carbines	1,841	1,337	0
Type 38 Rifles	5,761	7,405	3,888
Type 96 Light MGs	277	252	86
Type 92 Medium MGs	78	74	25
Type 89 Gren Dischr	255	252	0
81mm mortars	0	0	6
70mm infantry guns	18	18	0
37mm Type 94 AT guns	6	6	0
Type 41 Regt Guns	12	12	2
75mm Type 95 fld guns	24	24	1
105mm Type 91 how	12	0	0
30th Division			
Type 14 Pistols	462	504	0
Type 38 Carbines	1,483	1,716	5
Type 38 Rifles	5,217	4,332	0
Type 96 Light MGs	277	266	2
Type 92 Medium MGs	78	70	0
Type 89 Gren Dischr	255	243	0
70mm infantry guns	18	17	0
37mm Type 94 AT guns	22	20	0

	Authorized	Actual	
		Japanese	Captured
102nd Division			
Types 14 & 26 pistols	258	231	18
Type 38 Carbines	1,014	880	1,185
Type 38 Rifles	8,250	3,318	3,324
Types 11 & 96 Light MGs	299	181	78
Type 92 Medium MGs	64	55	38
Type 89 Gren Dischr	296	225	0
81mm mortars	8	0	14
70mm infantry guns	32	28	0
75mm fld guns	4	4	10
54th IMB			
Type 14 Pistols	101	153	0
Type 38 Carbines	319	101	0
Type 38 Rifles	2,292	1,875	60
Types 11 & 96 Light MGs	108	83	30
Type 92 Medium MGs	24	0	15
Type 89 Gren Dischr	111	52	0
81mm mortars	8	0	0
70mm infantry guns	12	4	0
75mm fld guns	0	0	2
55th IMB			
Type 14 Pistols	101	65	0

Table 2.16: Divisions & Brigades of 35th Army, 23 October 1944

	Authorized	Actual	
		Japanese	Captured
Type 41 Regt Guns	12	8	0
75mm Type 95 fld guns	12	8	0
105mm Type 91 how	15	6	0
100th Division			
Types 14 & 26 pistols	258	103	56
Type 38 Carbines	1,014	1,035	0
Type 38 Rifles	8,250	4,700	5,277
Types 11 & 96 Light MGs	299	291	123
Type 92 Medium MGs	64	55	33
Type 89 Gren Dischr	296	286	0
81mm mortars	8	0	10
70mm infantry guns	32	28	0
47mm AT guns	0	5	0
Type 41 Regt Guns	0	4	4
75mm fld guns	4	0	11

	Authorized	Actual	
		Japanese	Captured
Type 38 Carbines	319	267	0
Type 38 Rifles	2,292	908	283
Sniper rifles	0	139	0
Types 11 & 96 Light MGs	108	50	90
Type 92 Medium MGs	24	24	24
Type 89 Gren Dischr	111	30	0
81mm mortars	8	0	0
70mm infantry guns	12	3	0
Type 41 Regt Guns	0	3	0

78th Infantry Brigade (169th, 173rd, 174th, 355th IIBns)
103rd Division (ex-32nd IMB)
79th Infantry Brigade (175th, 176th, 178th, 356th IIBns)
80th Infantry Brigade (177th, 179th, 180th, 357th IIBns)
105th Division (ex-33rd IMB)
81st Infantry Brigade (181st, 182nd, 183rd, 185th IIBns)
82nd Infantry Brigade (184th, 186th, 358th, 359th IIBns)

These divisions were organized along the same lines as the earlier 100th Division, and like that formation had organic (albeit small) artillery units. As with the earlier division, the infantry battalions consisted of four rifle companies (each three platoons with three LMG and one grenade sections) and a machine gun company of three platoons. In most cases the MG company consisted of two MG platoons and an 81mm mortar platoon, each with four weapons. These divisions complemented the efforts of the 100th Division, which was based on Mindanao, with the 102nd Division (15,518 men) garrisoning the central Visayas, the 103rd Division (15,491 men) at Baguio on Luzon, and the 105th Division (16,101 men) in northern Luzon.

These divisions apparently relied heavily on captured rifles and machine guns, that was certainly the case with the two divisions in the southern Philippines under the 35th Army.[28]

Finally, to conserve manpower and to create a reserve of transportable divisions orders went out in mid-July 1944 to strip the fourth rifle company from each infantry battalion in three divisions in Manchuria (10, 25 and 71) and three in China (1, 11 and 12), saving 570 men per infantry regiment.

Homeland Defense: The Preliminary Moves

July of 1944 saw the activation of the last significant group of divisions of the type envisioned by the 1941 and 1944 mobilization plans. On 6 July the 3rd Guards, 44th, 72nd, 77th, 84th and 86th Divisions were officially activated, followed four days later by the 47th, 73rd, 81st and 93rd Divisions. All these units were activated in the homeland and all but the 47th remained there through the rest of the war, and all but the last two mentioned formed a single organizational group.

The infantry regiments of these new divisions were similar to the standard A1 type regiment, but with only three rifle companies per battalion, a leaner manpower allocation at most other levels and the use of draft horses instead of pack horses at lower levels. The artillery component was also reduced in these divisions, to a two-battalion regiment. In the 77th Division this was a mountain artillery regiment of 1,540 men with six batteries, each with three 75mm Type 94 mountain guns. In the other divisions fire support was provided by a 1,116-man field artillery regiment with six batteries each of three 105mm Type 91 field howitzers. Each division thus had eighteen artillery pieces for its fire support.

An innovation with this group of divisions was the replacement of the reconnaissance/cavalry regiment with a motorized anti-tank battalion with 67 motor vehicles and twelve 47mm Type 1 anti-tank guns.[29] The battalion consisted of three companies, each

28 Indeed, the artillery battalion of the 103rd Division, briefly an independent unit in the Philippines, included a battery of four ex-US half-track mounted 75mm guns, presumably captured in 1942, in addition to one battery of 75mm towed guns and two of 81mm mortars.

29 The 72nd and 81st Divisions were actually raised with cavalry regiments, but these were quickly

Table 2.17: Homeland Regular Infantry Regiments, 1944–45

	Personnel	Pistols	Rifles	Light Machine Guns	Medium Machine Guns	Grenade Launchers	37mm AT Guns	70mm Inf Howitzers	75mm Mtn Guns	Riding Horses	Pack Horses	Draft Horses	Motor Vehicles
In 3rd Gds, 44th, 72nd, 73rd, 77th, 84th & 86th Divisions													
Regiment Headquarters	132	0	23	0	0	1	0	0	0	*41	-	-	11
Mounted Platoon	38	1	28	3	0	0	0	0	0	38	3	0	0
Signal Company	133	0	120	0	0	0	0	0	0	3	18	1	0
Three Infantry Battalions, each													
Battalion Headquarters	103	0	23	0	0	1	0	0	0	11	0	39	0
Three Rifle Companies, each	180	0	144	9	0	9	0	0	0	0	0	0	0
Machine Gun Company	120	16	0	0	8	0	0	0	0	1	18	10	0
Infantry Gun Platoon	44	4	0	0	0	0	0	2	0	1	0	9	0
Infantry Gun Company	123	8	0	0	0	0	0	0	4	1	0	36	0
Anti-Tank Company	72	8	0	0	0	0	4	0	0	1	0	11	0
Pioneer Company	173	0	162	0	0	0	0	0	0	1	1	13	0
In 81st & 93rd Divisions													
Regiment Headquarters	206	0	44	0	0	1	0	0	0	28	0	98	0
Mounted Platoon	38	1	28	3	0	0	0	0	0	38	3	0	0
Signal Company	142	0	130	0	0	0	0	0	0	3	0	20	0
Three Infantry Battalions, each													
Battalion Headquarters	224	0	40	0	0	1	0	0	0	18	78	43	0
Four Rifle Companies	190	0	153	9	0	9	0	0	0	0	0	0	0
Machine Gun Company	200	24	0	0	12	0	0	0	0	1	54	0	0
Infantry Gun Platoon	69	4	0	0	0	0	0	2	0	1	0	21	0
Infantry Gun Company	155	8	0	0	0	0	0	0	4	5	0	36	0
Anti-Tank Company	106	9	0	0	0	0	4	0	0	3	0	17	0
Pioneer Company	173	0	162	0	0	0	0	0	0	1	1	13	0

* *total of all horses in regiment HQ*

converted to AT battalions. The others were raised *ab initio* with AT units.

of a headquarters, two firing platoons (each two squads) and an ammunition platoon. Completing the division were a headquarters, an engineer regiment (three companies and a materiel platoon), a signal unit, a transport regiment (one horsed and two motorized companies), an ordnance duty unit, a medical unit, two field hospitals and a veterinary unit.

As originally organized these divisions had the following strengths:

	Off	WO	MSG	SGT	PVT	Horses	Mtr Veh
Division Headquarters	42	2	31	30	252	64	35
Signal Unit	6	1	4	20	208	44	0
Three Infantry Regiments, each	95	15	31	285	2,493	415	0
Field Artillery Regiment	58	8	26	91	933	862	8
Anti-Tank Battalion	17	4	8	36	415	0	67
Engineer Regiment	27	3	12	80	839	128	10
Transport Regiment	22	0	25	25	677	296	76

On 6 February 1945 a new TO&E for the infantry regiment of these divisions was issued that added 173 men to yield a total of 98 officers, 16 warrant officers, 31 master sergeants, 296 sergeants and 2,651 privates. The general structure of the regiment was unchanged. A rifle company consisted of three platoons, each of four 13-man squads. Three of the squads were each armed with eleven rifles and a light machine gun, the fourth squad with ten rifles and three grenade dischargers. The battalion machine gun company had four platoons each with two medium machine guns, and the company carried a total of 51,840 rounds of ammunition. The battalion gun platoon had two squads and carried a total of 240 rounds of 70mm ammunition. The regimental gun and anti-tank companies were each divided into two 2-gun platoons and carried, respectively, 420 rounds of 75mm and 384 rounds of 37mm ammunition. The regimental signal company consisted of two platoons each of six squads. Two of the company's squads were telephone units, eight manned Mk 5 radios for the regimental net and two manned Mk 6 radios. The pioneer company had two platoons.

The two anomalous divisions, 81st and 93rd, were raised from school units and considered elite formations. Grouped into the 36th Army for the defense of the Kanto Plain, they essentially retained the divisional organization envisioned in the mobilization plans with only a few changes. Not only were the infantry regiments up to full normal strength (four rifle companies per battalion), but the artillery regiments were as well. The 81st Field Artillery Regiment consisted of two 621-man field artillery battalions, one 1,055-man mountain artillery battalion, a 124-man HQ and a 220-man regimental ammunition trains. Each of the field battalions was built around three 135-man batteries, each of four weapons (75mm Type 95 guns in one battalion, 105mm Type 91 howitzers in the other two). The mountain artillery battalion included three 203-man batteries, each with four 75mm Type 94 mountain guns. The 93rd Mountain Artillery Regiment had two 1,055-man battalions (each twelve Type 94 mountain guns) and one 1,136-man battalion (twelve 105mm Type 99 mountain howitzers). The division base of these divisions was almost identical to that of the other divisions of this group, except that the 93rd Division included a cavalry regiment of three rifle squadrons and a machine gun squadron.

	Off	WO	MSG	SGT	PVT	Horses	Mtr Veh
Division Headquarters (81 Div)	43	2	31	30	251	64	35
Division Headquarters (93 Div)	43	2	31	30	253	64	25
Signal Unit	6	1	4	20	208	44	0
Cavalry Regiment (93 Div)	25	5	23	71	548	783	0
Each Infantry Regiment (81 Div)	113	19	35	385	4,027	915	0
Each Infantry Regiment (93 Div)	114	19	35	381	3,994	870	0
Field Artillery Regiment (81 Div)	88	11	36	157	2,349	1,771	0
Mountain Artillery Regiment (93 Div)	94	13	38	187	3,732	2,178	0
Anti-Tank Battalion	17	4	8	36	415	0	69
Engineer Battalion	28	3	12	80	839	128	10
Transport Regiment	23	0	25	35	666	296	76
Medical Unit	30	3	30	30	436	59	0

China and Manchuria 1944

At the same time the 47th Division was activated for duty overseas and the 112th Division was activated in Manchuria. Because of their deployment they did not share the organization of the homeland divisions. Nor, of course, did two occupation divisions, 114th and 118th, organized in China at the same time.

The new 47th Division, activated in Japan, was shipped to China in September 1944 with the following organization:

	Off	WO	MSG	SGT	PVT	Total	Horses
Division Headquarters	37	2	29	30	299	397	64
Signal Unit	6	1	3	20	209	239	44
Reconnaissance Regiment	28	6	23	61	715	833	718
Three Infantry Regiments, each	98	14	35	289	3,400	3,836	1,079
Mountain Artillery Regiment	101	13	42	158	3,098	3,412	511
Engineer Regiment	27	3	12	80	839	961	125
Transport Regiment	35	5	28	68	878	1,014	472
Medical Unit	37	0	32	41	996	1,106	132
Two Field Hospitals, each	24	0	40	1	236	303	83
One Field Hospital	25	0	41	1	241	308	87
Ordnance Duty Unit	2	1	10	6	93	112	0
Total	640	73	405	1,334	18,040	20,492	5,556

With an authorized strength of over 20,000 men it was the largest division in the China theater. Each infantry regiment consisted of three infantry battalions, an infantry gun company, a signal company and a mounted platoon. An infantry battalion included three rifle companies (each 9 Type 11 LMGs and 9 grenade dischargers), a machine gun company (8 MGs) and an infantry gun company (four 70mm howitzers or 81mm mortars). The regimental gun company was equipped with two 75mm mountain guns and two 37mm anti-tank guns. The divisional mountain artillery regiment was made up of three battalions, each with three 3-gun batteries of 75mm guns. The cavalry regiment had two mounted squadrons and a 6-gun machine gun squadron.

The 112th Division was a triangular formation with the following strengths:

	Off	WO	MSG	SGT	PVT	Total	Horses
Division HQ	30	1	29	15	80	155	17
Signal Unit	5	1	5	17	179	207	40
Three Infantry Regiments, each	97	14	35	289	2,973	3,408	734
Artillery Unit	23	3	9	42	540	617	476
Engineer Unit	6	1	4	22	255	288	43
Transport Unit	35	5	28	68	878	1,014	472
Medical Unit	7	1	9	5	103	125	31
Ordnance Duty Unit	2	1	10	6	93	112	0
Total	399	55	199	1,042	11,047	12,742	3,281

In October 1944 a warrant officer, four sergeants and forty privates were added to the authorized strength of each infantry regiment, bringing them up to a total of 3,453; and on 7 December a further two company-grade officers, three sergeants and eighteen privates were added, increasing strength to 3,476. These gave the division an authorized strength on 1 January 1945 of 14,566. The most significant change, however, came in mid-July when the 3-battery artillery unit was upgraded to a 1,923-man regiment of three 3-battery battalions, and a 1,130-man raiding battalion added.

The 114th Division (not related to the earlier 114th Division deactivated in 1939) and the 118th Division were, like the earlier occupation divisions, formed by expansion of existing brigades.

114th Division (ex-3rd Independent Infantry Brigade)
- 83rd Infantry Brigade (199th, 200th, 201st, 202nd IIBns)
- 84th Infantry Brigade (381st, 382nd, 383rd, 384th IIBns)

118th Division (ex-9th Independent Infantry Brigade)
- 89th Infantry Brigade (223rd, 224th, 225th, 226th IIBns)
- 90th Infantry Brigade (392nd, 401st, 402nd, 403rd IIBns)

	Off	WO	MSG	SGT	PVT	Total	Horses
Division HQ	29	1	26	18	80	154	17
Signal Unit	15	4	4	36	270	329	44
Two Brigades, each							
Brigade HQ	6	0	2	8	34	50	3
4 Infantry Bns, each	44	13	14	117	1045	1233	108
Engineer Unit	6	2	5	14	150	177	6
Transport Unit	23	4	21	28	791	867	632
Field Hospital	26	0	33	9	284	352	45
Veterinary Unit	4	1	7	1	32	45	11

Each of the independent infantry battalions was composed of five rifle companies (each 9 Type 11 LMGs and 9 grenade dischargers), a machine gun company (8 MGs) and an infantry gun company (two 70mm howitzers and two 75mm mountain guns). The divisional transport unit had two draft companies.

In common with other China occupation divisions, the 114th and 118th Divisions were strengthened somewhat in early 1945. The engineer unit in both divisions was enlarged to a 901-man 3-company regiment, while fire support elements were added. In the case of the 114th Division this took the form of a 583-man mountain artillery battalion with three 4-gun batteries of 75mm guns added in July, and in the 118th Division a 577-man trench mortar battalion of three companies in February.

Two further occupation divisions were formed a few days later in China, the 115th and 177th Divisions. They were organized along the same lines as the 114th and 118th Divisions and were similarly reinforced, with mountain artillery battalions, in early 1945. In June 1945 the 117th Division was transferred to Manchuria where it was destroyed in the Soviet invasion.

- 115th Division (ex-7th IMB)
  - 85th Infantry Brigade (26th, 27th, 28th, 29th IIBns)
  - 86th Infantry Brigade (30th, 385th, 386th, 387th IIBns)
- 117th Division (ex-4th Independent Infantry Brigade)
  - 87th Infantry Brigade (203rd, 204th, 205th, 206th IIBns)
  - 88th Infantry Brigade (388, 389, 390, 391 IIBns)

These units had the following composition:

	Off	WO	MSG	SGT	PVT	Total	Horses
Division HQ	30	1	30	18	76	155	17
Signal Unit	15	4	4	36	270	329	44
Two Infantry Brigades, each							
Brigade HQ	6	0	2	8	34	50	3
Four Infantry Bns, each	44	13	14	117	1045	1,233	108
Engineer Unit	6	2	5	14	150	177	6
Transport Unit	23	4	21	28	754	830	632
Field Hospital	26	0	33	9	284	352	45
Veterinary Unit	4	1	7	1	32	45	11
Changes of 1 Feb 45:							
Engineer Unit changed to	24	3	14	77	783	901	119
Artillery Unit added	21	3	9	45	499	577	271

Mid-July 1944 also saw the formation of two further triangular divisions, the 66th Division in China and the 111th Division in Manchuria.

The 66th Division was notable was notable as a hybrid of the China-type organization modified by the incorporation of an anti-tank battalion needed to meet Western forces as part of the 10th Area Army on Formosa. The division was organized with the following components:

	Off	WO	MSG	SGT	PVT	Total	Horses
Division Headquarters	30	0	45	0	80	155	17
Signal Unit	15	4	5	36	270	330	44

	Off	WO	MSG	SGT	PVT	Total	Horses
Three Infantry Regiments, each	89	14	28	251	2,672	3,054	189
Mortar Unit	21	3	9	45	499	558	271
Anti-Tank Unit	17	4	8	36	415	480	0
Engineer Unit	4	2	4	15	148	173	0
Transport Unit	18	4	11	33	232	298	208
Medical Unit	19	17	30	34	598	698	0
Field Hospital	23	0	36	1	146	206	0
Ordnance Duty Unit	2	1	9	4	64	80	0

The infantry regiments were supposed to have a strength of 3,853 each but IGHQ reduced this by 5 company-grade officers, 30 sergeants and 749 privates, all of the infantry branch, on activation and this reduction was never restored. The mortar battalion, on the other hand, was strengthened, with the addition of 64 sergeants in August 1944, and then 30 more sergeants and 158 privates in January 1945.

The 111th Division was built around three 3,408-man infantry regiments (increased to 3,482 men in December), a 617-man artillery battalion, a 288-man engineer unit, a 1,104-man transport unit with 472 horses and 122 motor vehicles, and a 1,109-man medical unit.

This pace of force structure expansion could not be maintained and from August 1944 to the end of the year only three more divisions were formed, the 108th and 119th in Manchuria and the 120th in Korea.

The 108th and 120th divisions were organizationally near repeats of the 111th, and had a strength of 13,643. The bulk of these men were found in the three infantry regiments, each of which had a strength of 3,409 divided into three battalions, an infantry gun company, a mounted platoon and a signal company. The divisional artillery unit was actually a 617-man artillery battalion with 476 horses, equipped with four Type 95 field guns and eight Type 91 field howitzers. These were the first divisions to be organized in Manchuria whose initial organization actually reflected the severe shortages of artillery in that theater, although they were at full strength in personnel.

The 108th Division was reinforced by the 171st Cavalry Regiment in early 1945, and on 1 August a raiding battalion was formed and the artillery battalion expanded to a regiment of three battalions each of three 3-gun batteries of 75mm guns, probably Type 95s. The raiding battalion was formed by pulling men from other divisional units, including 280 from each infantry regiment, for a total of 1,000. By the outbreak of war against Russia in August 1945 the divisions boasted actual strengths of around 18,000 each.

The 119th Division was built around the elements left behind when the main body of the 23rd Division moved to the Philippines. In particular, it benefited from the motorized artillery regiment, which had stayed behind in its entirety, so that the division simply renamed the former 13th Field Artillery Regiment into the 119th and thus acquired three battalions of 75mm Type 90 field guns and a battalion of 15cm Type 96 howitzers.

The Draw-Down of the Kwantung Army

The three divisions were raised in an attempt to compensate for the continuing draw-down of forces under the Kwantung Army to reinforce active theaters. During February 1944 the 14th and 29th Divisions had been withdrawn for conversion to ocean divisions and the 27th Division for operations in China. In June the 28th Division (less the 36th Regiment)

Table 2.18: 1st Division (in Manchuria), from 26 October 1943

						Army			Manchurian		
	Officers	Warrant Officers	NCOs	Privates	Manchurians	Riding Horses	Pack Horses	Draft Horses	Pack Horses	Draft Horses	Main Weapons & Equipment
Division Headquarters	36	2	42	148	0	48	0	0	0	0	
Signal Unit	4	1	21	162	5	7	0	22	0	20	
Reconnaissance Regiment											
Regiment HQ	6	0	15	52	35	72	4	0	30	0	
Mounted Squadron	5	2	16	57	0	77	16	0	0	0	5 light MGs, 9 GLs
Three Infantry Regiments, each											
Regiment HQ	7	1	12	23	8	6	0	10	0	14	
Signal Company	3	1	10	79	5	3	16	0	0	2	
Three Infantry Battalions, each											
Battalion HQ	5	0	8	84	22	8	4	0	52	68	
4 Rifle Companies, each	4	1	14	99	0	0	0	0	0	0	9 light MGs, 9 GLs
Machine Gun Company	6	1	13	109	0	1	35	0	0	0	8 medium MGs
Infantry Gun Platoon	1	0	2	45	0	1	17	0	0	0	2 70mm howitzers
Artillery Battalion											
Battalion HQ	3	0	6	18	6	4	0	0	0	22	
Infantry Gun Company	3	1	10	106	0	5	0	38	0	0	4 75mm regtal guns
Anti-Tank Company	3	1	9	78	0	?	0	18	0	0	4 37mm AT guns

Table 2.18: 1st Division (in Manchuria), from 26 October 1943

						Army			Manchurian		
	Officers	Warrant Officers	NCOs	Privates	Manchurians	Riding Horses	Pack Horses	Draft Horses	Pack Horses	Draft Horses	Main Weapons & Equipment
Field Artillery Regiment											
Regiment HQ	8	0	18	97	0	39	8	17	0	6	
Three Light Battalions, each											
Battalion HQ	8	0	10	55	0	24	0	17	0	20	
3 Batteries, each	4	1	9	100	0	27	0	71	0	0	3 105mm howitzers
Heavy Battalion											
Battalion HQ	8	0	10	55	0	24	0	17	0	16	
2 Batteries, each	4	0	11	94	0	31	0	65	0	0	2 150mm howitzers
Engineer Regiment											
Regiment HQ	7	0	14	30	20	9	3	0	0	38	
Three Engineer Companies, each	4	1	12	116	0	5	0	24	0	0	4 light MGs
Materiel Platoon	1	0	2	42	0	3	0	0	0	36	
Transport Regiment											
Regiment HQ	6	0	10	15	0	0	0	0	0	0	
Two Transport Companies, each	3	0	10	46	0	0	0	0	0	0	16 trucks
Ordnance Repair Section	2	1	30	51	0	0	0	0	0	0	7 trucks
Decontamination Unit	4	1	15	73	0	0	0	0	0	0	5 trucks

Table 2.18: 1st Division (in Manchuria), from 26 October 1943

						Army			Manchurian		
	Officers	Warrant Officers	NCOs	Privates	Manchurians	Riding Horses	Pack Horses	Draft Horses	Pack Horses	Draft Horses	Main Weapons & Equipment
Medical Unit											
Unit HQ	7	0	8	56	0	8	0	7	0	25	
Stretcher-Bearer Company	1	4	9	93	0	1	0	0	0	0	
Veterinary Depot	2	0	6	30	0	3	0	0	0	0	5 trucks

and the 9th Division had been pulled out to reinforce the Ryukyus. July saw the departure of the 1st Division to the Philippines, the 10th Division to Formosa, and the 24th Division to the Homeland. In August the 8th Division also went to the Philippines, followed by the 23rd Division in October. Finally, the 12th Division also went to the Philippines in December.

The divisions that were transferred out of the Kwantung Army often bore little similarity to the proud formations of 1940–42. A primary reason was the continual administrative attrition suffered by these units as troops and equipment were pulled out to serve as replacements for active combat units. The second was the shortage of shipping that meant that each formation sent overseas had to first be stripped to its bare minimum in order to reduce shipping requirements.

The divisions in the Kwantung Army had been allowed to run down to peacetime strengths starting in December 1942. The strength of a motorized company in a reconnaissance regiment, for instance, fell from 150 men to 90.

An example of a reduced-strength division was the 1st Division as reorganized in October 1943 by the Kwantung Army. The infantry battalions in this division each included four rifle companies, but company strength was reduced drastically to four officers, a warrant officer, 14 NCOs and 99 privates (including 2 medics). Similarly, the machine gun company was restructured to four platoons each of two guns. Other elements of the infantry regiment were also reduced in strength, although without sacrificing firepower. The reconnaissance regiment was stripped of all but its HQ and mounted squadron, the motorized and maintenance elements being used to constitute new units elsewhere. The artillery regiment, although provided with powerful weapons (by Japanese standards) was reduced in establishment to two to three weapons per battery. The engineer regiment was reduced in strength throughout, while the transport regiment was reduced to two motor transport companies.

Particularly tempting targets were the massive all-pack A-2 type infantry regiments of the 9th, 11th and 25th Divisions, all still sitting relatively inert in Manchuria. They had each received authorization for an additional 45 men in late 1942, bringing their strengths up to 5,059, but it was all downhill after that. A decrement of 153 men was imposed in July 1943 on all the regiments, then varying amounts on the different units, totaling reductions of 2,179 and 201 horses for each of the regiments of the 11th Division, 931 for those of the 25th Division, and 1,515 and 284 horses for those of the 9th Division before it shipped out for Okinawa.

Further, by the time they had been nominated for transfer out of the Kwantung Army the 1st, 8th, 9th, 12th and 24th Divisions had all contributed an expeditionary unit to the Pacific theater in February 1944. This consisted of pulling one battalion out of each infantry and artillery regiment in the division, together with an engineer company and signal unit.

Building these divisions back up for operations in an active combat theater was not a simple matter. In the case of the once-proud 1st Division the most pressing need was to reconstitute the missing third battalion in each infantry regiment. That was accomplished by taking a rifle company plus additional rifle and machine gun personnel from each of the two remaining battalions to yield three battalions, each with three rifle companies and a machine gun company. At the same time the artillery battalion HQ in each infantry regiment was abolished, and the personnel distributed. Some recent inductees were

Table 2.19: 23rd Division, from 19 October 1944

	Officers	Enlisted	Pistols	Rifles	Light MGs	Medium MGs	Light Gren Lnchrs	Heavy Gren Lnchrs	37mm AT Guns	70mm Inf Howitzers	75mm Mountain Guns	105mm Howitzers	150mm Howitzers	Tankettes (MG)	Tankettes (37mm)	Cars	Trucks	Riding Horses	Pack Horses	Draft Horses
Division Headquarters	47	188	6	126	2	2	0	0	0	0	0	0	0	0	0	7	23	n/a	n/a	n/a
Reconnaissance Regiment																				
Regiment HQ	6	50	n/a	n/a	0	0	0	0	0	0	0	0	0	0	0	1	7	0	0	0
Signal Platoon	1	48	n/a	n/a	0	0	0	0	0	0	0	0	0	0	0	0	5	0	0	0
Two Motorised Squadrons, each	6	123	n/a	n/a	6	2	0	6	1	0	0	0	0	0	0	0	9	0	0	0
Two Tankette Squadrons, each	4	34	n/a	n/a	0	0	0	0	0	0	0	0	0	5	3	0	2	0	0	0
Total reconnaissance regiment	27	412	58	340	12	4	0	12	2	0	0	0	0	10	6	1	34	0	0	0
Three Infantry Regiments, each																				
Regiment HQ	11	86	n/a	n/a	0	0	0	0	0	0	0	0	0	0	0	0	11	6	0	0
Signal Company	4	129	n/a	n/a	0	0	0	0	0	0	0	0	0	0	0	0	0	3	19	1
Three Infantry Battalions, each																				
Battalion HQ	7	96	n/a	n/a	0	0	1	0	0	0	0	0	0	0	0	0	0	11	6	42
Three Rifle Companies, each	5	175	n/a	n/a	9	0	0	9	0	0	0	0	0	0	0	0	0	0	0	0
Machine Gun Company	6	114	n/a	n/a	0	8	0	0	0	0	0	0	0	0	0	0	0	1	18	10
Infantry Gun Platoon	1	43	n/a	n/a	0	0	0	0	0	2	0	0	0	0	0	0	0	1	0	9
Infantry Gun Company	4	119	n/a	n/a	0	0	0	0	0	0	4	0	0	0	0	0	0	1	1	36
Anti-Tank Company	4	68	n/a	n/a	0	0	0	0	4	0	0	0	0	0	0	0	0	1	1	11
Total infantry regiment	110	2736	76	1570	81	24	3	81	4	6	4	0	0	0	0	0	11	50	93	231

Table 2.19: 23rd Division, from 19 October 1944

	Officers	Enlisted	Pistols	Rifles	Light MGs	Medium MGs	Light Gren Lnchrs	Heavy Gren Lnchrs	37mm AT Guns	70mm Inf Howitzers	75mm Mountain Guns	105mm Howitzers	150mm Howitzers	Tankettes (MG)	Tankettes (37mm)	Cars	Trucks	Riding Horses	Pack Horses	Draft Horses
Field Artillery Regiment																				
Regiment HQ	13	146	n/a	n/a	2	0	0	0	0	0	0	0	0	0	0	0	0	44	11	12
Light Field Howitzer Battalion																				
Battalion HQ	11	172	n/a	n/a	0	0	0	0	0	0	0	0	0	0	0	0	0	34	0	81
Three Batteries, each	5	131	n/a	n/a	0	0	0	0	0	0	0	4	0	0	0	0	0	35	0	77
Service Battery	3	84	n/a	n/a	0	0	0	0	0	0	0	0	0	0	0	0	0	21	0	62
Heavy Field Howitzer Battalion																				
Battalion HQ	11	194	n/a	n/a	0	0	0	0	0	0	0	0	0	0	0	0	0	36	0	98
Three Batteries, each	5	187	n/a	n/a	0	0	0	0	0	0	0	0	4	0	0	0	0	42	0	121
Regimental Service Battery	4	160	n/a	n/a	2	0	0	0	0	0	0	0	0	0	0	1	27	0	0	0
Total artillery regiment	75	1823	138	153	4	0	0	0	0	0	0	12	12	0	0	1	27	392	11	939
Transport Regiment																				
Regiment HQ	8	53	n/a	n/a	0	0	0	0	0	0	0	0	0	0	0	1	2	0	0	0
Wagon Company	7	391	n/a	n/a	3	0	0	0	0	0	0	0	0	0	0	0	0	42	0	254
Two Motor Companies, each	4	141	n/a	n/a	3	0	0	0	0	0	0	0	0	0	0	0	37	0	0	0
Total transport regiment	23	726	27	293	9	0	0	0	0	0	0	0	0	0	0	1	76	42	0	254
Ordnance Duty Unit	4	77	0	17	0	0	0	0	0	0	0	0	0	0	0	1	5	0	0	0

Table 2.19: 23rd Division, from 19 October 1944

	Officers	Enlisted	Pistols	Rifles	Light MGs	Medium MGs	Light Gren Lnchrs	Heavy Gren Lnchrs	37mm AT Guns	70mm Inf Howitzers	75mm Mountain Guns	105mm Howitzers	150mm Howitzers	Tankettes (MG)	Tankettes (37mm)	Cars	Trucks	Riding Horses	Pack Horses	Draft Horses
Medical Unit																				
Unit HQ	21	211	n/a	n/a	0	0	0	0	0	0	0	0	0	0	0	0	0	13	0	59
Three Stretcher Companies, each	4	95	n/a	n/a	0	0	0	0	0	0	0	0	0	0	0	0	0	1	0	0
Ambulance Company	4	166	n/a	n/a	0	0	0	0	0	0	0	0	0	0	0	0	0	4	0	21
Total medical unit	37	662	30	17	0	0	0	0	0	0	0	0	0	0	0	0	0	20	0	80
Field Hospital	22	220	0	47	0	0	0	0	0	0	0	0	0	0	0	1	21	0	0	0
Veterinary Depot	5	47	0	9	0	0	0	0	0	0	0	0	0	0	0	0	2	5	0	5
Hygiene & Water Supply Company	11	185	2	51	0	0	0	0	0	0	0	0	0	0	0	1	20	0	0	0

Note: "officers" includes warrant officers.

A 75mm regimental gun of the Kwantung Army demonstrating its direct fire mode.

forthcoming from the depots in Japan and this permitted the rifle and MG companies to be built back up to a strength of 150 each. The withdrawal of units to form the expeditionary units had not changed the regiments' authorized strength, since this was regarded as a temporary expedient, with new battalions to be dispatched in replacement. When this did not happen IGHQ was finally forced, in July 1944, to issue a new authorized strength for the division's infantry regiments, a total of 2,200 personnel, slightly less than half of the 4,487 men authorized when those regiments were first mobilized, along with only 30 horses for transport.[30]

The division's field artillery regiment was reconfigured back to three light field and one heavy field artillery battalions, each of three batteries, but in this case with three weapons per battery. In the light battalions one of the batteries was equipped with Type (Meiji) 38 field guns and the other two with 105mm Type 91 howitzers, while in the heavy battalion all batteries used 150mm Type (Taisho) 4 howitzers. Most horses were stripped from the organization, and others lost at sea, so when the artillery landed at Leyte the guns had to be moved inland by borrowed trucks and tanks. In preparation for deployment the division's transport regiment was built back up to two horse-drawn companies and two motor companies, the latter each with 45 Chevrolet trucks. The horses, however, were left behind so in combat the first two companies functioned as labor troops. The engineer regiment was reconfigured back into three companies, but without any additional troops, to yield a total of 600 men (with 15 horses).

30 The same process was applied to the A-1 type infantry regiments of the Korea-based 19th Division, which had to rebuild in October 1944 for the Philippines after losing a battalion apiece to the expeditionary units in February of that year.

An example of a division that had not had to contribute an expeditionary unit to the Central Pacific was the 23rd Division. Although it had not been tapped for an expeditionary unit, normal administrative attrition combined with a lack of replacement personnel had reduced its strength considerably. The division was further "streamlined" in late 1944 to conserve shipping space in anticipation of its move to the Philippines and excess personnel transferred to the 8th Border Garrison Unit.

The revised 23rd Division, the original semi-motorized division in the Army, was barely a shell of its former self. Its motorized 13th Artillery Regiment was pulled out of the division and redesignated the 119th Field Artillery in October. A month later its place was taken by the horse-drawn 17th Field Artillery. The new regiment had only one battalion of light field howitzers and one of old 15cm Type 4 (1915) models with draft horses. The transport regiment traded in five motor transport companies for one horse-drawn wagon company.[31] The infantry battalions lost a rifle company apiece. The only aspect in which the division gained capability was a slight increase in the number of radios assigned, most significantly, the reconnaissance regiment now having eight MK 4 and four MK 5 radios.

Interestingly, the organization tables issued by the division HQ on 19 October 1944 (shown in Table 2.19) differ in many respects from the TO&E and authorized strengths issued for the division by IGHQ on 10 October. Under the IGHQ authorizations the strengths of the main components of the division were to be:

	Off	WO	MSG	SGT	PVT	Total	Horses
Each Infantry Regiment	94	15	31	281	2,489	2,910	413
Field Artillery Regiment	82	12	33	132	1,377	1,636	1,274
Engineer Regiment	24	3	13	74	784	898	119
Transport Regiment	23	0	25	35	666	749	296
Medical Regiment	22	4	5	61	347	439	35

The division had retained its unique reconnaissance regiment organization of two motorized rifle companies and two tankette companies, although by this time the motor transport company had been broken up and the vehicles directly assigned as needed. However, *en route* to Luzon the 2nd (motorized) and 3rd (tankette) companies were lost to submarine attack, leaving a regiment strength of 263 men.

Another reduced-strength division was the 10th, shipped to Formosa with infantry regiments reduced to 2,114 apiece and the artillery regiment to 1,700. Each of the infantry regiments now consisted of a 150-man headquarters group, three 509-man battalions, a 125-man signal company, an 80-man AT company, a 22-man separate AT platoon, a 110-man regimental gun company, and a 100-man pioneer company. Each of the battalions was made up of a 27-man headquarters, three 100-man rifle companies, a 136-man machine gun company and a 46-man infantry gun platoon. A rifle company had three platoons, each of three 6-man rifle squads and a 7-man grenadier squad. Its artillery regiment had excess weapons, but was almost completely devoid of transport and well understrength. Its third battalion, for instance, consisted of a 60-man HQ and three 84-man batteries, one battery having eight 75mm Type 95 field guns, and the other two each with seven 105mm howitzers. The battalion had telephones, but no radios.

31 This left one motor transport company with eight trucks for 105mm ammunition, nineteen for 150mm ammunition, three for fuel, five for baggage and two as reserve.

Similarly, the 8th Division was directed on 26 July to strip down to 12,740 men, including 2,500 in each infantry regiment, 200 in the reconnaissance regiment and 2,200 in the artillery regiment.

Local Reorganization 1944

Although the Type C division never became the standard envisioned in the 1944 mobilization plan, in fact several of the divisions in operational areas had been modifying their organizations to conform to local requirements in a manner that strongly suggested the Type C. The reconnaissance regiment of the 2nd Division had reorganized to three motorized companies (with organic motor vehicles), a tankette company (three platoons each three vehicles) and a signal platoon. In 1945 this was reconfigured as two motorized companies and two tankette companies (each two 3-tankette platoons). Divisional artillery had been reduced to one 75mm gun battalion and two 105mm howitzer battalions.

The 55th Division (in Burma) had been reduced to three rifle companies per battalion since 1943, and the 51st (in New Guinea) had been unofficially similarly reduced due to heavy losses the same year.[32] The 6th Division's infantry was reorganized in January 1944 with the MG company being broken up to provide a medium MG to each rifle platoon and the infantry gun platoons withdrawn from the battalions and centralized under the regimental gun company.

The 26th Division had lost its battalion gun platoons by mid-1944, but had gained a mounted platoon in each infantry regiment. This may have partially, but by no means completely, compensated for the lack of divisional reconnaissance elements. The division also had only six 75mm field gun batteries (and no battalion HQs) in its field artillery regiment, and only two engineer and three transport companies in those regiments. On the other hand, it received an additional allocation of twelve 37mm AT guns when it shipped out for the Philippines in July 1944. Unfortunately, only four of its artillery pieces successfully made it to the Philippines, so the regiment had to be brought up to strength locally almost from scratch.

Another example of local reorganization is the 54th Division, which on 18 October 1944 ordered each of its infantry regiments to form a colors platoon, three pioneer platoons and a mortar company, and the artillery regiment to form two more mortar companies. The regimental colors platoon was organized as a regular rifle platoon with three 11-man rifle squads and an 11-man grenadier squad, to be used for protection of the regimental HQ. A pioneer platoon was to consist of two 13-man squads specializing in assault preparations and anti-tank attacks, a 13-man squad specializing in earthwork construction and jungle penetration, and a 12-man squad specializing in the placing and removing of land mines. No information, unfortunately, has survived on the organization of the mortar companies. Presumably, these weapons were found much more useful in the jungles of Burma than artillery.

To counteract the continuing drawdown of the Kwantung Army eight new divisions were officially activated in Manchuria on 16 January 1945, each consisting of three infantry regiments, a field artillery regiment, a transport regiment and an engineer regiment, with a total authorized strength of 23,000 men. Unfortunately, these divisions did not represent any real increase in Kwantung Army strength since for the most part they were simply

32 Officially the 51st Division retained its original "B-type" organization through the war.

reorganizations of existing combat units:

121st Division: from left-behind elements of 28th Division
122nd Division: from miscellaneous troops
123rd Division: from 73rd Independent Mixed Brigade
124th Division: from left-behind elements of 111th Division
125th Division: from 13th Border Garrison Unit
126th Division: from 12th Border Garrison Unit
127th Division: from 9th Border Garrison Unit
128th Division: from left-behind elements of 120th Division

These divisions were raised with the following strengths:

	Off	WO	MSG	SGT	PVT	Total	Horses
Division Headquarters	30	1	29	15	80	155	17
Signal Unit	5	1	5	17	179	207	0
Three Infantry Regiments, each	100	14	35	289	2,972	3,410	734
Artillery Unit	23	3	9	42	540	617	476
Engineer Unit	6	1	4	22	255	288	43
Transport Unit (added Jan 45)	33	4	26	57	764	884	472
Veterinary Hospital	7	1	9	5	97	119	45
Ordnance Duty Unit (added May 45)	2	1	10	6	93	112	0
Total (original)	371	49	161	968	10,302	11,851	2,783

In July 1945 a 1,130-man "raiding" battalion was added to each division and the division artillery component expanded to 1,923 men with 1,466 horses.

Probably typical of this group, the 127th Division had on hand at the start of hostilities in Manchuria eighteen 75mm guns, 10 mountain artillery pieces and four 105mm howitzers in its artillery regiment. Each infantry regiment held 81 light machine guns, 18 medium machine guns, 48 grenade dischargers, 6 battalion guns (old 37mm models in a few cases) and six 37mm AT guns. The shortage of grenade dischargers appears to have been common. One division, the 125th, acquired a reconnaissance capability when they inherited the 57th Reconnaissance Regiment (one tankette company and one mounted company) in May from the 57th Division that was departing for Japan.

A relatively fortunate division was the 123rd, which was at or near strength in all its components and, indeed, overstrength in its infantry regiments. Nevertheless, rifle company strength varied from a low of 94 men in 3rd Co/270th Infantry to as many as 350 men in the 10th Co/269th Infantry, with an average strength of 194.

Consolidating in China 1945

Shortly thereafter, the China theater began reorganizing many of its smaller units into occupation divisions. The 131st, 132nd and 133rd Divisions were officially activated on 1 February 1945, followed by the 129th, 130th, and 161st on 12 April. Although the 19th IMB was disbanded to provide troops for the second batch, for the most part the source of personnel were field replacement units.

129th Division
- 91st Infantry Brigade (98th, 278th, 279th, 280th IIBns)
- 92nd Infantry Brigade (101st, 588th, 589th, 590th IIBns)

130th Division
- 93rd Infantry Brigade (97th, 99th, 100th, 277th IIBns)
- 94th Infantry Brigade (281st, 620th, 621st, 622nd IIBns)

131st Division
- 95th Infantry Brigade (591st, 592nd, 593rd, 594th IIBns)
- 96th Infantry Brigade (595th, 596th, 597th, 598th IIBns)

132nd Division
- 97th Infantry Brigade (599th, 600th, 601st, 602nd IIBns)
- 98th Infantry Brigade (603rd, 604th, 605th, 606th IIBns)

133rd Division
- 99th Infantry Brigade (607th, 608th, 609th, 610th IIBns)
- 100th Infantry Brigade (611th, 612th, 613th, 614th IIBns)

161st Division
- 101st Infantry Brigade (475th, 476th, 477th, 528th IIBns)
- 102nd Infantry Brigade (478th, 479th, 480th, 481st IIBns)

Each of the independent infantry battalions had a strength of 1,358 men and were organized along the lines of those of the 114th Division. Each brigade HQ had 117 men, giving a brigade strength of 5,549.

Division base elements included a 329-man HQ, a 901-man engineer regiment, a 95-man ordnance duty unit, a 230-man signal unit, two 495-man field hospitals, a 119-man veterinary unit, and a 239-man water purification & supply unit. Also included in each division was a transport regiment and an artillery regiment. The 131st, 132nd, and 133rd Divisions each had a transport regiment of three draft companies, while the 129th Division had a 583-man regiment of one draft and one motor company, the 130th Division had an 864-man regiment of two draft companies, and the 161st Division had two draft companies and one water transport company.[33]

The artillery component was equally disparate. The 129th and 130th Divisions each had a 777-man field artillery group (nominally a regiment) with two 75mm gun batteries, two 105mm howitzer batteries, and a 6-gun trench mortar battery. The 131st and 132nd Division Artillery Units were simply mountain artillery battalions with three 4-gun batteries of 75mm guns each. The 133rd Division Artillery Unit was a field artillery battalion with twelve 75mm guns, while the 161st Division Artillery Unit was similar but with 105mm howitzers.

These elements gave TO&E strengths that varied from 14,184 men and 2,148 horses in the 129th Division, to 17,210 men and 2,852 horses in the 131st and 132nd Divisions. Few new personnel would arrive in the China theater after the start of 1945 and this cannibalization of field replacement units represented a move borne of short-term desperation. This last reservoir exhausted, no further divisions would be formed in China.

33 The strength figures are those given to Allied forces at the end of the war by the GHQ, Japanese Expeditionary Army in China and translated by SINTIC in October 1945 as DT-157. IGHQ records show they authorized infantry battalion strength for these divisions at 1,549 with 130 horses. The lower local strength presumably reflects an attempt to distribute manpower shortages.

The attempt to stabilize the strength of forces in northern Asia continued in February 1945 with the activation of the 79th Division in northern Korea from the 19th Depot Division, and the 96th Division on Saishu Island off the Korean coast, both triangular divisions. The former was a full-strength division with an authorized strength of 18,471. In addition to the three 3,544-man infantry regiments it had a mountain artillery regiment (two battalions of 75mm), a cavalry regiment and the normal services, except medical units.[34] It was at full strength and its personnel were considered high quality due to the fact that they had served together as a depot division for some time. The latter division, on the other hand, had small 2,432-man infantry regiments, a trench mortar battalion for artillery and no transport unit, yielding a strength of only 8,745.

At the same time, the 306th Infantry Regiment was moved to Sakhalin, allowing the Karafuto Brigade to be strengthened and redesignated the 88th Division.

	Off	WO	MSG	SGT	PVT	Total	Horses
Division HQ	63	2	35	76	593	770	64
Signal Unit	29	4	15	99	999	1146	131
25th Infantry Regiment	135	22	40	398	3241	3836	424
125th Infantry Regiment	116	18	39	373	2961	3507	400
306th Infantry Regiment	143	35	42	421	4311	4952	871
88th Mountain Artillery Regiment	62	12	32	131	1595	1832	896
88th Engineer Regiment	30	4	15	100	1003	1152	142
88th Transport Regiment	60	0	67	81	1792	2000	1449
Ordnance Unit	2	1	10	6	93	112	0
Medical Unit	33	0	33	38	900	1004	53
Veterinary Unit	5	1	9	4	58	77	11

Similarly, two IMBs in the southern Kuriles were reinforced by two additional independent infantry battalions to form the 89th Division. This new division consisted of the 3rd (ex-43rd Ind) and 4th (ex-69th Ind) Mixed Brigades (the 1st and 2nd having been used to form the 109th Division) with some supporting troops:

	Off	WO	MSG	SGT	PVT	Total	Horses
89th Division HQ	85	2	71	44	575	777	64
Division signal unit	6	1	4	20	195	226	33
3rd Mixed Brigade							
Brigade HQ	21	0	10	23	244	298	8
Brigade signal unit	10	2	0	30	236	278	16
6 infantry battalions each	38	7	11	119	810	985	89
Brigade artillery unit	31	6	6	68	624	735	99
Brigade ordnance unit	3	0	14	4	133	154	0
4th Mixed Brigade							
Brigade HQ	21	0	10	23	244	298	8
Brigade signal unit	4	1	3	26	192	226	0

34 In July a 600-man rocket launcher battalion was attached to the 290th Infantry Regiment, but it actually had no rocket weapons. Other than that, the division was fully equipped.

4 infantry battalions each	32	6	6	101	669	814	29
Brigade artillery unit	22	6	5	43	389	465	112
Brigade labor unit	2	1	1	11	142	157	0
Division ordnance unit	2	1	10	6	93	112	0
Division medical unit	33	0	33	38	900	1004	0
Division field hospital	18	0	25	0	80	123	0
Other	10	1	19	21	212	263	0

Homeland Defense: the Coastal Divisions

By this time it was clear that the decisive battles were to be fought on the Japanese homeland. All efforts were hereafter geared towards the defense of that area.

On 28 February 1945 sixteen coastal defense divisions were ordered activated, of which eleven (140th, 144th–147th, 150th, 152nd, 154th–156th, and 160th) were officially activated immediately and five (142, 143rd, 151st, 153rd, and 157th) following in early April. Each of these divisions was built around three coastal infantry regiments and a mobile infantry regiment. The division was to hold 12km of beach with the coastal regiments on line while holding the mobile regiment for counter attack.

The basic unit of the coastal infantry regiment was the rifle platoon, which in this case consisted of three rifle squads (each eight men with one light MG and seven rifles), two machine gun squads (each six men with a medium MG), and a grenadier squad (ten men with three grenade dischargers and seven rifles). The rifle company was made up of three such platoons plus an infantry gun platoon (with two eight-man squads each with a 70mm infantry howitzer). Fire support for the coastal infantry battalion, such as it was, was provided by a three-platoon gun company that manned six 57mm guns salvaged from obsolete tanks and placed in fixed positions as anti-boat guns. The battalion was supported by a small HQ company with a transport platoon (15 one-horse carts) and a signal platoon.

At both the battalion and regimental level the coastal infantry was to be assisted by raiding units equipped, although rather sparingly, with submachine guns. A raiding company consisted of four platoons, each of three sections. A section had ten men and was armed with eight rifles, a submachine gun and a grenade discharger. The role of the raiding company was to infiltrate enemy lines and attack HQs, artillery positions and other high-priority targets.

Service support for the regiment consisted of a supply company of two platoons, a medical company of two platoons, and a signal company. The signal company was composed of a wire platoon (six squads) and a radio platoon (ten squads). The regiment was fairly well provided with a basic load of ammunition, including 140 rounds per gun of 70mm and 120 rounds per gun of 37mm ammunition, but was completely immobile. Without attached transport once in place the coastal infantry regiment could not move except by abandoning its heavy weapons, and its 57mm anti-boat guns would have to be left behind in any event.

In the division's mobile infantry the rifle company was more conventionally organized, consisting of three rifle platoons. Each of these platoons consisted of three 13-man rifle squads (each eleven rifles and a light MG) and a 19-man grenadier section (thirteen rifles and six grenade dischargers). The large number of Type 89 grenade dischargers was presumably designed to enhance their counter-attack capabilities. The mobile infantry regiment was fully provided with transport, although its communications net was not significantly better than that of the coastal infantry regiment. The mobile infantry regiment was provided with

Table 2.20: Coastal Defense Division (140–160 series), from February 1945

	Personnel	Pistols	Submachine Guns	Rifles	Light MGs	Medium MGs	Grenade Launchers	Flame Throwers	37mm AT Guns	47mm AT Guns	57mm Tank Guns	81mm Mortars	70mm Inf Howitzers	75mm Mountain Guns	75mm Field Guns	105mm Howitzers	200mm Rocket Launchers	Riding Horses	Pack Horses	Draft Horses	Wagons & Carts	Motor Vehicles
Division Headquarters	226	?	0	?	?	0	?	0	0	0	0	0	0	0	0	0	0	22	0	0	0	6
Signal Unit	235	0	0	220	0	0	0	0	0	0	0	0	0	0	0	0	0	0	0	10	9	0
Three Static Infantry Regiments, each																						
Regiment Headquarters	99	0	0	53	0	0	1	6	0	0	0	0	0	0	0	0	0	36	0	0	0	0
Signal Company	136	0	0	127	0	0	0	0	0	0	0	0	0	0	0	0	0	3	20	0	0	0
Supply Company	164	0	0	127	0	0	0	0	0	0	0	0	0	0	0	0	0	0	0	20	19	0
Medical Company	189	0	0	0	0	0	0	0	0	0	0	0	0	0	0	0	0	0	0	0	0	0
Anti-Tank Company	58	8	0	0	0	0	0	0	4	0	0	0	0	0	0	0	0	0	0	0	0	0
Three Infantry Battalions, each																						
Battalion HQ	99	0	0	59	0	0	1	0	0	0	0	0	0	0	0	0	0	1	0	15	15	0
Three Rifle Companies, each	182	?	0	?	9	6	9	0	0	0	0	0	2	0	0	0	0	0	0	0	0	0
Gun Company	98	?	0	?	0	0	0	0	0	0	6	0	0	0	0	0	0	0	0	0	0	0
Raiding Company	146	?	12	110	0	0	12	0	0	0	0	0	0	0	0	0	0	0	0	0	0	0
Raiding Battalion																						
Battalion HQ	99	0	0	59	0	0	1	0	0	0	0	0	0	0	0	0	0	1	0	15	15	0

Table 2.20: Coastal Defense Division (140–160 series), from February 1945

	Personnel	Pistols	Submachine Guns	Rifles	Light MGs	Medium MGs	Grenade Launchers	Flame Throwers	37mm AT Guns	47mm AT Guns	57mm Tank Guns	81mm Mortars	70mm Inf Howitzers	75mm Mountain Guns	75mm Field Guns	105mm Howitzers	200mm Rocket Launchers	Riding Horses	Pack Horses	Draft Horses	Wagons & Carts	Motor Vehicles
Three Raiding Companies, each	146	?	12	110	0	0	12	0	0	0	0	0	0	0	0	0	0	0	0	0	0	0
Mobile Infantry Regiment																						
Regiment HQ	167	0	0	39	0	0	1	0	0	0	0	0	0	0	0	0	0	21	0	72	?	0
Signal Company	143	0	0	130	0	0	0	0	0	0	0	0	0	0	0	0	0	3	20	0	0	0
Medical Company	189	0	0	0	0	0	0	0	0	0	0	0	0	0	0	0	0	0	0	0	0	0
Two Infantry Gun Companies, each	155	8	0	0	0	0	0	0	0	0	0	0	0	4	0	0	0	5	0	35	?	0
Two Infantry Battalions, each																						
Battalion HQ	197	0	0	40	0	0	1	0	0	0	0	0	0	0	0	0	0	19	70	33	?	0
Three Rifle Companies, each	207	0	0	159	9	0	18	0	0	0	0	0	0	0	0	0	0	0	0	0	0	0
Machine Gun Company	201	12	0	0	0	12	0	0	0	0	0	0	0	0	0	0	0	1	54	0	0	0
Mortar Company	180	12	0	0	0	0	0	0	0	0	0	6	0	0	0	0	0	9	0	55	?	0
Anti-Tank Battalion	480	0	0	142	0	0	0	0	0	12	0	0	0	0	0	0	0	0	0	0	0	15

Table 2.20: Coastal Defense Division (140–160 series), from February 1945

	Personnel	Pistols	Submachine Guns	Rifles	Light MGs	Medium MGs	Grenade Launchers	Flame Throwers	37mm AT Guns	47mm AT Guns	57mm Tank Guns	81mm Mortars	70mm Inf Howitzers	75mm Mountain Guns	75mm Field Guns	105mm Howitzers	200mm Rocket Launchers	Riding Horses	Pack Horses	Draft Horses	Wagons & Carts	Motor Vehicles
Field Artillery Battalion																						
Battalion HQ	89	0	0	0	0	0	0	0	0	0	0	0	0	0	0	0	0	8	0	0	0	0
Two Batteries, each	117	0	0	0	0	0	0	0	0	0	0	0	0	0	4	0	0	3	0	0	0	0
One Battery	117	0	0	0	0	0	0	0	0	0	0	0	0	0	0	4	0	3	0	0	0	0
Service Battery	68	0	0	0	0	0	0	0	0	0	0	0	0	0	0	0	0	2	0	66	?	0
or																						
Rocket Artillery Battalion																						
Battalion HQ	64	0	0	0	0	0	0	0	0	0	0	0	0	0	0	0	0	3	0	0	0	0
Three Batteries, each	188	0	0	0	0	0	0	0	0	0	0	0	0	0	0	0	12	3	0	55	?	0
Service Battery	64	0	0	0	0	0	0	0	0	0	0	0	0	0	0	0	0	2	0	41	?	0
Transport Unit																						
Unit HQ	74	6	0	27	0	0	0	0	0	0	0	0	0	0	0	0	0	0	0	0	0	?
Draft Company	286	4	0	44	1	0	0	0	0	0	0	0	0	0	0	0	0	55	0	181	170	0
Motorized Company	121	12	0	92	1	0	0	0	0	0	0	0	0	0	0	0	0	0	0	0	0	31
Ordnance Duty Unit	108	0	0	18	0	0	0	0	0	0	0	0	0	0	0	0	0	0	0	0	0	5
Field Hospital	208	0	0	0	0	0	0	0	0	0	0	0	0	0	0	0	0	0	0	0	0	5

Note: there are a total of 237 carts and wagons in the mobile infantry regiment.

30 telephones, eight Mark 5 radios and twelve Mark 6 radios, differing from the coastal regiment only in having four fewer Mark 6 units (although with only half the number of subordinate battalions).

Supporting the coastal division was a 304-man HQ, an anti-tank battalion, an artillery unit or rocket gun unit, a transport unit, a signal unit, an ordnance duty unit, and a field hospital.

The anti-tank battalion had three companies each with four 47mm Type 1 guns. Indirect fire support was provided in May with the addition of a rocket artillery battalion (in the 142nd, 143rd, 145th, 150th, 151st and 157th Divisions) or a field artillery battalion (in the others). A divisional field artillery unit was comprised of two batteries of 75mm guns and one of 105mm howitzers, while a rocket artillery unit had three companies each with twelve 200mm Type 4 rocket launchers.

Homeland Defense: the Mobile Divisions

With coastal defenses thus strengthened the IGHQ turned its attention to the need for mobile units for the maneuver role. The 11th, 25th, and 57th Divisions and the 1st Armored Division were all ordered home from Manchuria in March.

On 2 April it ordered the formation of eight mobile infantry divisions to complement the coastal divisions and strengthen the reserves available. Six of these (201st, 202nd, 205th, 206th, 209th and 212th) were activated immediately, while the remaining two (214th and 216th) followed at the end of the month.

The basic element of the mobile infantry division was the rifle squad, consisting of twelve men and armed with one Type 99 light MG and eleven rifles. Three of these rifle squads plus a grenadier squad (three Type 89 grenade dischargers) made up the rifle platoon. A rifle company had three rifle platoons and a small machine gun platoon with two squads. The infantry battalion was supported by two weapons companies, one with Type 92 medium MGs and one with mortars, each of two platoons. The mortar company was organized for the new 120mm weapon, but it would appear that not enough were built to go around, so the 81mm Type 99 mortar was substituted in many cases as the battalion support weapon.

Regimental support was provided by a large infantry gun company, a mounted platoon, and pioneer and signal companies. The infantry gun company was divided into two gun platoons (each two Type 41 mountain guns) and two mortar platoons (each two 120mm mortars). Once again, however, shortages of mortars plagued these divisions and some regiments were never able to form the mortar platoons. Although many divisions suffered these shortages of mortars, many were also strengthened by another new weapon that finally addressed a more serious shortcoming in the IJA inventory: the lack of effective anti-tank weapons. The year 1945 saw the introduction of a new 70mm AT rocket launcher. These were greeted enthusiastically and the 209th Division, at least, disbanded the pioneer companies in each of its regiments and replaced them with anti-tank companies, each of three platoons, each in turn equipped with ten of these rocket launchers.

Close-in fire support for these divisions was provided by trench mortar regiments, each of two battalions with eighteen 120mm mortars. These units, apparently having a higher priority than the battalion or regimental mortar units, were fully equipped. Each mortar company consisted of three platoons (each two weapons) and an ammunition platoon. The company had a total of 76 one-horse carts to carry the mortars and 500 rounds of ammunition. The battalion and regimental ammunition trains carried a further 720 rounds each. For communications the regiment had ten telephones and 20km of wire, but apparently no radios. Actual command of the mortar companies, then, must have been decentralized considerably in order for effective fire to be brought on targets.

Longer-ranged fire support was provided by a field or mountain artillery regiment. A field artillery regiment, such as those held by the 201st, 205th, 214th and 216th Divisions, consisted of three battalions with a total of 24 Type 38 improved 75mm guns and 12 Type 91 105mm howitzers. A mountain regiment, used by the 202nd, 206th, 209th and 212th Divisions, consisted of two battalions with a total of 24 Type 94 75mm mountain guns.

These divisions, formed with the best young troops available, were often referred to as the "elite" divisions of the homeland army and on these divisions rested much of the hope of IGHQ.

Homeland Defense: the Third Mobilization Wave

The third and final mobilization order for the homeland was issued on 23 May and called for the creation of eight more mobile divisions and eleven more coastal divisions. Four of the mobile divisions (224th, 230th, 231st and 234th) and six of the coastal divisions (303rd, 312th, 320th, 321st, 344th and 354th) were activated immediately. Three more mobile divisions (222nd, 225th and 229th) and two coastal divisions (308th and 351st) were activated in June. The final mobile division (221st) and three coastal divisions (316th, 322nd and 355th) were activated in July.

By this time, however, orders activating new divisions outstripped the Japanese ability to fill and, especially, equip them. The new mobile divisions were smaller and less powerful than their predecessors, having TO&E strengths of only 12,723. The combat units consisted of three infantry regiments and a trench mortar battalion. The anti-tank, artillery and machine cannon units of the earlier mobile divisions were completely absent in this second batch. The division base consisted of a HQ, signal unit, engineer regiment, and transport unit.

Similarly, the 300-series coastal divisions were substantially smaller than the earlier units. Combat components consisted of three infantry regiments and a rocket battalion in most divisions (an artillery battalion replaced the rocket battalion in the 321st and 322nd Divisions). The division base was made up of a HQ, signal unit, engineer unit, transport unit, and a field hospital. These elements gave the division a total of 11,926 men. As before, the coastal infantry regiment was a non-mobile formation with only 104 horses (including 65 in the supply company) for administrative resupply. The rocket battalion, however, was a semi-mobile formation with 58 pack horses allocated to each battery and 32 carts to the ammunition trains.

The manning and equipping of even these smaller divisions stretched the Army's capacity to the limit and most of these newer divisions were still incomplete at the surrender. As a result, no further divisions were formed in Japan beyond those ordered in the third mobilization.

Desperation in Manchuria

Divisions, however, were still being created in Manchuria through conversion and reorganization of existing units and a ruthless stripping of support units. The last wave to be mobilized by the Kwantung Army consisted of the 134th–139th and 148th and 149th

Table 2.21: Infantry Division (Mobile) (200–216 series), from April 1945

	Personnel	Pistols	Submachine Guns	Rifles	Light MGs	Medium MGs	Grenade Launchers	Flamethrowers	120mm Mortars	20mm AA Guns	47mm AT Guns	75mm Mountain Guns	75mm Field Guns	105mm Howitzers	Riding Horses	Pack Horses	Draft Horses	Wagons & Carts	Motor Vehicles
Division Headquarters	122	0	0	0	0	0	0	0	0	0	0	0	0	0	12	0	2	2	15
Signal Unit	300	0	0	283	0	0	0	0	0	0	0	0	0	0	2	16	18	17	0
Three Infantry Regiments, each																			
Regiment HQ	115	0	0	35	0	0	1	0	0	0	0	0	0	0	15	0	36	?	0
Signal Company	145	0	0	130	0	0	0	0	0	0	0	0	0	0	4	28	0	0	0
Mounted Platoon	27	0	0	25	0	0	0	0	0	0	0	0	0	0	27	0	0	0	0
Pioneer Company	186	0	0	176	0	0	0	10	0	0	0	0	0	0	1	0	18	?	0
Infantry Gun Company	262	?	0	0	0	0	0	0	4	0	0	4	0	0	5	0	90	?	0
Three Infantry Battalions, each																			
Battalion HQ	127	?	0	39	0	0	1	0	0	0	0	0	0	0	12	0	21	?	0
Four Rifle Companies, each	199	0	8	132	9	2	9	0	0	0	0	0	0	0	0	4	0	0	0
Machine Gun Company	138	16	0	0	0	8	0	0	0	0	0	0	0	0	0	43	0	0	0
Mortar Company	150	8	0	0	0	0	0	0	4	0	0	0	0	0	3	0	55	?	0
Mortar Regiment																			
Regiment HQ	81	9	0	2	0	0	0	0	0	0	0	0	0	0	10	0	18	15	9
Two Mortar Battalions, each																			
Battalion HQ	63	7	0	2	0	0	0	0	0	0	0	0	0	0	8	0	18	15	0

Table 2.21: Infantry Division (Mobile) (200–216 series), from April 1945

	Personnel	Pistols	Submachine Guns	Rifles	Light MGs	Medium MGs	Grenade Launchers	Flamethrowers	120mm Mortars	20mm AA Guns	47mm AT Guns	75mm Mountain Guns	75mm Field Guns	105mm Howitzers	Riding Horses	Pack Horses	Draft Horses	Wagons & Carts	Motor Vehicles
Three Mortar Companies	180	16	0	0	0	0	0	0	6	0	0	0	0	0	4	0	79	76	0
Ammunition Trains	115	7	0	0	0	0	0	0	0	0	0	0	0	0	1	0	80	76	0
Regimental Ammunition Trains	126	9	0	0	0	0	0	0	0	0	0	0	0	0	1	0	80	76	0
Field Artillery Regiment	2135	?	0	?	?	0	0	0	0	0	0	0	24	12	518	5	1161	?	?
or																			
Mountain Artillery Regiment	2496	?	0	?	?	0	0	0	0	0	0	24	0	0	316	991	180	?	0
Anti-Tank Battalion																			
Battalion HQ	61	0	0	24	0	0	0	0	0	0	0	0	0	0	0	0	0	0	3
Three Anti-Tank Companies, each	114	0	0	20	0	0	0	0	0	0	4	0	0	0	0	0	0	0	11
Ammunition Trains	80	1	0	42	0	0	0	0	0	0	0	0	0	0	0	0	0	0	11
Machine Cannon Battalion																			
Battalion HQ	46	6	0	8	0	0	0	0	0	0	0	0	0	0	0	0	0	0	5
Three Batteries, each	98	14	0	40	0	0	0	0	0	3	0	0	0	0	0	0	0	0	8
Engineer Unit																			
Unit HQ	70	1	0	11	0	0	0	0	0	0	0	0	0	0	9	0	19	13	0
Three Engineer Companies, each	233	0	0	221	0	0	0	0	0	0	0	0	0	0	1	0	10	9	0
Engineer Company	192	0	0	182	2	0	0	0	0	0	0	0	0	0	1	0	0	0	0

Table 2.21: Infantry Division (Mobile) (200–216 series), from April 1945

	Personnel	Pistols	Submachine Guns	Rifles	Light MGs	Medium MGs	Grenade Launchers	Flamethrowers	120mm Mortars	20mm AA Guns	47mm AT Guns	75mm Mountain Guns	75mm Field Guns	105mm Howitzers	Riding Horses	Pack Horses	Draft Horses	Wagons & Carts	Motor Vehicles
Materiel Platoon	38	0	0	35	0	0	0	0	0	0	0	0	0	0	1	0	13	12	4
Transport Unit																			
Unit HQ	58	4	0	20	0	0	0	0	0	0	0	0	0	0	0	0	0	0	7
Draft Company	254	3	0	35	0	0	0	0	0	0	0	0	0	0	52	0	175	167	0
Motorised Company	119	10	0	92	0	0	0	0	0	0	0	0	0	0	0	0	0	0	36
Ordnance Duty Unit	112	0	0	25	0	0	0	0	0	0	0	0	0	0	0	0	0	0	7
Field Hospital	277	0	0	0	0	0	0	0	0	0	0	0	0	0	9	0	70	?	0

Divisions, all officially activated on 10 July; and the 158th Division activated on 10 August.

These units had the following authorized strengths:

	Men	Horses
Division Headquarters	240	17
Signal Unit	239	46
Three Infantry Regiments, each	3,409	736
Raiding Battalion	1,130	0
Field Artillery Regiment	1,923	1,464
Engineer Regiment	964	125
Transport Regiment	1,180	708
Ordnance Duty Unit	112	0
Veterinary Hospital	119	45
Total	525 officers + 15,609 EM	4,613

Each of the infantry regiments was made up of a headquarters and company, a signal company, a regimental gun company (with mountain gun and AT platoons), a mounted platoon, and three infantry battalions. Each of these battalions consisted of a headquarters, three rifle companies (each three platoons), a machine gun company (four platoons) and an infantry gun platoon. Apparently plans were made to expand the regiments to four rifle companies apiece when circumstances permitted, but this never came about. The divisional raiding battalion was made up of three 350-man raiding companies.

The bulk of the personnel for these divisions (and a series of IMBs at the same time) came from conscription of Japanese citizens living in Manchuria and the organizational chaos engendered had still not been sorted out by the time of the Soviet attack. As a result, the divisions were in widely varying states of strength and readiness.

A relatively lucky division was the 135th, which was formed largely from units already existing, including the 77th IMB and the 3rd and 4th Border Garrison Units. Nevertheless, the division had only half its authorized number of light and medium machine guns, had no battalion guns, used ancient Type (Meiji) 31 mountain guns as regimental artillery, and its artillery regiment had only three 75mm cavalry guns per battery in two of its battalions, and four mortars per battery in the third battalion.

Other divisions were formed mostly with completely untrained conscripts, for whom even adequate supplies of rifles were unavailable. The 138th Division had no weapons other than rifles, while the 137th Division had only rifles and machine guns, and limited food stocks because the division had only three trucks for transport. The 148th Division had only 30 percent of its rifles and 10 percent of the machine guns, although they did have six 75mm Type (Meiji) 38 field guns, a 105mm howitzer and a 150mm howitzer for artillery. The 149th Division, on the other hand, did not have a single artillery piece.

In practice, none of these divisions were more than shadows of real divisions.

The Regular Army Brigades

Surprisingly, given the large number of mixed brigades fielded by the Army during the war, there were few brigades in the Army's permanent force structure. Through the late 1930s the only infantry brigade permanently organized was the Formosa Mixed Brigade, which consisted of the 1st and 2nd Formosa Infantry Regiments. Designed primarily to garrison

Table 2.22: Third-Wave Coastal and Mobile Infantry Divisions, from May 1945

Coastal Infantry Division	Mobile Infantry Division
HQ (403)	HQ (430)
Signal Unit (280)	Signal Unit (254)
Three Infantry Regiments, each (3,042)	Three Infantry Regiments, each (3,310)
Regiment HQ (132)	Regiment HQ (115)
Signal Company (133)	Signal Company (144)
Three Infantry Battalions, each (832)	Three Infantry Battalions, each (873)
Battalion HQ (1 GL)	Battalion HQ (1 GL)
3 Rifle Companies, each (9 LMG, 9 GL)	3 Rifle Companies, each (9 LMG, 9 GL)
Raiding/MG Company (6 MG, 9 GL)	Infantry Gun Company (4 70mm Inf How)
Infantry Gun Platoon (3 37mm AT)	Infantry Gun Company (233) (6 75mm Regtal guns)
Supply Platoon (131)	Pioneer Company (199)
Rocket Artillery Battalion (692)	Mortar Battalion (754)
Battalion HQ (64)	Battalion HQ (89)
3 Rocket Batteries, each (188) (12 20cm launchers)	3 Mortar Companies, each (180) (6 120mm mort)
Ammunition Trains (64)	Ammunition trains (125)
Engineer Unit (806)	Engineer Unit (886)
Unit HQ (53)	Unit HQ (90)
3 Companies, each (231)	3 Companies, each (252)
Materiel Platoon (60) (9 FT)	Materiel Platoon (40) (9 FT)
Transport Unit (360) (100 carts, 15 trucks)	Transport Unit
Field Hospital (232)	Unit HQ (61)
	Draft Transport Company (287) (120 carts)
Note: may have artillery unit in lieu of rocket unit:	Motor Transport Company (121) (45 trucks)
HQ (89)	
2 Mountain Batteries, each (117) (4 75mm mtn guns)	
1 Howitzer Battery (117) (4 105mm how)	
Ammunition trains (68)	
Total: 11,926 or 11,762 men	Total: 12,723 men

Taiwan, in 1937 the brigade was strengthened by the addition of a mountain artillery battalion (nominally a regiment) and a transport company. The brigade was further upgraded with additional service support units and participated in many campaigns on the Chinese mainland during 1937–40. On 30 November 1940 the 47th Infantry Regiment was added, combat support and service support units upgraded and the force redesignated the 48th Division.

The other permanent brigade was the Karafuto Mixed Brigade. Prewar the garrison of southern Sakhalin (in Japanese, Karafuto) island was undertaken by the 125th Infantry, a separate non-divisional regiment. In July 1941 the 7th Division, based in nearby Hokkaido, split off its 25th Infantry Regiment in the triangularization process and this unit formed the basis for the new Karafuto Brigade, along with an artillery unit, and engineer, signal, transport and medical elements. In May 1943 the 125th Regiment was mobilized and added to the brigade and other elements were enlarged, bringing its strength from 8,149 to 15,667. The brigade remained on garrison duty on Sakhalin Island until February 1945 when it was expanded and reorganized to form the 88th Division.

The two brigades were similar in structure consisting of one or two "A1" type infantry regiments, a mountain artillery "regiment" of three batteries, an engineer regiment of two companies and service support units. The main difference between the two brigades was that while the Karafuto Brigade included a conventional transport unit with 304 Type 39 carts, the Formosa Brigade's transport unit was simply a headquarters element to administer locally-requisitioned and contracted transport services. Both brigades were deficient in fire support and rather ungainly due to their binary configuration. This was less important in the case of the Karafuto Brigade, which deployed one regiment on the Soviet–Japanese border on the island and the other in the south of the island to guard against flank attacks from the sea.

Two more brigades were formed later by the Army, the 65th and 68th. The 65th Brigade was formed in early 1941 by redesignation of the 65th Independent Infantry Group with a HQ, an engineer unit, a signal unit and the 122nd, 141st and 142nd Infantry Regiments. The infantry regiments were at a low strength and had only two battalions apiece (each of 3 rifle companies and an MG company) plus a regimental gun company, giving the brigade a strength of only 247 officers and 6,471 enlisted. The brigade fought in the Philippines and remained there on occupation duty. In December 1942 the 122nd and 142nd Regiments were dropped from the organization, although the rest of the organization was little changed.[35] The sole remaining infantry regiment received its third battalion and an AT company (although losing its transport horses) in late 1943 for its move to New Britain, yielding the following strengths:

	Off	WO	MSG	SGT	PVT	Total	Horses
Brigade Headquarters	17	0	13	8	76	114	0
Signal Unit	2	0	0	7	58	67	0
141st Infantry Regiment	64	11	25	181	1,602	1,883	235
141st Infantry Regiment (from Nov 43)	87	15	31	254	2,284	2,671	0

35 The 142nd Regiment remained in the Philippines on garrison duty until December 1943 when it was broken up and its battalions redesignated the 181st and 182nd Independent Infantry Battalions as part of the 33rd IMB.

Engineer Unit	3	1	0	12	122	138	0
Field Hospital	11	0	18	2	118	149	0

The 68th Brigade was formed in Manchuria on 19 June 1944 by mobilizing the Kungchuling Military School Training Regiment. The brigade had the following strengths:

	Off	WO	MSG	SGT	PVT	Total	Horses
Brigade Headquarters	19	0	16	9	126	170	30
Signal Unit	5	1	5	17	179	207	40
126th Inf Regt	72	17	27	228	2,376	2,721	616
Artillery Unit	26	4	18	53	637	738	572
Engineer Unit	6	1	4	22	235	288	43
Medical Unit	11	1	13	14	214	252	0
Total	139	24	83	343	3,767	4,376	1,301
changes of 22 Sept 44:							
126th Inf Regt	92	23	26	308	3,205	3,654	0
Artillery Unit	52	8	16	108	1,236	1,420	0
Total	185	34	80	478	5,195	5,992	70

The brigade was built around the 126th Infantry Regiment, which followed the standard "A1" type organization except that it had only two battalions instead of three, and only three rifle companies per battalion instead of four. Fire support was provided by the brigade artillery unit, which initially consisted of a 228-man HQ, a 131-man field gun battery (four horse-drawn Type 95 field guns), two 119-man howitzer batteries (each three horse-drawn Type 91 howitzers), and a 122-man service battery. In June of 1944 the Brigade was transferred to the Formosa Army where the 126th Regiment received its third battalion in September and the artillery unit was brought up to regimental strength. This latter involved adding a second battalion with three 3-gun batteries of Type 91 howitzers. Shortly thereafter it was sent to the Philippines where it was destroyed.

The Independent Infantry Groups

When the five reservist divisions (101, 106, 108, 109 and 114) were brought back from China and disbanded in late 1939 their constituent units did not disappear, but were merely assigned to their parent depot divisions. These, along with further reservist infantry regiments, were used to form seven independent infantry groups between August 1940 and June 1941.

61st Independent Infantry Group: 101st, 149th & 157th Infantry Regiments
62nd Independent Infantry Group: 129th, 130th & 158th Infantry Regiments
63rd Independent Infantry Group: 118th, 135th & 136th Infantry Regiments
64th Independent Infantry Group: 106th, 153rd & 168th Infantry Regiments
65th Independent Infantry Group: 122nd, 141st & 142nd Infantry Regiments
66th Independent Infantry Group: 123rd, 145th & 147th Infantry Regiments
67th Independent Infantry Group: 105th, 125th & 131st Infantry Regiments

Each of these groups had an authorized strength of 232 officers and 6,283 enlisted,

Table 2.23: Regular Army Mixed Brigades, from October 1940

	Personnel	Pistols	Carbines	Rifles	Light MGs	Light Gren Lnchrs	75mm Type 94 Mtn Guns	105mm Type 91 Howitzers	Riding Horses	Pack Horses	Draft Horses	Wagons & Carts	Cars	Trucks
Karafuto Mixed Brigade														
Brigade Headquarters	227	0	27	89	1	0	0	0	24	0	25	24	5	7
Signal Unit	192	1	143	0	0	0	0	0	7	0	29	28	0	0
Two Type "A1" Infantry Regiments														
see separate table														
Karafuto Mountain Artillery Regiment														
Regiment HQ	192	14	4	*200	2	1	0	0	44	17	48	55	0	0
Two Mountain Batteries, each	193	12	0	0	0	0	4	0	18	88	0	0	0	0
Field Battery	130	12	0	0	0	0	0	4	12	0	38	7	0	0
Ammunition Trains	129	9	0	0	1	0	0	0	16	6	61	58	0	0
Engineer Unit														
Unit HQ	78	1	2	6	0	0	0	0	10	0	38	37	0	0
Two Engineer Companies, each	242	0	0	231	0	0	0	0	7	16	0	0	0	0
Materiel Platoon	34	0	0	33	0	0	0	0	1	0	17	16	0	0
Transport Unit	472	3	66	100	0	0	0	0	71	0	320	304	0	0
Medical Unit														
Unit HQ	163	6	4	*50	0	0	0	0	14	0	49	47	0	0

Table 2.23: Regular Army Mixed Brigades, from October 1940

	Personnel	Pistols	Carbines	Rifles	Light MGs	Light Gren Lnchrs	75mm Type 94 Mtn Guns	105mm Type 91 Howitzers	Riding Horses	Pack Horses	Draft Horses	Wagons & Carts	Cars	Trucks
Medical Unit														
Unit HQ	163	6	4	*50	0	0	0	0	14	0	49	47	0	0
Two Stretcher Companies, each	184	21	0	0	0	0	0	0	1	0	0	0	0	0
Vehicle Company	182	3	12	0	0	0	0	0	3	0	24	n/a	0	0
Formosa Mixed Brigade														
Brigade Headquarters	147	0	2	59	0	0	0	0	19	0	24	23	0	0
Signal Unit	227	1	176	0	0	0	0	0	7	0	31	30	0	0
Two Type "A1" Infantry Regiments														
see separate table														
Formosa Mountain Artillery Regiment														
Regiment HQ	189	15	4	*270	0	0	0	0	46	32	60	57	0	0
Three Mountain Batteries, each	196	12	0	0	0	0	4	0	19	73	0	0	0	0
Ammunition Trains	155	9	0	0	0	0	0	0	17	6	78	74	0	0
Engineer Regiment														
Regiment HQ	78	1	2	6	0	0	0	0	10	0	38	37	0	0

Table 2.23: Regular Army Mixed Brigades, from October 1940

	Personnel	Pistols	Carbines	Rifles	Light MGs	Light Gren Lnchrs	75mm Type 94 Mtn Guns	105mm Type 91 Howitzers	Riding Horses	Pack Horses	Draft Horses	Wagons & Carts	Cars	Trucks
Two Stretcher Companies, each	184	21	0	0	0	0	0	0	1	0	0	0	0	0
Vehicle Company	182	3	12	0	0	0	0	0	3	0	24	n/a	0	0
Formosa Mixed Brigade														
Brigade Headquarters	147	0	2	59	0	0	0	0	19	0	24	23	0	0
Signal Unit	227	1	176	0	0	0	0	0	7	0	31	30	0	0
Two Type "A1" Infantry Regiments														
see separate table														
Formosa Mountain Artillery Regiment														
Regiment HQ	189	15	4	*270	0	0	0	0	46	32	60	57	0	0
Three Mountain Batteries, each	196	12	0	0	0	0	4	0	19	73	0	0	0	0
Ammunition Trains	155	9	0	0	0	0	0	0	17	6	78	74	0	0
Engineer Regiment														
Regiment HQ	78	1	2	6	0	0	0	0	10	0	38	37	0	0
1st Engineer Company	242	0	0	231	0	0	0	0	5	16	0	0	0	0
2nd Engineer Company	220	0	0	215	0	0	0	0	0	0	0	0	0	0

Table 2.23: Regular Army Mixed Brigades, from October 1940

	Personnel	Pistols	Carbines	Rifles	Light MGs	Light Gren Lnchrs	75mm Type 94 Mtn Guns	105mm Type 91 Howitzers	Riding Horses	Pack Horses	Draft Horses	Wagons & Carts	Cars	Trucks
Materiel Platoon	17	0	0	0	0	0	0	0	1	0	10	9	0	0
Transport Unit	71	3	53	0	0	0	0	0	55	0	0	0	0	0
Medical Unit														
Unit HQ	188	7	4	*50	0	0	0	0	18	0	65	61	0	0
Two Stretcher Companies, each	167	20	0	0	0	0	0	0	1	0	0	0	0	0
Vehicle Company	222	3	19	0	0	0	0	0	4	0	28	n/a	0	0

* *includes undistributed rifles for unit self-defense*

although it seems likely that actual strength while in depots at the homeland averaged about half that, with infantry regiments having 750–1500 men apiece.

Lacking combat support and service support elements these groups were not considered field formations, but in mid-1941 the 65th Group was reinforced with small service support units and redesignated the 65th Brigade. The other groups disappeared during 1942–43 as their infantry units were drawn off to form divisions, without ever having left the homeland. Thus, the 61st Group provided the basis for the 61st Division, the 62nd Group for the 42nd Division, the 63rd Group for the 43rd Division, the 64th Group for the 49th Division, the 66th Group for the 46th Division, and the 67th Group for the 47th Division.

The First Independent Mixed Brigades

Far more important than these four permanent brigades were the large numbers of "independent mixed brigades" raised for wartime duties. Initially, these were raised for service in China where their rather low ratio of artillery-to-infantry would not be a serious handicap, but their use quickly spread to the Pacific and then to the Kwantung Army and Japan itself.

Operations in Manchuria in the early 1930s often featured "mixed brigades," which were task forces usually consisting of an infantry regiment and a field artillery battalion drawn from divisions in the homeland for temporary duty with the Kwantung Army. Mixed brigades deployed to China in 1937 usually also represented a portion of a division formed by adding the bulk of the divisional artillery regiment and a portion of the other divisional services to one of the division's infantry brigades. Thus, the 2nd Division's 3rd Brigade was transformed into the 3rd Mixed Brigade through the addition of the 2nd Field Artillery Regiment and one company of the 2nd Engineer Regiment.

The first *independent* mixed brigade, designated the 1st, was formed in 1934, followed shortly by the 11th, consisting of the 11th and 12th Independent Infantry Regiments, the 12th Independent Mountain Artillery Regiment, 11th Independent Field Artillery Regiment, as well as cavalry, engineer and transport companies.[36] Apparently the components were still on peacetime strength, for brigade strength only totaled 4,095 men.

The 11th IMB, however, was not to be a prototype for the future. The use of formations at peacetime strength was not efficient and instead more compact war-strength components would be used. The new generation of mixed brigades first showed up in 1937.

The 2nd through 5th IMBs (an early 1st IMB initially followed the pattern of the 21st) were organized in late 1937 and officially activated in February 1938. Each of these brigades had the following components:

	Off	WO	MSG	SGT	PVT	Total	Horses
Brigade Headquarters	17	0	15	10	25	67	26
Signal Unit	7	2	3	21	142	175	27
Five Infantry Battalions, each	27	8	9	70	696	810	40
Artillery Unit	20	4	14	35	348	421	258
Engineer Unit	7	2	5	14	150	178	6
Total	186	48	82	430	4,145	4,891	517

36 This initial 11th IMB was expanded into the 26th Division in September 1937.

Being designed for garrison use the infantry battalions of these brigades initially loosely followed the organization of the reserve and conscript infantry in China. A battalion consisted of four rifle companies, a machine gun company and an infantry gun company. The rifle company had three rifle platoons each of about 40 men with two light MG squads (each eight men with a Type 11 LMG), a rifle squad (seven men with rifles) and a grenadier squad (eight men with two Type 89 grenade dischargers). The machine gun company seems to have had two platoons each of two 12-man gun squads (each with a medium machine gun and a pack horse) and a 12-man ammunition squad. The infantry gun company had two platoons each with two weapons, one platoon with 70mm Type 92 battalion guns and the other with 75mm Type 41 regimental guns.

The brigade artillery unit was usually a mountain artillery battalion with two batteries of 75mm field or mountain guns, although the 3rd IMB had one field battery and two mountain batteries for a total strength of 620 men. The engineer unit usually consisted of two small companies. These elements combined to give the IMBs a strength of about 4,900 men with 517 horses. They were deployed mainly in the garrison role, and indeed the absence of logistical transport units at any level rendered them suitable for little else.

The performance of these IMBs in the garrison role in China was apparently considered satisfactory for in early 1939 nine more IMBs (6th–14th) were activated. These were organizationally identical to the earlier formations. From mid-1939 to late 1940 six more brigades were activated (15th–20th IMBs), still using the same organization. These initial twenty brigades form a distinct group and may be thought of as the first-generation independent mixed brigades.

1st IMB (Ind Inf Bns 72, 73, 74, 75, 76)
2nd IMB (Ind Inf Bns 1, 2, 3, 4, 5)
3rd IMB (Ind Inf Bns 6, 7, 8, 9, 10)
4th IMB (Ind Inf Bns 11, 12, 13, 14, 15)
5th IMB (Ind Inf Bns 16, 17, 18, 19, 20)
6th IMB (Ind Inf Bns 21, 22, 23, 24, 25)
7th IMB (Ind Inf Bns 26, 27, 28, 29, 30)
8th IMB (Ind Inf Bns 31, 32, 33, 34, 35)
9th IMB (Ind Inf Bns 36, 37, 38, 39, 40)
10th IMB (Ind Inf Bns 41, 42, 43, 44, 45)
11th IMB (Ind Inf Bns 46, 47, 48, 49, 50)
12th IMB (Ind Inf Bns 51, 52, 53, 54, 55)
13th IMB (Ind Inf Bns 56, 57, 58, 59, 60)
14th IMB (Ind Inf Bns 61, 62, 63, 64, 65)
15th IMB (Ind Inf Bns 77, 78, 79, 80, 81)
16th IMB (Ind Inf Bns 82, 83, 84, 85, 86)
17th IMB (Ind Inf Bns 87, 88, 89, 90, 91)
18th IMB (Ind Inf Bns 92, 93, 94, 95, 96)
19th IMB (Ind Inf Bns 97, 98, 99, 100, 101)
20th IMB (Ind Inf Bns 102, 103, 104, 105, 106)

In March 1941 each of the independent mixed brigades received an additional authorization of ten privates per rifle company, bringing battalion strength up to 850 men.

Of more significance was the elimination in May of the artillery and engineer units in most of the IMBs. Apparently it was felt that with the brigades scattered about in small outposts, usually platoon-size, on garrison duty the centralized artillery and engineer assets were of little use. Only the 17th IMB is known to have retained its engineer and artillery units. In any event, the bulk of the first-generation IMBs were used to create occupation divisions during 1943–44.

For the most part the IMBs did not engage in campaigning but the gradual withdrawal of many of the first-line divisions from China forced one of them into a more mobile role. To that end the 8th Brigade received augmentation in the form of 10 officers, 4 warrant officers, 30 sergeants and 340 privates with 244 horses added to the brigade HQ for use as a brigade transport unit on 1 May 1943. At about the same time the 3rd Brigade, in remote Shansi province, received a motor transport company of 100 men in three platoons with a total of 70 Ford 2-ton trucks. Each man in that company was provided with a Type (Meiji) 38 rifle, while the company held four light machine guns (2 Japanese and 2 captured Chinese).

The surviving first-generation IMBs in China were reorganized in March 1945 to conform, with a few changes, to the organization tables used by the 81st–92nd IMBs then being raised. The main difference was that the first-generation IMBs now held five smaller rifle companies in each infantry battalion instead of the four large companies in the new units. A battalion thus held 45 light machine guns, 45 grenade dischargers, eight heavy MGs, two 70mm howitzers and two 75mm mountain guns, a considerable increase in automatic weapons firepower over the prior organization. For fire support the 1st, 2nd, 3rd and 5th IMBs were given mortar battalions, each with three 6-gun batteries, while the 8th, 9th and 17th IMBs used artillery battalions, each with one 4-gun battery of mountain guns and one 4-gun battery of field guns. Brigade headquarters strength varied slightly from one unit to the next, but typical of the 1945 organization was the 1st Brigade with the following strengths:

	Off	WO	MSG	SGT	PVT	Total	Horses
Brigade Headquarters	24	1	23	15	106	169	21
Signal Unit	7	2	3	21	142	175	29
Five Infantry Battalions, each	45	13	14	128	1,349	1,549	130
Artillery Unit	21	3	9	43	506	582	288
Engineer Unit	17	2	9	49	454	531	52
Total	294	73	114	768	7,953	9,202	1,040

The break in this organizational pattern came with the second 21st IMB which, unlike the others, was not intended for service in China, but rather for the occupation of French Indochina. This unit, reformed in February–March 1941, essentially represented a combat team built around the 170th Infantry Regiment (split off from the 104th Division when it converted to triangular configuration). The regiment consisted of three battalions (each four rifle companies, an MG company and an infantry gun platoon), a signal company, infantry gun company and anti-tank company. Supporting the regiment was the brigade artillery battalion,[37] and tank, AA, engineer, signal and motor transport companies, and a

37 The artillery battalion was a 428-man medium formation consisting of one battery of 105mm Type (Taisho) 14 field guns, one of 150mm Type 96 howitzers, both drawn by 6-ton tractors, and a service battery that included a transport platoon with 19 trucks and

field hospital. In November 1942 the regiment's second battalion reinforced by one platoon from each of the regimental gun and AT companies was sent to garrison Wake Island, while the rest of the brigade, less its horses, was sent to Rabaul and the motor transport company remained in Indochina. The brigade suffered heavy casualties in New Guinea before retiring to New Britain. It was disbanded in June 1943.

An IMB-equivalent formed at the same time was the Hong Kong Defense Force, activated on 19 January 1942, with the 67th, 68th, and 69th Independent Infantry Battalions, a brigade artillery unit and a hospital. The artillery consisted of an AA battery (six guns), a trench mortar battery and two field artillery batteries. With an authorized strength of 106 officers and 3,028 enlisted the unit survived unchanged until the end of the war.

IMB Conversions in China

No further IMBs were formed for over a year. Indeed, six of the brigades (10th, 11th, 14th, 16th, 18th and 20th) were expanded and reorganized into occupation divisions, reducing the number of IMBs in the Army's force structure to fifteen. One new IMB (the 22nd) was raised in late 1942 in China and another (the 23rd) in January 1943 on Taiwan, but these were the only conventional IMBs activated in China between mid-1939 and early 1944.

22nd IMB (Ind Inf Bns 66, 70, 71, 125, 126, 127)
23rd IMB (Ind Inf Bns 128, 129, 130)[38]

Except for the varying numbers of line battalions these were organized along the same lines as their predecessors.

In the meantime, other IMBs had followed the example of the six earlier units and had been converted to occupation divisions. In May 1943 the 4th, 12th, and 13th IMBs were expanded and converted, and in June the 15th IMB was converted. Thus, by mid-1943 half of the original China-type IMBs had been reorganized into divisions.

IMBs for the South

The IMB was gaining popularity as an organizational form in the southern areas, however. In November 1943 IGHQ ordered the 24th IMB formed in Burma, along with the 25th and 26th IMBs in Sumatra and the 27th and 28th IMBs in Java, the 29th IMB in Thailand, the 30th–33rd IMBs in the Philippines, and the 34th IMB in French Indochina.

24th IMB (Ind Inf Bns 138, 139, 140, 141)
25th IMB (Ind Inf Bns 142, 143, 144, 145)
26th IMB (Ind Inf Bns 146, 147, 148, 149)
27th IMB (Ind Inf Bns 150, 151, 152, 153)
28th IMB (Ind Inf Bns 154, 155, 156, 157)
29th IMB (Ind Inf Bns 158, 159, 160, 161, 162)
30th IMB (Ind Inf Bns 163, 164, 165, 166, 167, 168)
31st IMB (Ind Inf Bns 169, 170, 171, 172, 173, 174)

four more 6-ton tractors.

38 Independent Infantry Battalion 70 was transferred to this brigade in mid-1943 and Battalions 247 and 248 were added in February 1944 by IGHQ order.

32nd IMB (Ind Inf Bns 175, 176, 177, 178, 179, 180)
33rd IMB (Ind Inf Bns 181, 182, 183, 184, 185, 186)
34th IMB (Ind Inf Bns 187, 188, 189, 190)[39]

These units had the following strengths:

	Off	WO	MSG	SGT	PVT	Total	Horses
Brigade Headquarters	19	2	23	3	151	198	4
Signal Unit	5	2	6	18	147	178	17
Each Infantry Battalion							
Battalion HQ	8	0	17	5	63	93	7
Four Rifle Companies, each	4	2	2	15	153	176	0
Heavy Weapons Company	4	2	2	13	113	134	13
Artillery Unit							
Unit HQ	8	0	9	5	48	70	12
Two Batteries, each	4	2	1	9	129	145	58
or Artillery Unit	*4*	*2*	*7*	*7*	*111*	*131*	*73*
Engineer Unit	4	2	6	13	155	180	11
Total (6-battalion brigade)	212	70	208	525	5,487	6,502	280

These were organized along a common pattern in spite of the varying number of infantry battalions per brigade. The brigade HQ included a 79-man ordnance duty unit and a 67-man medical unit. The rifle companies had the conventional organization with nine light MGs and nine grenade dischargers, while the heavy weapons company operated a mixture of medium MGs, 70mm howitzers, 37mm AT guns and captured weapons depending on availability. To move the weapons and supplies short distances the battalion had four pack horses in its headquarters and twelve in the heavy weapons company.

In the case of the 30th–33rd IMBs the fire support element was a 131-man battery with four 75mm guns. In the others it was a two-battery 360-man battalion. In the 24th–26th IMBs both batteries were equipped with 90mm or 150mm mortars, in the 27th–29th IMBs one battery was equipped with 75mm mountain guns and the other with mortars, while the 34th IMB had two mountain gun batteries. The organization was identical in both cases, with a battery having 54 pack horses for transport. The rest of the brigade of this series consisted of a 180-man engineer company and a 178-man signal company. Once again, the brigade had no transport unit, effectively restricting it to the garrison role.

The creation of these IMBs did not really represent a substantial addition to the Army's force structure. For the most part they were simply conversions of existing Independent Garrison Units, which were essentially static infantry regiments. Sacrificed in this mobilization were Independent Garrison Units 10 (for the 30th IMB), 11 (for the 31st), 13 (for the 27th), 14 (for the 28th), 15th (for the 25th), and 17 (for the 32nd). The other IMBs of this series were raised using personnel of the 3rd and 6th Field Replacement Units; the 142nd Infantry Regiment, and various lines of communications units. Thus, although the IMBs were certainly more capable formations than the garrison units they replaced, the improvement to the Japanese situation was considerably less than would be supposed

39 In November 1944 the 187th and 188th Battalions were transferred to the 72nd Ind Inf Bde, and in April 1945 the 672nd to 675th Battalions were formed and added.

by a simple list of activation of these units. The 30th IMB of this series barely existing at all, being almost immediately expanded and converted to the 100th Division, and in June 1944 the other three Philippine-based IMBs followed suit.

These formations were followed by three more IMBs raised in January 1944: the 35th in the Andaman Island, the 36th in the southern Nicobars and the 37th in the northern Nicobars. These, too, were drawn primarily from existing units already present at those locations: the 35th from the 1st Southwestern Garrison Unit, the 36th from the 2nd Southwestern Garrison Unit, and the 37th from other miscellaneous troops *in situ*.

35th IMB (Ind Inf Bns 251, 252, 253, 254, 255, 256)
36th IMB (Ind Inf Bns 258, 259, 260)
37th IMB (Ind Inf Bns 262, 263, 264)[40]

These brigades had the following strengths:

	Off	WO	MSG	SGT	PVT	Total	Horses
Brigade Headquarters	19	1	22	10	111	163	4
Signal Unit	5	2	4	20	147	198	17
Each Infantry Battalion	28	10	9	96	815	958	20
Artillery Unit	16	4	7	27	306	360	128
Engineer Unit	4	2	4	15	155	180	11

Each of the infantry battalions consisted of four rifle companies and a heavy weapons company. The artillery unit varied; in the 35th it had one field gun battery and one of mountain guns, plus four captured field guns and four 15cm naval guns; in the 36th it had three field gun batteries (Type 38 and Type 90); and in the 37th it had one infantry mortar battery and one with captured field artillery.

The Southeastern Theater and the Homeland, Spring 1944

The southeastern theater raised its first IMBs in the late spring of 1944. As Army units were isolated on islands in the Solomons/New Britain area IMB headquarters were established to control them. Thus, the 38th IMB was activated in July 1944 on Bougainville to command the 81st Infantry Regiment plus an artillery battalion, engineer company and signal company. The infantry regiment, taken from the 17th Division, had a new theoretical strength figure issued at that time of 4,089 men, but apparently lost all its transport horses.

The 40th IMB, with a strength of 8,000 men, was built around the 230th Infantry Regiment (formerly of the 38th Division), the remnants of the 1st Independent Mixed Regiment, an artillery battalion, engineer unit, signal company and transport unit to garrison New Ireland. The 39th IMB was formed from 3,000 miscellaneous personnel stranded in Rabaul and was divided into the 4th and 5th Mixed Regiments, an artillery unit, signal unit, transport unit, and shipping engineer unit. The 38th IMB was officially added to the 17th Army and the 39th and 40th to the 8th Area Army on 26 June 1944.

40 The gaps in the numerical sequence of the infantry battalions was caused by plans to reinforce each of these IMBs with another battalion, and in July 1944 these battalions were dispatched from Manchuria, the 257th for the 35th IMB, the 261st for the 36th IMB, and the 265th for the 37th IMB, although these were removed again four months later.

Two more non-standard IMBs, along with two on the earlier standard pattern, were raised at the same time around the empire. The two standard-type brigades were the 43rd and 45th, the former activated in May 1944 to command units on the southern Kuriles island of Etorofu, while the latter was formed on Shikoku before being shipped to Ishigaki Island in the Ryukyus.

43rd IMB (Ind Inf Bns 294, 295, 296, 297)[41]
45th IMB (Ind Inf Bns 271, 298, 299, 300, 301)

The 43rd Brigade consisted of a 298-man HQ, four (later six) 985-man infantry battalions, each with 89 horses for transport, an engineer unit and an artillery unit for a total of 4,719 men. The 45th IMB components were a 59-man brigade HQ, five 683-man infantry battalions, and a 173-man engineer unit, being notable for having no organic artillery, although the 8th Heavy Artillery Regiment (former Funauki Fortress HAR) was attached.

The two non-standard IMBs were the 44th and 46th. The 44th Brigade was raised in Kyushu in June 1944 and was immediately dispatched to Okinawa. The brigade had 4,657 men divided into a 60-man HQ, two 2,045-man "infantry units," a 332-man artillery unit and a 173-man engineer unit. Each of the infantry units consisted of three small battalions (each three rifle and one MG companies and a 70mm howitzer platoon), an anti-tank company (four 37mm guns) and an infantry gun company (four 75mm mountain guns). The brigade artillery unit had two batteries, each with four 105mm Type 91 horse-drawn howitzers. The engineer unit was made up of three 50-man platoons and a HQ section. The transport carrying the bulk of the brigade to Okinawa was torpedoed *en route* and only about 600 men made it to the island.

On arrival the brigade was reorganized effective 27 September 1944. The surviving infantrymen were concentrated in the 2nd Infantry Unit and the 1st Infantry Unit was disbanded. Replacement personnel and locally-conscripted Okinawans brought the 2nd Unit back up to a strength of 2,500 men with three battalions, organized similarly to its original configuration. To replace the missing 1st Infantry Unit the 15th Independent Mixed Regiment was flown out from Japan in September and replacement personnel brought in to reconstitute the artillery, engineer and HQ elements. The 44th IMB thus faced the US invasion consisting of the 15th IMR, the 2nd Infantry Unit, and an artillery battalion. It was destroyed on the island after heavy fighting.

The 46th IMB, similar to the 44th, was raised on Formosa in May 1944. Its existence was brief, forming the basis for the 66th Division two months later.

IMBs in the Central Pacific

A massive reorganization of forces in April 1944 by the 31st Army responsible for the central Pacific brought into being a new generation of independent mixed brigades. In order to clarify and simplify the rather cumbersome lines of command on the larger Pacific islands the garrison forces were reorganized into independent mixed brigades and regiments. Six new brigades resulted from this reorganization: the 47th IMB (on Saipan, from the 1st Expeditionary Unit), the 48th (on Guam, from part of the 6th Expeditionary Unit),

41 Independent Infantry Battalions 419 and 420 were added to this brigade on 29 July 1945. With these additions brigade strength rose to 6,513.

the 49th (on Yap, from the 4th Expeditionary Unit), the 50th (on Woleai, from the 7th Expeditionary Unit) the 51st (on Truk, from the 8th Expeditionary Unit), and the 52nd (on Ponape, from the 3rd South Seas Detachment). These were shortly followed by the 53rd IMB (in the Palau Islands, from various troops present).

47th IMB (Ind Inf Bns 315, 316, 317, 318)
48th IMB (Ind Inf Bns 319, 320, 321, 322)
49th IMB (Ind Inf Bns 323, 324, 325, 326, 327, 328, 329, 330)
50th IMB (Ind Inf Bns 331, 332, 333, 334, 335)
51st IMB (Ind Inf Bns 336, 337, 338, 339, 340, 341)
52nd IMB (Ind Inf Bns 342, 343, 344, 345)
53rd IMB (Ind Inf Bns 364, 347, 348, 349, 350, 351, 352)

These brigades adopted a homogeneous, modular organization that required considerable reshuffling of existing units. The infantry battalions were identical to those in the 109th Division (organized at the same time), consisting of three rifle companies, a machine gun company, an infantry gun company and a pioneer platoon, but minus 39 trains personnel in the battalion HQ.

As was the case with the 109th Division the mobilization order issued by IGHQ in Tokyo differed from the organizational orders issued by the 31st Army on the same date. The data shown in Table 2.15 reflects the 31st Army order. Tokyo authorized slightly lower strengths for the brigade components as follows:

	Off	WO	MSG	SGT	PVT	Total
Brigade Headquarters	19	0	31	4	95	149
Infantry Battalion	25	9	6	79	460	579
Artillery Battalion	19	6	5	37	348	415
Anti-Aircraft Battery	4	1	3	13	149	170
Anti-Aircraft Battalion	24	4	6	58	619	711
Tank Company	5	2	6	24	55	92
Engineer Company	6	2	3	21	189	221

For the most part the differences between the two sets of strength figures were simply reduced manpower in the Tokyo tables, not affecting organization of major items of equipment. The one exception was in the infantry battalions, where the IGHQ strengths did not include the 67-man pioneer platoon. Since the brigades were not moved after their formation, the pioneer platoons became superfluous once initial field positions were established, and were apparently abolished. Under the IGHQ tables the rifle company had a total strength of four officers, two warrant officers and 108 NCOs and privates. The company was built around the standard three rifle platoons, each with three seven-man rifle squads and an 8-man grenadier squad. The machine gun company was also little changed, consisting of three platoons each of a lieutenant and four seven-man MG squads. The infantry gun company was reduced to a strength of four officers and 78 other ranks. These changes reduced battalion strength of 34 officers and warrant officers and 545 enlisted.

Each of these IMBs also included an artillery battalion. Since these battalions were static units with no organic transport, gun crew size (and hence battalion size) remained the same regardless of the type or caliber of the weapons used. These were 75mm mountain

Table 2.24: Central Pacific Type IMBs

	Officers	Warrant Officers	NCOs	Privates	Pistols	Rifles	Light MGs	Medium MGs	Heavy Gren Lnchrs	70mm Infantry Howitzers	37mm AT Guns	Flamethrowers	75mm AA Guns	Artillery Pieces	Light Tanks		Searchlights	Riding Horses		Motorcycles	Cars	Trucks
47th Ind Mixed Brigade																						
Brigade HQ	17	2	29	108	0	84	0	0	0	0	0	0	0	0	0		0	0		2	1	10
Signal Unit	3	1	16	111	0	118	0	0	0	0	0	0	0	0	0		0	0		0	0	0
Four Infantry Battalions, each	29	9	98	560	32	318	27	12	27	2	2	0	0	0	0		0	0		0	0	0
Artillery Battalion	18	6	41	375	40	0	0	0	0	0	0	0	0	12	0		0	16		0	0	0
Engineer Company	5	2	23	189	0	205	0	0	0	0	0	6	0	0	0		0	0		0	0	0
48th Ind Mixed Brigade																						
Brigade HQ	17	2	29	108	0	84	0	0	0	0	0	0	0	0	0		0	0		2	1	10
Signal Unit	3	1	16	111	0	118	0	0	0	0	0	0	0	0	0		0	0		0	0	0
Four Infantry Battalions, each	29	9	98	560	32	318	27	12	27	2	2	0	0	0	0		0	0		0	0	0
Artillery Battalion	18	6	41	375	40	0	0	0	0	0	0	0	0	12	0		0	16		0	0	0
Engineer Company	5	2	23	189	0	205	0	0	0	0	0	6	0	0	0		0	0		0	0	0
49th Ind Mixed Brigade																						
Brigade HQ	17	2	29	108	0	84	0	0	0	0	0	0	0	0	0		0	0		2	1	10

Table 2.24: Central Pacific Type IMBs

	Officers	Warrant Officers	NCOs	Privates	Pistols	Rifles	Light MGs	Medium MGs	Heavy Gren Lnchrs	70mm Infantry Howitzers	37mm AT Guns	Flamethrowers	75mm AA Guns	Artillery Pieces	Light Tanks		Searchlights	Riding Horses		Motorcycles	Cars	Trucks
Signal Unit	3	1	16	111	0	118	0	0	0	0	0	0	0	0	0		0	0		0	0	0
Eight Infantry Battalions, each	29	9	98	560	32	318	27	12	27	2	2	0	0	0	0		0	0		0	0	0
Artillery Battalion	18	6	41	375	40	0	0	0	0	0	0	0	0	12	0		0	16		0	0	0
Anti-Aircraft Battery	4	1	16	148	13	40	0	0	0	0	0	0	6	0	0		0	0		0	0	0
Engineer Company	5	2	23	189	0	205	0	0	0	0	0	6	0	0	0		0	0		0	0	0
50th Ind Mixed Brigade																						
Brigade HQ	17	2	29	108	0	84	0	0	0	0	0	0	0	0	0		0	0		2	1	10
Signal Unit	3	1	16	111	0	118	0	0	0	0	0	0	0	0	0		0	0		0	0	0
Five Infantry Battalions, each	29	9	98	560	32	318	27	12	27	2	2	0	0	0	0		0	0		0	0	0
Artillery Battalion	18	6	41	375	40	0	0	0	0	0	0	0	0	12	0		0	16		0	0	0
Tank Company	5	3	27	55	24	48	0	0	0	0	0	0	0	0	11		0	0		0	0	4
Anti-Aircraft Battery	4	1	16	148	13	40	0	0	0	0	0	0	6	0	0		0	0		0	0	0
Engineer Company	5	2	23	189	0	205	0	0	0	0	0	6	0	0	0		0	0		0	0	0
51st Ind Mixed Brigade																						
Brigade HQ	17	2	29	108	0	84	0	0	0	0	0	0	0	0	0		0	0		2	1	10

Table 2.24: Central Pacific Type IMBs

	Officers	Warrant Officers	NCOs	Privates	Pistols	Rifles	Light MGs	Medium MGs	Heavy Gren Lnchrs	70mm Infantry Howitzers	37mm AT Guns	Flamethrowers	75mm AA Guns	Artillery Pieces	Light Tanks		Searchlights	Riding Horses		Motorcycles	Cars	Trucks
Signal Unit	3	1	16	111	0	118	0	0	0	0	0	0	0	0	0		0	0		0	0	0
Six Infantry Battalions, each	29	9	98	560	32	318	27	12	27	2	2	0	0	0	0		0	0		0	0	0
Two Artillery Battalions, each	18	6	41	375	40	0	0	0	0	0	0	0	0	12	0		0	16		0	0	0
Anti-Aircraft Battalion	23	4	55	618	58	93	0	0	0	0	0	0	18	0	0		6	0		2	0	18
Engineer Company	5	2	23	189	0	205	0	0	0	0	0	0	0	0	0		0	0		0	0	0
52nd Ind Mixed Brigade																						
Brigade HQ	17	2	29	108	0	84	0	0	0	0	0	0	0	0	0		0	0		2	1	10
Signal Unit	3	1	16	111	0	118	0	0	0	0	0	0	0	0	0		0	0		0	0	0
Four Infantry Battalions, each	29	9	98	560	32	318	27	12	27	2	2	0	0	0	0		0	0		0	0	0
Artillery Battalion	18	6	41	375	40	0	0	0	0	0	0	0	0	12	0		0	16		0	0	0
Tank Company	5	3	27	55	24	48	0	0	0	0	0	0	0	0	11		0	0		0	0	4
Engineer Company	5	2	23	189	0	205	0	0	0	0	0	0	0	0	0		0	0		0	0	0
53rd Ind Mixed Brigade																						
Brigade HQ	17	2	29	108	0	84	0	0	0	0	0	0	0	0	0		0	0		2	1	10
Signal Unit	3	1	16	111	0	118	0	0	0	0	0	0	0	0	0		0	0		0	0	0

Table 2.24: Central Pacific Type IMBs

	Officers	Warrant Officers	NCOs	Privates	Pistols	Rifles	Light MGs	Medium MGs	Heavy Gren Lnchrs	70mm Infantry Howitzers	37mm AT Guns	Flamethrowers	75mm AA Guns	Artillery Pieces	Light Tanks		Searchlights	Riding Horses		Motorcycles	Cars	Trucks
Seven Infantry Battalions, each	29	9	98	560	32	318	27	12	27	2	2	0	0	0	0		0	0		0	0	0
Artillery Battalion	18	6	41	375	40	0	0	0	0	0	0	0	0	12	0		0	16		0	0	0
Engineer Company	5	2	23	189	0	205	0	0	0	0	0	0	0	0	0		0	0		0	0	0

guns in the case of the 47th IMB (Type 94s) and the 51st IMB (Type 41s), and a mix of one battery of 75mm field guns and two batteries of 105mm howitzers in the other IMBs. It should be noted, however, that once in static positions the artillery units were often given responsibility for other weapons so that in fact, for instance, the 47th IMB artillery unit on Saipan actually had one battery of eight 75mm guns and two batteries of seven 105mm howitzers each.

Some of the IMBs also had tank companies, each with 11 light tanks; and anti-aircraft batteries (six 75mm guns) or battalions (three gun batteries and one searchlight battery). In addition, each IMB had an engineer company with four 45-man platoons and a materiel platoon.

Most, and perhaps all, of the expeditionary units had received additional weapons in the form of about 20 Type 2 anti-tank grenade launchers and 50 Type 100 rifle grenade launchers per infantry battalion. Presumably these supplemental weapons remained with the units when they reorganized into IMBs.

The most notable shortcoming of this series of IMBs was their complete lack of mobility. Barring the attachment of outside transport elements the brigades, and especially the artillery, were simply incapable of movement. This weakness was exacerbated by a very weak radio link. A brigade signal unit included eight Type 94 Mk 5 radios, but these were the only wireless sets in the brigade (except those in the light tanks, where present). All other communications were by telephone and runner. The wire net presumably served fairly well as long as the brigade was in a static mode, but once it had to move under fire communications must have been tenuous at best.

Southern Army IMBs – the Second Wave

In June 1944 Southern Army ordered the formation of another bloc of IMBs. The 54th IMB was raised for the Zamboanga Peninsula of Mindanao, the 55th IMB for the Sulu Archipelago (particularly Jolo), the 56th IMB for Borneo, the 57th IMB for the Celebes, and the 58th IMB for Luzon. The 58th IMB was a conversion of a pre-existing unit (the 12th Expeditionary Unit), but the rest of the brigades represented actual accretions to strength of Southern Army, the personnel having been shipped from Japan for local organization.

54th IMB (Ind Inf Bns 360, 361, 362)
55th IMB (Ind Inf Bns 363, 364, 365)
56th IMB (Ind Inf Bns 366, 367, 368, 369, 370, 371)
57th IMB (Ind Inf Bns 372, 373, 374, 375, 376, 377)
58th IMB (Ind Inf Bns 378, 379, 380)

These units adopted almost precisely the organizational structure of the 24th–34th IMBs raised earlier in that theater.

	Off	WO	MSG	SGT	PVT	Total	Horses
Brigade Headquarters	22	1	22	10	111	166	4
Signal Unit	5	2	4	20	147	178	17
Each Infantry Battalion	30	11	10	103	843	997	20
Artillery Unit	16	4	7	27	306	360	128
Engineer Unit	4	2	4	15	155	180	11

Each of the infantry battalions was divided into four 176-man rifle companies (each 9 light MGs and 9 grenade dischargers) and a 134-man weapons company with medium MGs, 37mm AT guns and 70mm howitzers. The artillery unit consisted of two batteries of 75mm mountain guns. These brigades were more mobile than those raised by 31st Army in the central Pacific, having the responsibility for larger land areas, but only slightly. The brigade headquarters was provided with 18 motor vehicles to supplement the meager allotment of horses.

IMBs for the Philippines

In July 1944 the Kwantung Army was ordered to raise two more four-battalion mixed brigades (59 and 60) for service in the Pacific and these were shipped out almost immediately to the Ryukyus. A third brigade was raised *in situ* in the Philippines at the same time to garrison the Batan and Babuyan islands north of Luzon. A fourth brigade (the 62nd) was activated simultaneously for service in China.

59th IMB (Ind Inf Bns 393, 394, 395, 396)
60th IMB (Ind Inf Bns 397, 398, 399, 400)
61st IMB (Ind Inf Bns 405, 406, 407, 408, 409)[42]
62nd IMB (Ind Inf Bns 410, 411, 412, 413, 414)[43]

The two brigades raised by the Kwantung Army had the following organization initially:

	Off	WO	MSG	SGT	PVT	Total	Horses
Brigade Headquarters	19	0	16	9	126	170	30
Signal Unit	5	1	5	17	179	207	40
Each Infantry Battalion	31	8	9	92	869	1,009	147
Artillery Unit	23	4	10	58	771	866	486
Engineer Unit	6	1	4	22	255	288	43
Total	177	38	71	474	4,807	5,567	1,187

These units, highly mobile by IMB standards, retained this organization only a short time. On 26 July, two weeks after their formation, their strengths were dramatically reduced, presumably to conserve shipping space. IGHQ dictated new manpower ceilings for each unit, without specifying ranks, of 150 (with 3 horses) for the HQ, 200 (with 5 horses) for the signal unit, 600 (with 4 horses) for each infantry battalion, 400 (with 35 horses) for the artillery unit, and 280 (with 8 horses) for the engineer unit.

These two brigades, the 59th and 60th, served on smaller islands at the southern end of the Ryukyus, with most component units being 5–10 percent understrength at the end of the war.

The 61st Brigade was completely different from the prior two, having a 103-man HQ,

42 Independent Infantry Battalion 302 was added on 26 August 1944.

43 The brigade was significantly strengthened on 1 February 1945 with the 625th Independent Infantry Battalion, 10th Machine Gun Battalion, 10th Independent Medium Artillery Battalion, 54th Independent Field AA Battery, and 64th Independent Field Machine Cannon Battery for a new strength of 9,947 men. By the end of the war the locally-authorized strength had risen to 10,456, although no new units were added.

562-man infantry battalions, a 348-man artillery unit, a 174-man engineer unit and a 156-man signal unit. The 62nd Brigade was generally similar to the early China-type IMBs. Each of the infantry battalions had four rifle companies (each with nine light MGs and nine grenade dischargers), a machine gun company (eight MGs), and a gun company (two 37mm AT guns, two 70mm howitzers and two 75mm regimental guns). The brigade's artillery unit consisted of one battery of field guns and one of mountain guns, while the engineer unit was made up of two companies. Each of its infantry battalions had only 15 horses, while the battalion-size artillery unit had 65 horses, the engineer unit six and the signal unit ten, something of an improvement over the immobile Pacific-type, but just barely.

IMBs for the Homeland: a First Group via Redesignation

There then followed a series of non-standard IMBs raised for the defense of the homeland. In these cases the IMBs HQs were essentially activated to command existing units for the defense of specific areas, mostly in the Tokyo Bay area. These units were:

64th IMB: HQ (114 men), 21st & 22nd Independent Mixed Regiments (each 2230), 6th Heavy Artillery Regiment (997)

65th IMB: 27th Independent Mixed Regiment, 1st Special Garrison Battalion (B)[44]

66th IMB: 18th Independent Mixed Regiment; 427th Independent Infantry Battalion; Niijima Detachment; 27th ind. AT Company; 22nd ind. Mountain Artillery Battalion; 5th ind. Machine Gun Battalion; 15th, 16th & 17 Special Garrison Companies

67th IMB: 16th Independent Mixed Regiment; 425th and 426th Independent Infantry; Battalions; 16th ind. Machine Gun Battalion; 15th ind. AT Battalion; 24th ind. AT Company; 12th ind. Field Artillery Battalion; 23rd ind. Mountain Artillery Battalion; 100th ind. Medium; Artillery Battalion; 52nd Field AA Battery; 50th Field Machine Cannon Company; 41st & 42nd Special Machine Cannon Units; 63rd Engineer Battalion (A); 5th Special Garrison Battalion (B)[45]

69th IMB: Ind Infantry Battalions 421, 422, 423 & 424

The 64th IMB, essentially a coastal defense unit, was built around the 6th Heavy Artillery Regiment, a coastal artillery formation divided into one 2-battery and one 3-battery battalion with a total of four 24cm and two 15cm weapons, along with seven field guns. Infantry consisted of two independent mixed regiments, each consisting of three small battalions, an AT company (four 37mm guns), and a mountain gun company (four 75mm Type 41). Each of the 600-man battalions had three rifle companies (each six light

44 Independent Infantry Battalions 668, 669 and 670 were added to this brigade on 6 February 1945. Each of these had 898 men in a headquarters, three rifle companies (each nine light MGs and six grenade dischargers), a machine gun company (eight medium MGs) and an anti-tank company (four 37mm AT guns).

45 Independent Heavy Artillery Battery 4 was added to this brigade in January 1945, and Independent Infantry Battalions 668 and 669 were added in March.

MGs and nine grenade dischargers) and a machine gun company (four medium MGs).

The elements above also gave the 66th IMB a TO&E strength of 4,323 men and the 67th IMB of 8,227 men. Their small personnel strengths relative to the large list of component units was due largely to the somewhat diminutive size of their constituent elements. The independent infantry battalions of these brigades were of the smaller type, each having a strength of 564 men, while the mountain artillery battalions had strengths of only 415 men each.

In August 1944 the 8th Independent Mixed Regiment in the Kuriles was augmented by two additional infantry battalions and redesignated the 69th IMB to consist of a 298-man headquarters, four 814-man infantry battalions and an artillery unit.

Southern Army and Kwantung Army

In September 1944 the Southern Army raised three more IMBs: the 70th in French Indochina (later moved to Malaya), the 71st in Borneo, and the 72nd in Burma. The Indochina and Borneo units represented actual accretions to strength, since the personnel for these were shipped from Japan. This was made possible by the assessment that the defenses of the Philippines, which had previously had first call on all personnel, were judged to be at full strength. The 72nd IMB had two battalions drawn from the 53rd Division and two reorganized from the 61st Infantry Regiment.

69th IMB (Ind Inf Bns 421, 422, 423, 424)
70th IMB (Ind Inf Bns 428, 429, 430, 431)[46]
71st IMB (Ind Inf Bns 538, 539, 540, 541)
72nd IMB (Ind Inf Bns 187, 188, 542, 543)

All these brigades followed the normal Southern Area Army organizational pattern of including an artillery battalion and engineer company for support. The 70th IMB, in addition, included a tank company.

	Off	WO	MSG	SGT	PVT	Total	Horses
Brigade Headquarters	19	1	24	10	111	165	4
Signal Unit	5	2	4	20	147	178	17
Each Infantry Battalion	30	11	10	103	843	997	20
Tank Unit (70th IMB only)	4	2	5	22	92	125	0
Artillery Unit	16	4	7	27	306	360	128
Engineer Unit	4	2	4	15	155	180	11
Total	164	53	79	484	4,091	4,871	200

The hapless 72nd Brigade was ordered organized on 17 December but when its organization was nominally completed in late January 1945 it had an actual strength of only 850 men. Reinforced with other provisional units, it reached a strength of around 2,000 men before being ground down around Moulmein in 1945. The 71st still lacked two infantry battalions and the engineer unit at the end of the war. The other brigades appear to have reached nominal strength, although sometimes through the incorporation

46 Independent Infantry Battalions 676 and 677 were added 14 April 1945.

of untrained personnel.

In October the Kwantung Army formed the 73rd IMB from left-behind elements of the 1st Division but this lasted only a short time, being expanded to the 123rd Division in February 1945. The 10th Area Army formed two IMBs in January 1945, the 75th and 76th, each with independent infantry battalions, an engineer unit and the 12th and 13th, respectively, heavy artillery regiments.[47] These essentially static units were organized with the following strengths:

	Off	WO	MSG	SGT	PVT	Total	Horses
Brigade Headquarters	6	0	1	14	71	92	26
Each Infantry Battalion	30	5	7	87	680	809	0
12th Hvy Art Regt (75th IMB)	25	2	12	59	505	603	3
13th Hvy Art Regt (76th IMB)	28	4	14	88	808	942	3
Engineer Unit	5	1	2	21	220	249	0

In February four additional IMBs were formed in Manchuria: the 77th (from the 3rd Cavalry Brigade), the 78th (from elements of the 71st Division), the 79th (from units in the Yehho-Mutanchiang area), and the 80th (from left-behind elements of the 23rd Division).

75th IMB (Ind Inf Bns 560, 561, 562, 563 & 564)
76th IMB (Ind Inf Bns 565, 566 & 567)
77th IMB (Ind Inf Bns 568, 569, 570, 571 & 572)
78th IMB (Ind Inf Bns 573, 574, 575, 576 & 577)
79th IMB (Ind Inf Bns 578, 579, 580, 581 & 582)
80th IMB (Ind Inf Bns 583, 584, 585, 586 & 587)

The four February brigades had the following composition:

	Off	WO	MSG	SGT	PVT	Total	Horses
Brigade HQ	20	0	15	9	124	168	30
Signal Unit	5	1	5	17	179	207	40
Each Infantry Battalion	34	0	32	75	678	819	119
Artillery Unit	21	0	12	45	499	577	271
Engineer Unit	6	0	5	22	255	288	43
Transport Unit	25	0	24	45	514	608	216
Total (5-battalion brigade)	247	1	221	513	4,961	5,943	1,195

The 77th and 78th brigades were disbanded on 10 July 1945 to provide personnel for the new divisions of the 130-series.

The Last Gasp in China

In March 1945 the China theater raised twelve more IMBs. All were for the southern portion of the country under the 6th Area Army, with the 20th Army receiving four IMBs, the 13th Army four, the 34th Army three and the 11th Army one. The brigades

47 These were former fortress heavy artillery regiments, often field artillery battalions in all but name.

were organized from replacement personnel mobilized in Japan and shipped to China. Although the officers were for the most part trained the enlisted men had received little, if any, training before arriving in China.

81st IMB (Ind Inf Bns 484, 485, 486, 487, 488)
82nd IMB (Ind Inf Bns 489, 490, 491, 492, 493)
83rd IMB (Ind Inf Bns 494, 495, 496, 497, 498)
84th IMB (Ind Inf Bns 499, 500, 501, 502, 503)
85th IMB (Ind Inf Bns 504, 505, 506, 507, 508)
86th IMB (Ind Inf Bns 509, 510, 511, 512, 513)
87th IMB (Ind Inf Bns 514, 515, 516, 517, 518)
88th IMB (Ind Inf Bns 519, 520, 521, 522, 523)
89th IMB (Ind Inf Bns 524, 525, 526, 527, 528)
90th IMB (Ind Inf Bns 626, 627, 628, 629, 630)
91st IMB (Ind Inf Bns 631, 632, 633, 634, 635)
92nd IMB (Ind Inf Bns 615, 616, 617, 618, 619)

These brigades were identically organized (except for the 89th IMB), being authorized the following strengths by IGHQ:

	Off	WO	MSG	SGT	PVT	Total	Horses
Brigade HQ	26	1	22	19	202	270	8
Signal Unit	4	1	3	26	192	226	25
Each Infantry Battalion	45	13	14	128	1349	1549	130
Artillery Unit	21	3	6	39	543	612	423
Engineer Unit	17	2	9	49	449	526	29
Total (5-battalion brigade)	312	72	110	773	8131	9381	1135

The China theater, however, actually raised them with a lower strength of 8,441 men mainly by using only four rifle companies per battalion instead of five. This resulted in the organization seen in Table 2.25.

The rifle companies followed the standard pattern, being composed of three rifle platoons, each with three LMG and one grenadier squads. With a strength of 230 men, however, they were larger than most of their contemporaries. The 130-man machine gun company had four platoons, each of two gun squads. The 100-man infantry gun company had one platoon with two 70mm infantry howitzers and one platoon with 75mm Type 41 mountain guns. Characteristic of these independent infantry battalions was the absence of any battalion ammunition trains. The 16 rounds carried in the rifle company for each grenade discharger represented the total battalion holdings, as did the 280 rounds of 70mm and 105 rounds of 75mm HE shells held by the gun company. The machine gun company carried 6,480 rounds for each of its weapons. Since the IMBs lacked an organic transport unit, no additional ammunition was carried at the brigade level, either.

Although there were some variations, the brigade artillery unit generally consisted of two mortar companies and a mountain gun company. A mortar company consisted of a HQ platoon, two firing platoons, and a trains platoon. Each firing platoon had three squads, each manning an 81mm Type 97 mortar, with the squad and the trains platoon carrying

Table 2.25: 80-Series Independent Mixed Brigades in China, March 1945

	Officers	Warrant Officers	NCOs	Privates	Pistols	Carbines	Rifles	Light MGs	Medium MGs	Type 89 Gren Lnchrs	81mm Mortars	70mm Inf Howitzers	75mm Type 41 Mountain Guns	75mm Type 94 Mountain Guns	Riding Horses	Pack Horses	Draft Horses	Carts	Cars	Trucks
Brigade Headquarters																				
Headquarters	16	0	20	71	0	30	33	1	0	1	0	0	0	0	5	0	0	0	1	3
Ordnance Duty Unit	1	1	5	30	0	41	0	0	0	0	0	0	0	0	1	0	0	0	0	0
Medical Section	6	0	7	63	0	0	45	0	0	0	0	0	0	0	2	0	0	0	0	0
Signal Unit	4	1	29	192	0	?	0	0	0	0	0	0	0	0	1	10	14	14	0	0
Five Infantry Battalions, each																				
Battalion HQ	?	?	?	?	0	0	25	0	0	1	0	0	0	0	9	3	12	10	0	0
Four Rifle Companies, each	?	?	?	?	0	0	196	9	0	9	0	0	0	0	0	0	0	0	0	0
Machine Gun Company	?	?	?	?	16	0	0	0	8	0	0	0	0	0	1	34	0	0	0	0
Infantry Gun Company	?	?	?	?	8	0	0	0	0	0	0	2	2	0	4	51	0	0	0	0
Signal Unit	?	?	?	?	0	0	71	0	0	0	0	0	0	0	1	6	2	2	0	0
Artillery Unit																				
Battalion HQ	8	0	19	27	7	0	0	0	0	0	0	0	0	0	5	0	2	0	0	0
Two Mortar Companies, each	4	1	13	157	14	0	15	0	0	0	6	0	0	0	6	0	79	71	0	0
Mountain Gun Battery	4	1	11	167	14	0	15	0	0	0	0	0	0	4	16	83	0	0	0	0

Table 2.25: 80-Series Independent Mixed Brigades in China, March 1945

	Officers	Warrant Officers	NCOs	Privates	Pistols	Carbines	Rifles	Light MGs	Medium MGs	Type 89 Gren Lnchrs	81mm Mortars	70mm Inf Howitzers	75mm Type 41 Mountain Guns	75mm Type 94 Mountain Guns	Riding Horses	Pack Horses	Draft Horses	Carts	Cars	Trucks
Engineer Unit																				
Battalion HQ	6	1	15	55	5	0	0	0	0	0	0	0	0	0	7	12	0	0	0	0
Two Companies, each	5	1	22	227	2	16	218	1	0	0	0	0	0	0	5	18	0	0	0	0

67 rounds of HE ammunition for each weapon. The mountain gun company, normally equipped with 75mm Type 94 guns, was unusual in that many of the companies were apparently authorized only three guns instead of the usual four. Somewhat anomalously, the mortars (which were well adapted to pack transport) used carts for transport, while the mountain guns used the more manpower-intensive pack horse system.

The brigade engineer unit consisted of two companies, each of four platoons and an equipment squad. The equipment squad had 16 pack horses to carry tools and demolition equipment. The engineer platoons were commanded by lieutenants and composed of four squads.

The brigade signal unit was composed of one wire platoon and two radio platoons. The wire platoon had four squads, while the radio platoons each had eight radio squads. In addition to the 16 radios held by the signal unit, the brigade HQ itself had five Type 94 Mk 5 radios, while each infantry battalion HQ had a further six for internal distribution. The artillery unit had Type 94 Marks 3, 5 and 6 for its internal net. The 80-series China-type IMB was thus considerably better equipped in terms of signal equipment than its Central Pacific type counterpart.

Surprisingly, however, the China-type IMBs did not have organic transport, which must have reduced their flexibility considerably.

The 89th IMB was supplemented by many attached units, including the 9th Ind Machine Gun Battalion, the 53rd AA Unit, the 63rd Machine Cannon Unit, a transport unit and other service elements that brought authorized strength up to 11,116. These 80-series represented the last IMBs raised in the China theater.

Homeland Defense: the Second IMB Group

The struggle to prepare the homeland for defense began to be reflected in the mobilization of IMBs in mid-February 1945, when eight brigades were ordered activated.[48]

95th IMB (Ind Inf Bns 651, 652, 653, 654)
96th IMB (Ind Inf Bns 655, 656, 657, 658, 659, 660)
97th IMB (Ind Inf Bns 661, 662, 663)
98th IMB (Ind Inf Bns 664, 665, 666, 667)
100th IMB (Ind Mixed Regt 30; Hvy Art Regt 16)
101st IMB (Ind Inf Bns 456, 457, 458, 459)
102nd IMB (Ind Inf Bns 464, 465, 466, 467)
103rd IMB (Ind Inf Bns 468, 469, 470)

Of these units, three (100th, 102nd, 103rd) were raised in Taiwan for the defense of that island, while the 101st IMB (with a strength of 5,696) represented a reorganization of the 31st and 32nd Independent Garrison Units on Hokkaido. The remaining brigades were raised for the rest of the Homeland, with one each going to the 11th, 12th, 13th and 16th Area Armies.

The homeland brigades were built around a variable number of standardized infantry battalions and had the following components:

48 There were no IMBs numbered 93 and 94.

	Off	WO	MSG	SGT	PVT	Total	Horses
Brigade Headquarters	17	2	20	21	177	237	0
Each Infantry Battalion	27	5	6	92	768	898	2
Engineer Unit	5	1	2	21	220	249	0

The infantry battalions each consisted of a 77-man HQ, three 187-man rifle companies, a 138-man machine gun company and a 122-man mortar company. The rifle companies were of the standard pattern, with three platoons each of three 13-man rifle squads (each a light MG and twelve rifles) and a 13-man grenadier squad (with three rifles and three grenade dischargers). The machine gun company was made up of four 2-gun platoons, while the mortar company had two platoons each with two 81mm mortars. Each battalion was provided with six telephones and twelve reels of wire, but no radios. The brigades were each supported by a 237-man HQ (that included a 50-man radio platoon with five radios and a 19-man motor transport section with a car and five trucks) and a 249-man engineer unit. The IMBs in Japan proper were activated without an artillery component, although in the case of the 98th IMB the 15th Heavy Artillery Regiment, a 1,221-man formation, was attached later on. The 95th IMB included the 33rd Heavy Artillery Battalion of 584 men, reduced in February to 356. Under both strength figures the battalion was an immobile coastal formation with only three riding horses for transport.

The three brigades on Formosa each had a unique organization. The 100th IMB was simply the Takao Fortress (HQ and Heavy Artillery Regiment) redesignated and reinforced by the 30th Independent Mixed Regiment and a guard battalion. It had the following strengths:

	Off	WO	MSG	SGT	PVT	Total	Horses
Brigade Headquarters	12	0	6	9	82	116	0
30th Ind Mixed Regt	82	14	24	262	2,037	2,417	0
51st Guard Battalion	17	3	3	68	546	637	2
16th (ex-Takao Fort) Hvy Art Regt	27	2	14	76	591	710	3

The 102nd Brigade consisted of the four infantry battalions, two artillery units and a signal unit for a total strength of 241 officers and 5,815 enlisted. The 103rd IMB had three infantry battalions, an artillery unit, a transport unit, a medical unit, a signal unit and an ordnance duty unit for a total of 219 officers and 5,232 enlisted.

The unique 101st IMB had the following components:

	Off	WO	MSG	SGT	PVT	Total	Horses
Brigade Headquarters	20	0	10	23	244	297	8
Signal Unit	10	2	4	26	236	278	16
Four Infantry Battalions, each	37	7	10	119	810	983	7
Artillery Unit	31	6	6	68	627	708	99
Engineer Unit	7	2	4	22	232	499	17
Field Hospital	3	1	13	4	133	154	0
Total	219	39	77	619	4,712	5,666	168

These units were followed by the 105th IMB. On 5 January 1945 the Rangoon Independent Mixed Brigade had been organized under the 28th Army in Burma for the defense of that city. On 10 March the brigade was redesignated the 105th IMB, consisting of the 451st–453rd Independent Infantry Battalions and an artillery unit. Each of the infantry battalions had a TO&E strength of 580 men and was divided into four rifle companies, a heavy weapons company (with MGs and 70mm howitzers) and a labor platoon. The artillery unit had two batteries each with four 81mm mortars. In April, the artillery batteries were re-equipped with three captured 25pdr guns each.[49]

The same month saw the raising of the 107th and 109th IMBs in Japan for Homeland defense, the 108th IMB for Korea and the 112th IMB in Taiwan.

107th IMB (Ind Inf Bns 636, 637, 638, 639, 640, 641)
108th IMB (Ind Inf Bns 642, 643, 644, 645, 646, 647)
109th IMB (Ind Inf Bns 678, 679, 680, 681, 682, 683, 684)
112th IMB (Mixed Regts 32, 33)[50]

The two Homeland IMBs were both raised under the 16th Area Army but had different organizations. In the case of the 107th and 108th IMBs the infantry battalions had strengths of 805 men and were supported by a 572-man artillery unit, a 178-man engineer unit and a 175-man signal unit. These elements, plus HQ and administrative units, gave the brigade a total strength of 7,343. Each of the infantry battalions was made up of four rifle companies (each six light MGs and nine grenade dischargers), a machine gun company (eight medium MGs) and an infantry gun company (two 70mm infantry howitzers and two 75mm Type 41). The artillery unit consisted of two mortar batteries (each six 120mm mortars) and a field gun battery (two 75mm improved Type 38). The brigade garrisoned Fukue Shima at the southern end of the Goto Retto chain.

The 109th IMB had two types of infantry battalions, with the 678th–680th having strengths of 590 each and made up of four rifle companies (each 12 light MGs and 12 grenade dischargers), while the 681st–684th had strengths of 705 men and made up of three rifle companies and a machine gun company. These were supported by a 188-man engineer unit, a 163-man transport unit, a 70-man medical unit, and a 244-man HQ. Attached units reinforced the brigade: the 420-man 203rd Special Garrison Battalion (with 12 light MGs and 12 grenade dischargers), the 85-man 55th Special Machine Cannon Unit, and the 492-man 26th Ind Field Artillery Battalion (twelve improved Type 38 field guns).

The Last Wave of IMBs for the Homeland, May 1945

On 23 May the last homeland mobilization was ordered, this including 15 IMBs (113th–126th and 128th) for the defense of Japan itself and one (127th) for Korea.

113th IMB (Ind Inf Bns 685, 686, 687, 688, 689)
114th IMB (Ind Inf Bns 690, 691, 692, 693, 694, 695)

49 The IGHQ files credit the 105th IMB with a 171-man HQ, three 997-man infantry battalions and a 59-man mortar artillery unit (later replaced by a field artillery unit). It seems unlikely, however, that it actually ever reached this strength.

50 These were former independent mixed regiments that lost the "independent" appellation when incorporated into the IMB.

115th IMB (Ind Inf Bns 696, 697, 698, 699, 700, 701)
116th IMB (Ind Inf Bns 702, 703, 704, 705, 706)
117th IMB (Ind Inf Bns 707, 708, 709, 710, 711, 712)
118th IMB (Ind Inf Bns 713, 714, 715, 716, 717)
119th IMB (Ind Inf Bns 718, 719, 720, 721, 722)
120th IMB (Ind Inf Bns 723, 724, 725, 726, 727)
121st IMB (Ind Inf Bns 728, 729, 730, 731, 732)
122nd IMB (Ind Inf Bns 733, 734, 735, 736, 737)
123rd IMB (Ind Inf Bns 738, 739, 740, 741, 742, 743)
124th IMB (Ind Inf Bns 744, 745, 746, 747, 748)
125th IMB (Ind Inf Bns 749, 750, 751, 752, 753, 754)
126th IMB (Ind Inf Bns 755, 756, 757, 758, 759)[51]
127th IMB (Ind Inf Bns 760, 761, 762, 763, 764)
128th IMB (Ind Inf Bns 768, 769, 770, 771, 772, 773)

The basic structure of the 113th to 126th brigades was as follows:

	Off	WO	MSG	SGT	PVT	Total	Horses
Brigade Headquarters	12	1	17	10	150	190	10
Signal Unit	5	1	4	20	194	224	15
Each Infantry Battalion	30	23	8	101	733	895	23
Artillery Unit (typical)	21	3	6	42	473	545	80
Engineer Unit	23	2	9	55	486	565	16

This group of brigades was built around a variable number of 895-man infantry battalions. Each of these battalions consisted of a HQ (with a grenade discharger), four 183-man rifle companies (each 135 rifles, nine light MGs and nine grenade dischargers), and a 133-man machine gun company (eight medium MGs and two 70mm Type 92). For communications the battalion had two Mk5 and six Mk6 radios, along with eight telephones and sixteen rolls of light wire. An artillery battalion provided fire support. In the 113th–116th, 119th, 121st, 123rd and 125th IMBs the artillery unit was a conventional formation of three batteries with a total of eight 75mm field guns and four 105mm howitzers. The effectiveness of these pieces must have been severely limited by the fact that the battalion had only three radios and only eight wagons to carry ammunition and equipment. In most of the other IMBs of this group the artillery unit had 820 men and five batteries each manning four ex-naval guns of undetermined type. That of the 118th Brigade had 856 men, but only two 135-man batteries had been organized, and they had no heavy weapons.

The IMB engineer unit consisted of two 3-platoon companies and was provided with 14 Type 39 carts, along with six Type 100 flamethrowers, three light demolition mortars, four Type 91 pneumatic rafts and five Type 95 collapsible boats. The engineers were provided with six telephones and eighteen reels of wire, along with two Mk 5 radios. The brigade signal unit was provided with 4 switchboards, 24 telephones and 120 reels of wire

51 In addition three unaffiliated independent infantry battalions were attached: the 433rd and 434th to the 126th IMB, and the 435th to the 125th IMB.

split between two 71-man wire platoons, along with six Mk 3 high-power and eight Mk 5 portable transceivers for a 51-man radio platoon, all transported by 20 Type 39 carts. The brigade headquarters consisted of a 105-man command group with ten horses, a car, two trucks and five Mk 5 radios; a 40-man ordnance duty detachment and a 45-man HQ guard detachment with two light MGs.

In addition, the 125th Brigade (ex-3rd Amphibious Brigade) included a 190-man medical unit and a 66-man motor transport unit. It may have been planned to strengthen the anti-tank capabilities of the brigades with close range weapons. In early August the 118th Brigade received an attached rocket unit of 1 officer and 150 enlisted that was to have been equipped with 70mm rocket launchers, but no weapons had been received by the time of the surrender.

The two anomalous brigades were:

	Off	WO	MSG	SGT	PVT	Total	Horses
127th IMB							
Brigade Headquarters	17	1	18	19	170	225	10
Signal Unit	4	1	3	26	190	224	15
Each Infantry Battalion	28	5	7	106	751	897	21
Artillery Unit	27	10	8	67	708	820	60
Engineer Unit	16	4	12	52	488	572	20
128th IMB							
Brigade Headquarters	19	1	22	10	111	163	4
Signal Unit	5	2	4	20	147	178	17
Each Infantry Battalion	32	11	8	103	843	997	20
Engineer Unit	5	2	3	15	155	180	11

In some cases these brigades represented actual accretions to strength, but often they were simply reorganizations of existing units, as in the 118th IMB (ex-Hoyo Fortress Unit) and 122nd IMB (ex-Nagasaki Fortress Unit). In these cases, the fortress heavy artillery "regiments" were given numerical designations and kept within the brigade, as in the 650-man 18th Heavy Artillery Regiment (ex-Hoyo Fortress Heavy Artillery Regiment) in the 118th IMB and the 387-man 17th Heavy Artillery Regiment (ex-Nagasaki Fortress Heavy Artillery Regiment) in the 122nd IMB. Similarly, the 125th IMB was simply a redesignation and reorganization of the former 3rd Amphibious Brigade.

In any event few of these brigades were fully formed, and apparently none fully equipped, before the surrender.

The 129th IMB was formed on 16 July 1945 with the 807th, 808th and 809th Independent Infantry Battalions (each 800 men), an artillery battalion, an anti-tank company and other small units, for a strength of 112 officers and 3,660 enlisted to augment the defense of the Kuriles.

The final group of independent mixed brigades were raised by the Kwantung Army. These units, the 130th to 136th IMBs, were largely drawn from personnel and units already in Manchuria and were officially activated on 10 July 1945. Units abolished to fill out these brigades included the 1st, 2nd, and 10th Border Guard Units (to form the 132nd IMB); the 5th, 6th and 7th Border Guard Units (to form the 135th and 136th IMBs) and the 74th LoC Garrison Unit (to form the 131st IMB).

130th IMB (Ind Inf Bns 775, 776, 777, 778)
131st IMB (Ind Inf Bns 779, 780, 781, 782)
132nd IMB (Ind Inf Bns 783, 784, 785, 786)
133rd IMB (Ind Inf Bns 787, 788, 789, 790)
134th IMB (Ind Inf Bns 791, 792, 793, 794)
135th IMB (Ind Inf Bns 795, 796, 797, 798)
136th IMB (Ind Inf Bns 799, 800, 801, 802)

These units had the following composition:

	Men	Horses
Brigade HQ	183	30
Signal Unit	207	40
Four Infantry Battalions, each	718	119
Raiding Battalion (3 raiding companies)	1,130	0
Artillery Unit (3 batteries)	577	271
Engineer Unit (3 platoons)	288	43
Transport Unit (3 companies)	608	239

The infantry battalions had a standard organization of a headquarters, four rifle companies, a machine gun company and an infantry gun platoon The artillery unit was made up of three batteries armed with whatever weapons were available. In the case of the 135th IMB that artillery unit nominally had two batteries of 105mm howitzers and one of 150mm howitzers salvaged from fortifications, but in fact had to content itself with 12 trench mortars until after hostilities started.

The Independent Infantry Brigades

Indicative of the nature of combat in the China theater, as weakly armed as the IMBs were, there remained roles for which the limited combined-arms capabilities of these units were not required. Duties such as local guard missions and the pursuit of small guerrilla bands and bandit groups could be entrusted to pure infantry units and in December 1943 a new type of unit, the independent infantry brigade, was brought into being.

Ten such units were raised on 10 December 1943 and officially established on 1 February 1944. The 2nd Independent Infantry Brigade was allotted to the North China Area Army; the 3rd and 10th Brigades to the 1st Army; the 1st, 4th and 9th Brigades to the 12th Army; the 5th and 7th Brigades to the 11th Army; the 6th Brigade to the 13th Army and the 8th Brigade to the 23rd Army. These units were formed in part by taking personnel from existing units (primarily infantry elements of divisions), but more often from semi-trained personnel shipped over from Japan.

1st Ind Inf Bde (Ind Inf Bns 191, 192, 193, 194)
2nd Ind Inf Bde (Ind Inf Bns 195, 196, 197, 198)
3rd Ind Inf Bde (Ind Inf Bns 199, 200, 201, 202)
4th Ind Inf Bde (Ind Inf Bns 203, 204, 205, 206)
5th Ind Inf Bde (Ind Inf Bns 207, 208, 209, 210)
6th Ind Inf Bde (Ind Inf Bns 211, 212, 213, 214)

7th Ind Inf Bde (Ind Inf Bns 215, 216, 217, 218)
8th Ind Inf Bde (Ind Inf Bns 219, 220, 221, 222)
9th Ind Inf Bde (Ind Inf Bns 223, 224, 225, 226)
10th Ind Inf Bde (Ind Inf Bns 227, 228, 229, 230)

Four more independent Infantry Brigades quickly followed these in late 1943 and early 1944:

11th Ind Inf Bde (Ind Inf Bns 231, 232, 233, 234) [March 1944]
12th Ind Inf Bde (Ind Inf Bns 235, 236, 237, 238) [December 1943]
13th Ind Inf Bde (Ind Inf Bns 239, 240, 241, 242) [December 1943]
14th Ind Inf Bde (Ind Inf Bns 243, 244, 245, 246) [March 1944]

Each of these brigades consisted of the following elements:

	Off	WO	MSG	SGT	PVT	Total	Horses
Brigade Headquarters	20	0	21	15	110	166	10
Signal Unit	3	1	4	10	93	113	11
Each Infantry Battalion	38	10	13	132	1234	1,427	72
Total	175	41	77	553	5,139	5,985	309

The TO&E called for each battalion to be made up of a HQ, five 226-man rifle companies, a mountain gun company and a signal unit. A rifle company was to consist of a 40-man HQ and three 62-man rifle platoons. In keeping with the limited need for heavy weapons the rifle platoons followed the pre-1940 organization with two light MGs and three grenade dischargers each. In practice, however, at least some of the infantry battalions reconfigured their platoons to the more modern organization of a headquarters (11 men), three rifle squads (each 13 men with a light MG) and a grenadier squad (twelve men with three grenade dischargers). The large personnel allotment made possible further changes to company organization. In the 210th Independent Infantry Battalion, for instance, a water-cooled machine gun and an 81mm mortar were added to each rifle company when operations required, these weapons manned by rifle company personnel.

About 30 of the horses in the battalion were located in the mountain gun company, which manned three 75mm Type 41 regimental guns. The battalion baggage trains held the remaining 40 horses. Four of the horses towed Type 39 carts, while the remainder were pack horses.

Unless reinforced, the independent infantry brigade had no medium machine guns, mortars or battalion 70mm howitzers. The 75mm regimental guns were presumably included because of their long range, a necessary characteristic when the battalions were spread thinly on security duties.

Given their limited armament and mobility the scope for employment of the independent infantry brigades was extremely narrow. Although all these units remained on the field, unchanged, until the surrender no further independent infantry brigades were formed.

The Amphibious Brigades

There can, perhaps, be no more certain indication of the unpreparedness of the Japanese Army for the Pacific War than the fact it was not until late 1943, almost two years after Pearl Harbor, that they formed their first amphibious unit. The organization of amphibious brigades was announced by Organization Order A-106 of 16 November 1943. This document ordered the creation of a 1st Amphibious Brigade as soon as possible and 2nd, 3rd and 4th Amphibious Brigades in March 1944.

The 1st Amphibious Brigade was organized in November 1943 in Manchuria from the 3rd Independent Garrison Unit. In mid-December the brigade embarked for the Pacific and arrived at Truk. There it passed under the command of the 4th Fleet. In preparation for the move, the Navy Section of IGHQ had sent a directive a month earlier to the 4th Fleet commander stating that the role of the Brigade was "to wait in readiness in the Marshall Islands area to engage in mobile counterattacks in said area." To this end barracks were ordered built on Kwajalein, which was to serve as the base for this mobile striking force, but before the Brigade left Truk the increasing threat of American attack in that sector made immediate reinforcement of certain points urgent and resulted in a change of mission. As a result, 2nd Battalion of the Brigade was assigned to the defense of Wotje and Maloelap, while one company of 3rd Battalion was assigned to Kwajalein. The rest of the Brigade was sent to Eniwetok, where it was destroyed in the US invasion.

The brigade was almost identical to the amphibious regiment of the ocean-type division discussed earlier. The only organic difference between the amphibious brigade and the amphibious-type infantry regiment was that the former included a slightly different mix of technical personnel in the brigade/regiment HQ that lowered strength from 172 men to 121 men. As such it suffered from the same absences of transport and field artillery. As events transpired, the shortage of transport did not factor in the brigade's fate, for it was assigned to the garrison of a coral atoll, and no significant tactical movement was required. This was undoubtedly just as well, for the brigade was allotted quite a bit of ammunition and certainly could not move more than a small fraction if it had had to displace over ground. The allotment of 81mm mortar rounds, for instance, was 35,643 (almost ten per man for the entire brigade), while 7,445 rounds of ammunition was provided for the 75mm regimental mountain guns.

The only non-organic difference between the amphibious brigade and regiment was that the brigade had an attached (but apparently not organic) sea transport unit similar to that of the ocean division. The 1st Amphibious Brigade's sea transport unit did not accompany its parent brigade to the central Pacific, but instead wound up in the Palau Islands, where it was reorganized, along with other units, into the 53rd Independent Mixed Brigade in May 1944.

While the 1st Amphibious Brigade was being moved to its new location in the central Pacific, an identical 2nd Amphibious Brigade was being raised in Manchuria for a similar fire-brigade role in the south Pacific. Formed using the HQ of the 29th Infantry Group and the 70th and 150th Line of Communications Garrison Units as a nucleus, the 2nd Amphibious Brigade was assigned to the Southern Army on 11 April 1944. By May, the Brigade had moved to the Philippines, apparently losing its tank company in the process, and in June it moved to Sorong on the Vogelkop Peninsula of western New Guinea. An attempt was made to land the brigade on contested Biak Island while the fighting was going on there, but the convoy turned back before this could be accomplished. The Brigade

Table 2.26: Independent Mixed Regiment, from October 1940

	Personnel	Pistols	Carbines	Rifles	Light Machine Guns	Medium Machine Guns	Light Gren Lnchrs	Heavy Gren Lnchrs	70mm Inf Howitzers	75mm Guns	Riding Horses	Pack Horses	Draft Horses	Carts	Wagons	Trucks
Regiment Headquarters	157	1	6	97	0	0	0	0	0	0	9	12	18	17	0	2
Cavalry Squadron	172	3	0	147	4	0	0	4	0	0	175	15	0	0	0	0
Engineer Company	291	1	2	251	0	0	0	0	0	0	8	32	0	0	0	0
Two or Three Infantry Battalions, each																
Battalion Headquarters	95	1	3	*86	0	0	1	0	0	0	7	0	0	0	0	0
Four Rifle Companies, each	194	0	0	169	6	0	0	6	0	0	0	0	0	0	0	0
Machine Gun Company	82	8	0	0	0	4	0	0	0	0	1	9	6	5	0	0
Infantry Gun Platoon	62	4	0	0	0	0	0	0	2	0	1	13	0	0	0	0
Mountain Artillery Battery	230	13	26	60	0	0	0	0	0	4	21	114	0	0	0	0
or																
Field Artillery Battery	153	11	1	50	0	0	0	0	0	4	35	0	49	11	8	0

* *includes 70 rifles as undistributed battalion reserve*

finished the war isolated and ravaged by disease and malnutrition at Sorong.

In May 1944 the two final amphibious brigades were actually raised, the 3rd and 4th, both identical to the first two brigades. They were formed in the Kuriles for local defense, but in April 1945 the 4th Brigade was ordered to Honshu and on 1 May the 3rd Brigade to Kyushu, although leaving their sea transport units behind in the Kuriles where they would be of more use. The 3rd Brigade was reorganized and reinforced into the 125th Independent Mixed Brigade on 23 May 1945, while the 4th Brigade was assigned to the defense of Tokyo, where it was subordinated to the 1st Armored Division. Since the 4th Brigade still had no organic transport, its usefulness to an armored division was probably limited.

The Independent Mixed Regiments

Although only a few specified brigades were part of the Army's permanent force structure, the mobilization plans did provide for an unspecified number of slightly smaller combined-arms units, the independent mixed regiment. Under the 1941 mobilization plan (published in October 1940) an IMR was to consist of a HQ, two or three infantry battalions, a cavalry squadron, a field or mountain artillery battery and an engineer company. Apparently, the IMRs were not envisioned as front-line combat units at this early juncture, for they were provided with a somewhat lower than normal ratio of medium machine guns and still used the old pattern rifle company organization of three rifle platoons each with two light MGs and four rifle squads. Before being shipped off overseas, however, the rifle companies were reorganized onto the more modern basis (three platoons each with three rifle and one grenadier squads), the machine gun company expanded to 8–10 weapons, and the infantry gun platoon enlarged to a small company by adding a 2-gun 37mm anti-tank platoon.

For the most part the IMRs had rather short lives, often forming the basis for larger units. Although the 2nd through 7th IMRs appear to have been formed by December 1941, few saw combat in the early phases of the war. The 2nd IMR was sent to China under the 13th Army in September 1941 and seems to have remained there until it was demobilized in mid-1943. The 4th IMR was assigned to the 25th Army for the occupation of French Indochina in September still in the configuration envisioned by the mobilization plan. In April of 1942 the 4th Regiment (less its 3rd Battalion, cavalry squadron and field artillery battery) was transferred to Borneo for occupation duties. The 4th IMR was reorganized into the 40th and 41st Independent Garrison Infantry Battalions in July 1942.

Two new IMRs were ordered activated in September 1943, the 1st and the 5th. The 1st was originally intended for duty with the Chichi Jima Fortress, but in December it departed instead for New Ireland in the south Pacific. Immediately prior to its departure the regiment was reorganized along the lines adopted by the 5th IMR, consisting of a 213-man HQ, three 938-man battalions (but with infantry gun platoons rather than companies), a 566-man field artillery battalion, a 292-man engineer company and a 276-man medical unit. The 2nd Battalion of the regiment was sent to the Admiralties where it was destroyed, while the rest of the regiment served out the war near Kavieng.

The 5th IMR, which was distinguished from other IMRs in having no cavalry squadron but with a full field artillery battalion of improved Type (Meiji) 38s, was sent to the Central Pacific in late 1942, providing an infantry battalion each to Wake, Ponape and Marcus Islands. Each of its infantry battalions consisted of three rifle companies, a machine gun company and an infantry gun company, while its artillery battalion had three batteries. It was deactivated in May 1944 as part of the 31st Army reorganization.

Table 2.27: Central Pacific Independent Mixed Regiments, from May 1944

	Officers	Warrant Officers	NCOs	Other Ranks	Pistols	Rifles	Light MGs	Medium MGs	Heavy Gren Lnchrs	37mm AT Guns	70mm Inf Howitzers	75mm Type 94 Mountain Guns	75mm Type 38 Field Guns	105mm Type 91 Howitzers	75mm Type 88 AA Guns	Light Tanks	Motorcycles	Trucks
9th & 11th Ind Mixed Regiments																		
Regiment Headquarters	10	1	19	110	0	28	0	0	1	0	0	0	0	0	0	0	1	8
Signal Unit	1	1	10	61	0	60	0	0	0	0	0	0	0	0	0	0	0	0
Three Infantry Battalions, each																		
Battalion HQ	7	0	12	16	0	14	0	0	1	0	0	0	0	0	0	0	0	0
Three Rifle Companies, each	4	2	16	93	0	77	9	0	9	0	0	0	0	0	0	0	0	0
Machine Gun Company	4	2	19	99	24	0	0	12	0	0	0	0	0	0	0	0	0	0
Infantry Gun Company	4	1	9	71	12	0	0	0	0	2	2	0	0	0	0	0	0	0
Pioneer Platoon	1	0	7	59	0	73	0	0	0	0	0	0	0	0	0	0	0	0
Artillery Battalion																		
Battalion HQ	6	0	12	45	8	0	0	0	0	0	0	0	0	0	0	0	0	0
Three Batteries, each	4	2	10	110	11	0	0	0	0	0	0	*4	0	*4	0	0	0	0
Anti-Aircraft Battery	4	1	14	148	12	0	0	0	0	0	0	0	0	0	6	0	0	0
Engineer Company	5	2	21	189	0	205	0	0	0	0	0	0	0	0	0	0	0	0

Table 2.27: Central Pacific Independent Mixed Regiments, from May 1944

	Officers	Warrant Officers	NCOs	Other Ranks	Pistols	Rifles	Light MGs	Medium MGs	Heavy Gren Lnchrs	37mm AT Guns	70mm Inf Howitzers	75mm Type 94 Mountain Guns	75mm Type 38 Field Guns	105mm Type 91 Howitzers	75mm Type 88 AA Guns	Light Tanks	Motorcycles	Trucks
10th Ind Mixed Regiment																		
Regiment Headquarters	10	1	19	110	0	28	0	0	1	0	0	0	0	0	0	0	1	8
Signal Unit	1	1	10	61	0	60	0	0	0	0	0	0	0	0	0	0	0	0
Three Infantry Battalions, each																		
as above																		
Artillery Battalion																		
Battalion HQ	6	0	12	45	8	0	0	0	0	0	0	0	0	0	0	0	0	0
Three Batteries, each	4	2	10	110	11	0	0	0	0	0	0	4	0	0	0	0	0	0
Engineer Company	5	2	21	189	0	205	0	0	0	0	0	0	0	0	0	0	0	0
12th Ind Mixed Regiment																		
Regiment Headquarters	10	1	19	110	0	28	0	0	1	0	0	0	0	0	0	0	1	8
Signal Unit	1	1	10	61	0	60	0	0	0	0	0	0	0	0	0	0	0	0
Three Infantry Battalions, each																		
as above																		

Table 2.27: Central Pacific Independent Mixed Regiments, from May 1944

	Officers	Warrant Officers	NCOs	Other Ranks	Pistols	Rifles	Light MGs	Medium MGs	Heavy Gren Lnchrs	37mm AT Guns	70mm Inf Howitzers	75mm Type 94 Mountain Guns	75mm Type 38 Field Guns	105mm Type 91 Howitzers	75mm Type 88 AA Guns	Light Tanks	Motorcycles	Trucks
Artillery Battery	4	2	9	110	11	0	0	0	0	0	0	0	4	0	0	0	0	0
Tank Company	5	3	27	55	24	0	0	0	0	0	0	0	0	0	0	11	1	3
13th Ind Mixed Regiment																		
Regiment Headquarters	10	1	20	104	0	28	0	0	1	0	0	0	0	0	0	0	1	8
Signal Unit	1	1	10	61	0	60	0	0	0	0	0	0	0	0	0	0	0	0
Three Infantry Battalions, each																		
as above																		
Artillery Battery	4	2	9	110	11	0	0	0	0	0	0	0	4	0	0	0	0	0
Tank Company	5	3	27	55	24	0	0	0	0	0	0	0	0	0	0	11	1	3
14th Ind Mixed Regiment																		
Regiment Headquarters	10	1	20	104	0	28	0	0	1	0	0	0	0	0	0	0	1	8
Signal Unit	1	1	10	61	0	60	0	0	0	0	0	0	0	0	0	0	0	0
Two Infantry Battalions, each																		
as above																		

Table 2.27: Central Pacific Independent Mixed Regiments, from May 1944

	Officers	Warrant Officers	NCOs	Other Ranks	Pistols	Rifles	Light MGs	Medium MGs	Heavy Gren Lnchrs	37mm AT Guns	70mm Inf Howitzers	75mm Type 94 Mountain Guns	75mm Type 38 Field Guns	105mm Type 91 Howitzers	75mm Type 88 AA Guns	Light Tanks	Motorcycles	Trucks
Artillery Battery	4	2	9	110	11	0	0	0	0	0	0	4	0	0	0	0	0	0
Two Tank Companies, each	5	3	27	55	24	0	0	0	0	0	0	0	0	0	0	11	1	3

* *75mm Mountain Guns in the 9th IMR, 105mm Howitzers in the 11th IMR*

Note: also 19 riding horses in the 9th–12th IMRs and two riding horses in the 13th and 14th IMRs.

No further IMRs were raised until May 1944, when the 8th Independent Mixed Regiment was activated as a HQ to command troops resident in the southern Kuriles. The regiment had a life of about three months, being reorganized into the 69th IMB in August.

A New Generation of IMRs for the Central Pacific

The reorganization of Central Pacific forces under the 31st Army in May 1944 not only spelled the end of the 5th IMR, but brought into being six new IMRs. These units, activated to rationalize command and control of existing units on the various islands were the 9th IMR (Pagan), 10th IMR (Guam), 11th IMR (Puluwat), 12th IMR (Marcus), 13th IMR (Wake) and, although not in the 31st Army's chain of command, the 14th IMR on Bougainville.

Although the regiments differed from one another in size and composition due to the availability of troops and equipment and the sizes of the areas to be defended, they used standard organizational building blocks to achieve a great deal of commonality. They used the same TO&E "building blocks" as the Central Pacific-type IMBs discussed earlier, the main difference between an IMB and an IMR under this reorganization being that the former had four or more infantry battalions, while the latter had only two or three. As was the case with the IMBs, it would seem that shortly after initial activation the infantry battalions of the IMRs were reorganized with the pioneer platoon being disbanded. It is also possible that the 13th IMR on Wake added a second tank company shortly after activation to make full use of the 18 Type 95 light tanks present.

The Third Generation IMRs

These Central Pacific-type IMRs were followed by a ten-unit series raised in Japan for widely dispersed destinations, these being the 15th to 18th and 21st in June 1944, and the 22nd, 23rd, and 25th to 27th in July.[52]

Each of these was authorized 83 officers, 15 warrant officers and 2,132 enlisted, along with a sparse allocation of only 29 horses for transport. A regiment consisted of 110 men in its HQ, 590 in each of three infantry battalions, 102 men in the infantry gun company (four 75mm regimental guns), 64 men in an anti-tank company (four 37mm guns) and 184 men in an engineer company. Each of the infantry battalions was composed of three small rifle companies (each 9 light MGs and 9 grenade dischargers), a machine gun company (4 medium MGs) and a 70mm howitzer platoon.

The 15th IMR was slated for duty with the 32nd Army with a reduced authorized strength of 2,180, the 16th IMR for the Formosa Army, the 17th IMR for the Ogasawara (Bonins) Group, and the 18th IMR for the 14th Army. It would appear that the 16th and 18th IMRs were never actually dispatched, however.

The 15th IMR was flown to Okinawa in June 1944 and subordinated to the depleted 44th IMB. Okinawan Boeitai were added to the regiment on arrival: 200 men to each infantry battalion and 20 men each to the engineer, AT and regimental gun companies. The regiment was destroyed on Okinawa in 1945. The 17th IMR was sent to Chichi Jima, where it served out the war, except for its 3rd Battalion, which had the misfortune to be caught on Iwo Jima.

The 21st and 22nd IMRs were raised for the 32nd Army, the 23rd IMR for Kyushu,

52 There were no 19th, 20th, 24th, 28th and 29th IMRs.

the 25th for the Borneo Garrison Army, 26th in Manila and 27th in Japan. Some units of this series were incorporated into non-standard independent mixed brigades, 16th IMR into the 67th IMB, the 18th IMR into the 66th, and the 21st and 22nd into the 64th.

The 1945 Group of IMRs

In early 1945 three independent mixed regiments were raised for the Formosa Army (later 10th Area Army): the 30th IMR in January with a strength of 3,000 men, followed in February by the 2,116-man 32nd and 33rd IMRs. None of these lasted very long, the 30th being disbanded to form the 100th IMB in mid-February and the 32nd and 33rd being incorporated into the 112th IMB.

The 31st IMR was raised in March with an authorized strength of 2,262 men and 481 horses for the 23rd Army around Canton. The 34th and 35th IMRs, organized like the 33rd, were activated in March as well, from assets present in the 8th Area Army in the SW Pacific. The 36th and 37th IMRs, each authorized 3,199 men and 159 horses, were formed for Tokyo's defense in May, along with the 43rd IMR with 1,500 men divided into a HQ, two infantry battalions, an anti-tank company (four 37mm guns) and an infantry gun company (four 75mm mountain guns).

An organizationally homogeneous group of three IMRs was raised in June, each with 3,442 men: the 38th IMR in 15th Area Army in Japan, and the 39th and 40th IMRs in Korea where they stayed until the end of the war. The 41st IMR was also formed in June from units remaining on Matsuwa Island in the Kuriles. This unit had a strength of 2,380 men in two infantry battalions (each four rifle companies and an infantry gun company) and an artillery battalion (two artillery batteries and an AA battery) plus a signal company. The 42nd, and final, IMR was officially activated in late June for the 10th Area Army on Formosa with a strength of 2,230.

Separate Infantry Units

When the 124th Infantry Regiment was split from its parent 18th Division as part of the triangularization process it was used to form the nucleus of the separate Kawaguchi Detachment, which swept through the Dutch East Indies starting in mid-December 1941. It was then ordered to participate under the 17th Army and to that end was reinforced in May 1942 to a strength of 3,913 men built around three infantry battalions, each of a 111-man HQ, four 168-man rifle companies, a 136-man 8-gun MG company, a 77-man 2-gun infantry gun platoon, and a 37-man 2-gun mortar platoon. Supporting those battalions was a 153-man signal company, a 165-man regimental gun company, a 125-man AT company (4 guns), a 150-man MG company (6 guns), and a 153-man signal company. The remnants of the regiment were evacuated from Guadalcanal and in March 1943 they were assigned to the 31st Division in Burma, and rebuilt with the structure of that division.

A few unaffiliated infantry units were formed near the end of the war for homeland defense. The largest was the 364th Infantry Regiment, which was attached to the 86th Division and seems to have shared its organizational format. That division also exercised tactical command of the 765th, 766th and 767th Independent Infantry Battalions, each of 897 men. The 16th Area Army held the 433rd, 434th, and 435th Battalions, each of 1,036 men, as direct commands. Other independent infantry battalions held in the homeland included the 28th, 290th, 460th and 649th.

South Seas Detachments

In addition to the various independent brigades and regiments the Army raised a confusingly wide variety of other, quasi-permanent, infantry units of similar size for duties in the Pacific. The formation of these units was brought about by a number of factors, the most important of which were the dispersed nature of the fighting on the many islands in the southwest and central Pacific; the piecemeal requests for reinforcements issued by the Navy, which was in charge of most of the Pacific theater and underestimation of the US ability to counterattack, forcing hurried decisions and improvisation.

For the initial advance into the SW Pacific the Army formed the "South Seas Detachment" out of a mathematically near-perfect one-third share of the 55th Division.

The Detachment consisted of the 55th Infantry Group HQ (180 men), the 144th Infantry Regiment (3,500), the 3rd Squadron (less one platoon, but reinforced by one 37mm gun) of the 55th Cavalry Regiment (130), the 1st Battalion of the 55th Mountain Artillery Regiment (900), the 1st Company (plus part of the materiel platoon) of the 55th Engineer Regiment (300), the 2nd (Motorized) Company of the 55th Transport Regiment (145), one field hospital (140) and one-third of the division's signal unit and other services for a total of 5,586 men. Attached was one company of the 47th Field AA Battalion, one company of the 10th Independent Engineer Regiment and some smaller units. On 10 December 1941 the Detachment captured Guam and on 23 January 1942 it captured Rabaul. After capturing Salamaua and Lae, and after reinforcement by most of the rest of the 47th Field AA Battalion, the Detachment landed at Buna and attempted to cross the Owen Stanley mountains to capture Port Morseby. It suffered heavy losses and on 17 June 1943 was deactivated and the remaining divisional troops returned to the 55th Division in Burma.[53]

While the original South Seas Detachment represented simply a combat group detached from a division, the group of detachments raised next were regarded as near-permanent organizations with their own TO&Es. Faced with increasing Allied pressure in the Pacific the Army ordered the raising "as soon as possible" of six new south seas detachments on 16 November 1943. The 1st South Seas Detachment was to be raised by the 8th Area Army out of the 122nd Infantry Regiment and the tankette company of the 38th Infantry Group; the 2nd and 3rd South Seas Detachments were to be raised by the Kwantung Army (with the 4th Independent Garrison Unit providing the nucleus for the 2nd SSD, and the 142nd Infantry Regiment for the 3rd SSD); and the 4th, 5th and 6th South Seas Detachments were to be raised by the Western Military District in the homeland.[54]

Only one of these, the 6th South Seas Detachment at Palau, was to remain under Army control. The other five detachments were placed under the Fourth Fleet and on 23 November 1943 the Navy GHQ issued orders to assign the 1st South Seas Detachment to the Marshall Islands, the 2nd SSD to Kusaie, the 3rd SSD to Ponape, the 4th SSD to Mortlock, and the 5th SSD to Woleai.

The South Seas Detachments used common TO&E building blocks to create flexible

53 Non-divisional units were used to form the 5th Ind Medium Artillery Battalion and the 42nd Ind Field AA Battery.

54 The General Staff's Unit Organization Tables and Unit Historical Tables give activation dates for the 1st through 3rd South Seas Detachments as 16 November 1943, and for the 4th through 6th Detachments as 16 February 1943. It seems likely that the latter were clerical errors with actual organization in December 1943, although they may have begun raising earlier as different units.

organizations that could be adapted to their mission. Assigned strengths were as follows:

	Off	WO	NCO	PVT	Total
Battalion Headquarters	6	0	13	15	34
Rifle Company	3	1	16	82	102
Machine Gun Company	5	1	16	54	76
Infantry Gun Company	5	1	12	94	112
Tank Company	3	1	15	40	59
Engineer Company	4	1	12	96	113
Detachment HQ	11	1	23	74	109

Aside from some attachments to detachment HQs (mostly medical personnel, but including some administrative functions as well) these strengths were common to all detachments. Of the detachments the 4th and 5th SSDs were the smallest, consisting of a detachment HQ (reinforced by 3 officers, 5 NCOs and 8 privates), three rifle companies, a machine gun company, an infantry gun company and a tank company. At the other end of the spectrum were the 2nd and 3rd SSDs, each consisting of a detachment HQ (plus 4 officers, 8 NCOs and 20 privates), a tank company, an engineer company, and three battalions each made up of a battalion HQ, three rifle companies, a machine gun company, and an infantry gun company. Between these two extremes were the 1st and 6th SSDs, each made up of a detachment HQ (plus 4 officers, 7 NCOs and 15 privates), a tank company and two battalions similar to those of the 2nd and 3rd SSDs.

The rifle companies, although small, followed the general Japanese pattern of three platoons (each with three light MGs and three grenade dischargers) reinforced by an anti-tank platoon with two 20mm AT rifles. A machine gun company had three platoons each with two medium MGs. An infantry gun company had two platoons, one with four 81mm Type 97 mortars and one with four 37mm Type 94 AT guns. Exceptions appear to have been the infantry gun company of the 4th SSD, in which the 37mm AT guns were replaced by 47mm Type 1 models, and the 81mm mortars by 70mm howitzers; and that of the 6th SSD which held four 37mm AT guns and two battalion guns. A tank company had nine light tanks, except in the 3rd SSD, which inexplicably was assigned 18 tanks. As was the case with most other formations tailored for the Pacific, the South Seas Detachments were almost completely immobile once landed. The only transport assets in most cases was a section of a sergeant and 15 privates in the detachment HQ to operate a car and ten trucks plus requisitioned vehicles and wagons.

Two of the South Seas Detachments were deactivated in the May 1944 reorganization ordered by the 31st Army. The 3rd SSD was reinforced by an infantry battalion and an artillery battalion to become the 52nd IMB, while the 5th SSD received three infantry and an artillery battalion to become the 50th IMB. In addition, the 6th SSD on Palau provided 230 men to the new 53rd IMB before being moved to New Guinea, where the HQ and 1st Battalion were destroyed at Hollandia. Three survived to the end of the war, the 1st SSD on Mille and Jaluit, the 2nd on Kusai and the 4th on Nomoi.

The Expeditionary Units

As the US threat across the central Pacific increased, further measures were desperately needed. In reaction IGHQ ordered the Kwantung Army to form eight "expeditionary units" on 21 February 1944. In order to minimize disruption of unit cohesion a policy was implemented that envisioned forming one expeditionary unit from each of eight divisions, with each division to contribute its infantry group HQ, one battalion from each of its infantry regiments, one battalion from its artillery regiment and one company from its engineer regiment. Thus, each expeditionary unit would become a miniature clone of its parent division.

In practice, of course, the differing requirements for defending the various islands in the Pacific made complete uniformity impossible. Thus, the 6th EU was actually two divisions' worth of contributions, and extra units had to be found for the 1st EU as well, but overall the original objective of retaining unit integrity was adhered to as much as possible. Once their organization was complete they were dispatched to the Pacific: the 1st EU to Ponape (but diverted to Saipan), the 2nd EU to Nomoi, the 3rd EU to Eniwetok (but diverted to Puluwat), the 4th EU to Yap, the 5th EU to Pagan, the 6th EU to Guam, the 7th EU to Woleai, and the 8th EU to Truk, arriving in mid-March to mid-April.

EU	Parent Div(s)	Inf Bns	Fld Art Bns	Mtn Art Bns	Eng Cos
1st	25th	4	0	1	1
2nd	9th	2	0	1	1
3rd	8th	3	1	0	1
4th	12th	3	1	0	1
5th	71st	2	0	1	1
6th	1st & 11th	6	1	1	2
7th	24th	2	1	0	1
8th	19th	3	0	1	1

Since the parent divisions were expected to reconstitute the units detached for service, some personnel from the elements earmarked for the expeditionary units were left behind to form a cadre. Also a severe shortage of shipping space and a reduced requirement for land mobility contributed to a decision to create a new, leaner set of TO&Es for units tapped for expeditionary unit duties.

These new tables called for eliminating 70 riflemen, 10 light MG crew and 10 grenadier slots in each rifle company, reducing company strength to 4 officers, a warrant officer, sixteen NCOs and 82 privates. The company's organizational structure and heavy weapons armament, however, remained the same. In the machine gun companies the entire 16-man AT rifle section was abolished, as was the 34-man ammunition platoon. In addition, all 58 horses were eliminated, as were the 29 men who handled them, and 44 members of machine gun crews, bringing company strength down to about 100 men. No reduction was made in the number of machine guns held (12), but all anti-tank rifles were left behind. The infantry gun and mortar platoons were each reduced by 31 men, along with all 22 horses, but retained the full quota of heavy weapons. In the battalion HQ the veterinary officer and his NCO assistant were dropped from the organization, along with 174 (in A1 type units) or 239 (in A2 type units) transport personnel and all horses. The infantry battalions of the expeditionary units were thus stripped of all transport assets, as well as all heavy weapons crewmen not strictly necessary to operate the weapons. The only reinforcement they received, in contrast, was an equally immobile anti-tank platoon of one officer, two NCOs and 30 privates with two anti-tank guns. To partially compensate for the lack of horses, each battalion was authorized thirty "rear cars," small road-bound passenger vehicles. Each battalion also received an additional allocation of 300 rifles so that each member of the 565-man battalion could be armed. The changes to the basic organization are summarized as follows:

	Original		Changes	
	Men	Horses	Men	Horses
Type A-1 Infantry Battalion				
Battalion Headquarters	187	133	-176	-133
Rifle Company	190	0	-90	0
Machine Gun Company	223	59	-123	-59
Infantry Gun Platoon	69	22	-31	-22
Anti-Tank Platoon	0	0	+31	0
Type A-2 Infantry Battalion				
Battalion Headquarters	289	181	-242	-181
Rifle Company	190	0	-90	0
Machine Gun Company	223	59	-123	-59
Mortar Platoon	69	21	-31	-21
Anti-Tank Platoon	0	0	+31	0

The artillery battalions of the expeditionary units were similarly stripped of all transport assets. The field artillery battalions were shorn of all horses, along with veterinary and transport personnel, amounting to 122 men in each battalion HQ, 52 men in each battery, and 87 men in the service batteries, and usually configured as one battery of 75mm Type 38 improved or Type 95 field guns and two of 105mm Type 91 howitzers. For transport each battalion received five trucks. The mountain artillery battalions lost 181 in the battalion HQ, 106 men in each battery and 198 men in the service battery. All horses were also deleted from these organizations, but in partial compensation they were augmented by 10 drivers with 5 trucks and 20 rear cars.

The most striking departure for the artillery component was in the 1st EU, which had been assigned the 3rd Independent Mountain Artillery Regiment for its fire support element. The regiment had been reduced to a single battalion by that time, and its horses and associated personnel were removed, but 269 men along with twelve 15cm Type 4 howitzers and an equal number of trucks were added. It took all 12 of its Type 94 mountain guns and the Type 4 howitzers to Saipan, where the EU was further reinforced by the III/10 Field Artillery, with four 75mm Type 95 guns and eight 105mm Type 91 howitzers.

The EU's engineer companies were also reduced in size, losing 66 men and 22 horses each. They were augmented however, by 20 rear cars, 4 flamethrowers, an air compressor, a generator, and 30 collapsible boats. An expeditionary unit also included in its headquarters a signal unit drawn from the divisional signal element. This signal unit consisted of an officer, a wire/optical section of 10 NCOs and 40 privates, and a radio section of 10 NCOs and 100 privates. It manned 24 telephones, one Mk 3A radio, 8 Mk 5 radios, and 12 Mk 6 radios. These assets were distributed within the expeditionary unit as required.

An exception was the 5th Expeditionary Unit, which was based on a "special" division, the 71st. It apparently had not infantry gun platoon in its infantry gun battalions, but the end result was otherwise almost the same. Its organization was:

	Off	WO	NCO	PVT	Total
Expeditionary Unit HQ	13	1	42	171	227
Two Infantry Battalions, each					
Battalion HQ	5	0	9	0	14
Four Rifle Companies	4	1	15	83	103
Machine Gun Company	5	1	20	30	106
Anti-Tank Platoon	1	0	2	30	33
Mountain Artillery Battalion					
Battalion HQ	5	0	7	20	32
Three Batteries, each	5	1	16	97	119
Anti-Aircraft Battery	4f	1	18	110	133
Engineer Company	5	3	18	160	186
Total	96	18	315	1,636	2,065

The expeditionary units received an initial allocation of extra weapons on their departure from Manchuria. The most notable of these were Type 100 rifle grenade launchers and Type 2 anti-tank grenade launchers. The provision of 20 Type 2 launchers per infantry battalion would have improved their anti-tank capabilities but for the fact that they only received 5–10 grenades per launcher, including for training.

Table 2.28: Initial Allocations of Extra Weapons to Expeditionary Units

	Type 14 Pistols	Type 99 Light MGs	Type 92 Heavy MGs	Type 89 Gren Dischargers	Type 100 Grenade Launchers	Type 2 AT Grenade Launchers	47mm AT Guns
1st Exped Unit	20	20	8	30	200	80	8
2nd Exped Unit	10	10	4	15	100	40	4
3rd Exped Unit	14	15	6	20	150	60	6
4th Exped Unit	14	15	6	20	150	60	6
5th Exped Unit	10	10	4	15	100	40	4
6th Exped Unit	30	30	12	50	300	120	12
7th Exped Unit	10	10	4	15	100	40	4
8th Exped Unit	15	15	6	20	150	60	6

On their arrival in the Pacific the expeditionary units were reinforced with additional personnel and weapons found locally and brought in from the homeland, allowing the AT platoons to be expanded to infantry gun companies. The rifle and machine gun companies were conventionally organized (albeit smaller than usual) when they left Manchuria. By May 1944, however, these forces had been considerably reinforced, particularly with grenade dischargers, infantry guns and medium machine guns. The infantry gun/mortar platoons of the infantry battalions uniformly received double allocations of weapons, so that the infantry gun platoons (in A1 type units) actually manned four 70mm howitzers and the mortar platoons (in A2 type units) eight 81mm mortars. The rifle companies also received additional weapons, particularly in grenade dischargers, which generally doubled in number. Similarly, the machine gun companies usually received an additional 6–12 weapons. All these additional weapons were supplied without an increase in personnel, rendering the expeditionary units even more immobile than before.

Following shortly on the creation of these expeditionary units IGHQ ordered the activation of the 9th EU (for Yap) on 1 April 1944 and the 10th and 11th EUs (for the Philippines and the Kuriles, respectively) on 4 April. These were also to be raised by the Kwantung Army, this time through the conversion of independent garrison infantry battalions.

9th EU: 12th, 14th & 28th ind. Garrison Infantry Bns
10th EU: 22nd, 23rd & 30th ind. Garrison Infantry Bns
11th EU: 2nd, 4th & 29th ind. Garrison Infantry Bns

These units differed from the earlier series in that they did not have any supporting artillery or engineers, and largely retained their earlier garrison unit TO&Es intact. The only changes made to the battalions were the deletion of 200 veterinary and transport personnel along with most of the battalion's horses, and the addition of an anti-tank platoon identical to those in the earlier EUs. Each battalion then consisted of four rifle companies, a battalion gun platoon (two 70mm howitzers), a regimental gun platoon (two 75mm mountain guns), an anti-tank platoon (two 47mm AT guns), and a signal platoon. Each rifle company had a strength of 144 men and was divided into a HQ, two rifle platoons and a machine gun platoon. The rifle platoon had three 9-man rifle squads (each with a light MG) and a 9-man grenadier squad (with four grenade dischargers) as well as the platoon leader and his runner. In some cases the number of light machine guns was doubled on leaving for the Pacific. The machine gun platoon was made up of the platoon leader and two 10-man squads each with a Type 92 medium MG.

For transport each battalion had five Type 39 one-horse carts, while the unit HQ had ten trucks as well as ten more carts. These second-generation EUs were fairly well provided with communications, with each battalion having four Mk 5 radios, four Mk 6 radios and eight telephones; and the unit HQ no fewer than 21 Mk 5 radios, two Mk 9 radios, two Mk 3 radios and four telephones. Total strength of each of these expeditionary units was 126 officers and 4,002 enlisted.

Finally, on 13 June 1944, IGHQ ordered the activation of the 12th, and last, Expeditionary Unit in Japan out of the 47th Division, with three infantry battalions, a mountain artillery battalion, and an engineer company, for service in the Philippines. In this case, full units were taken from the division, leaving no cadre behind since the division was at its mobilization base at the time, except that the number of rifle companies was reduced to three per battalion and a 2-gun AT platoon and a 2-gun regimental gun platoon were added to each infantry battalion.

Unlike the South Seas Detachments, which were raised largely *de novo* and had organic domestic designations (i.e., 2nd Battalion/3rd South Seas Detachment), it was apparently envisioned initially that the Expeditionary Units would eventually be reunited with their parent formations.[55] Thus, the components of the EUs retained their original designations;

55 Non-divisional elements were used to form the 5th Ind Medium Artillery Battalion and the 42nd Ind Field AA Company.

for example the 3rd Expeditionary Unit consisted of the 8th Infantry Group HQ; I/5, I/17 and III/31 Infantry Battalions; the III/8 Field Artillery Battalion; and the III/8 Engineer Company.

The hopes of reuniting the EUs with their parent formations, however, proved futile and when the 31st Army took command in the Central Pacific they acknowledged this by completely reorganizing the expeditionary units into independent mixed brigades and regiments. The four small rifle companies of each EU's infantry battalions were generally consolidated into three larger ones for the new independent infantry battalions, while the MG, infantry gun, artillery and engineer units were generally simply redesignated.

The 1st Expeditionary Unit on Saipan was reorganized into 47th IMB, the 2nd and 8th EUs on Truk and Nomoi were reorganized into the 51st IMB, the 3rd EU on Puluwat became the 11th IMR, the 4th EU on Yap became, with reinforcements, the 49th IMB, while the 5th EU on Pagan became the 9th IMR. The large 6th Expeditionary Unit was split in two, with units drawn from the 1st Division forming the 48th IMB and units drawn from the 11th Division forming the 10th IMR. The 7th EU on Woleai became the 50th IMB and the scattered 9th EU was split up among the 47th, 49th and 50th IMBs. The 10th EU finally ended up on Halmahera where it finished the war as the sole surviving expeditionary unit, while the 11th EU was quickly incorporated into the 1st Kuriles Garrison Unit. The 12th EU was reorganized into the 58th IMB on its arrival in the Philippines. Thus, none of the expeditionary units participated in combat before being reorganized into other types of units.

The Garrison Infantry

The subject of the garrison and guard infantry is a complex one, in part because the Allied translators were inconsistent in translating the Japanese terms, leading to a great deal of confusion. The basic units were as follows:

Independent Garrison Commands (*Dokoritsu shubitai shireibu*): the original garrison formations, raised in Manchuria, originally to guard the railway. Each composed of a number of independent garrison infantry battalions (*Dokoritsu shubitai hohei daitai*).

Independent Garrison Units (*Dokoritsu Shubitai*): similar to the independent garrison commands, above, and following in the same numerical sequence but with a slightly different designation and formed for duties in the south Pacific.

Border Garrisons (*Kokkyō shubitai*): formed to man the border defenses of Manchuria. Composed of infantry "units" and artillery "units" of varying size.

Garrison Units (*Shubitai*): named units (e.g., South Seas, North Seas, Southwestern) formed to garrison outlying possessions.

Guard Commands (*Keibi shirebu*): numbered units formed in Japan in 1944 and composed of a variable number of guard battalions (*Keibi daitai*), guard infantry units (*Keibi hohei tai*) and supporting arms and services.

Guard Units (*Keibi Tai*): named units formed in Manchuria of approximately battalion strength.

Independent Guard Units (*Dokoritsu Keibi Tai*): formed in China in early 1945 and made up of a variable number of independent guard infantry battalions (*Dokoritsu Keibi Hohei daitai*).

Guard Brigades (*Keibi ryodan*): formed in February 1945 to guard Tokyo, each consisting of six guard infantry battalions (*Keibi hohei daitai*) and supporting units.

Guard Battalions (*Keibi Daitai*): in addition to forming the basis for some of the Guard Commands, these units were also incorporated into the divisional districts in the homeland.

The first of the non-line infantry was the garrison infantry. These were initially formed to guard the railway lines in Manchuria granted to Japan under the 1905 Portsmouth Treaty. In 1907 these were formed into seven independent garrison infantry battalions deployed at important stations along the line with a central HQ at Kungchuling. These railway guards answered to the commander of the Kwantung Army, not to the civil governor, and were thus in the forefront of Japan's imperial drive. Because they were almost constantly employed in anti-bandit and extensive security and retaliatory operations they quickly developed a reputation as tough campaigners and were more highly regarded by the Chinese in the 1930s than the rather more sedentary infantry divisions used for occupation duties.

In spite of their usefulness, the "railway guards" were not spared completely in the Army reductions of the 1920s. By 1931 the guards had been reorganized into the "Independent Garrison Command," commanded by a lieutenant general and numbering 4,000 men. The IGC consisted of six garrison battalions, totaling 24 companies, dispersed along the railway lines. The only strength figure for this period to have survived regards the 2nd Battalion (based at Mukden) which had 25 officers and 674 enlisted.

As the Kwantung Army moved in to occupy Manchuria directly the garrison units were expanded. By late 1937 there were three independent garrison commands based at Mukden, Chilin and Harbin, and two more were added at the end of the year. Each controlled six independent garrison infantry battalions:

1st IGC (Ind Garr Inf Bns 1–6)
2nd IGC (Ind Garr Inf Bns 7–12)
3rd IGC (Ind Garr Inf Bns 13–18)
4th IGC (Ind Garr Inf Bns 19–24)
5th IGC (Ind Garr Inf Bns 25–30)

To increase flexibility three more independent garrison command headquarters (7, 8 and 9) were added in August 1939, followed by the 9th IGC headquarters in late 1940. There was no expansion in the number of independent garrison infantry battalions, however, and the new commands were filled out by transferring existing battalions from the original commands.

On mobilization to war strength in 1941 the IGCs had the following strengths:

1st IGC	5,491
2nd & 5th IGCs, each	4,125
3rd & 9th IGCs, each	3,315
4th & 8th IGCs, each	2,417
6th IGC	1,622
7th IGC	3,972
10th IGC	3,833

The organization tables in effect for the independent garrison infantry battalions on their mobilization in July 1941 provided for the following strength:

	Fld Off	Co Off	WO	MSG	SGT	PVT	Horses
Line Branch	2	22	8	0	70	672	-
Medical	0	2	0	2	0	12	-
Technical	0	0	0	2	0	0	-
Veterinary	0	0	0	1	0	0	-
Intendance	0	1	0	4	0	0	-
Total	2	25	8	9	70	684	40

Although the official organization never changed during the war, in practice the organization of the independent garrison infantry battalions varied with time and place. That of the 16th IGIB was probably typical: in 1940 it consisted of a HQ, three rifle companies, a machine gun platoon and an infantry gun platoon. Each of the rifle companies was made up of a HQ, 3 rifle platoons and a machine gun platoon. By August 1941 the Battalion had been reorganized slightly to give a strength of 31 officers, 80 WOs and NCOs, and 791 other ranks (more than a hundred men above authorized strength), divided into a HQ (106 men), four rifle companies (each 162–166 men), a battalion gun platoon (42 men) and a mountain gun platoon (42 men), plus 68 men on detached duty with the IGU HQ. In late 1941 the infantry gun and mountain gun platoons were apparently dropped from the organization. Similarly, the 8th IGIB had 810 men in 4 companies (each of 3 rifle and 1 MG platoons). It formed a gun company of two regimental and two battalion guns by drawing personnel from the line companies, but abolished this in November 1942.

The conquest of large areas in the SW Pacific created a need for further garrison units. To fill this requirement IGHQ ordered the transfer in January–March 1942 of the 10th IGU headquarters from Manchuria and activation of the 11th IGU (Philippines), 12th IGU (Malaya), and 13th and 14th IGUs (Java), followed in September by the 15th and 16th IGUs (Sumatra).

IGU	Ind Garr Inf Bns	Strength
10	31, 32, 33, 34 & 35	3,833
11	36, 37, 38, 39	2,971
12	43, 44, 45, 46, 47	3,708
13	48, 49, 50, 51	2,971
14	52, 53, 54 & 55	2,971
15	56, 57, 58 & 59	2,971
16	60 & 61	1,497

In addition to the units shown the two IGUs in Java were reinforced by attached independent mountain artillery battalions, the 51st in the 13th IGU and the 52nd in the 14th IGU.

Further, the 40th–42nd Independent Garrison Infantry Battalions were raised as separate formations, each with 737 men. The 40th and 41st Battalions (each with 549 men) were created out of the 5th IMR in Borneo and the 42nd IGIB activated for the 15th Army.

An independent garrison infantry battalion for these south Pacific IGUs consisted of an HQ, a signal platoon, four rifle companies, a regimental gun platoon, a battalion gun platoon, and an AT platoon. The rifle companies each had 144 men and consisted of two rifle platoons (each three 9-man rifle squads and a 9-man grenadier squad) and a machine gun platoon (two 10-man squads). Heavy weapons for the company consisted of six light MGs, two medium MGs, and eight grenade dischargers. The supporting gun platoons each had two weapons, 75mm Type 41 mountain guns, 70mm Type 92 howitzers or 37mm Type 94 AT guns. The battalion signal platoon was provided with four Mk 5 radios, four Mk 6 radios, eight telephones and 28 rolls of wire. The battalion was largely immobile, and the HQ included all the transport assets, ten Type 39 carts.

In late 1942 the 17th IGU was formed to garrison Manila and in April 1943 the 18th IGU was activated to guard Singapore.

IGU	Ind Garr Inf Bns	Strength
17	62, 63 & 64	3,041
18	65 & 66	n/a

A little later, in August 1943, a pair of other garrison formations were formed outside of the normal numerical sequence. The 1st Southwest Garrison Unit (78 officers and 1,888 enlisted in two infantry battalions, an artillery battalion and a signal company) and the 2nd Southwest Garrison Unit (36 officers and 898 enlisted in two rifle companies and a field artillery battery) were formed at Port Blair and Car Nicobar, respectively. The personnel for these units were drawn from the elements of the 2nd Guards Division already present in the Andamans and Nicobars.

One named garrison unit was also formed by the Kwantung Army, the Aershan Garrison Unit. This consisted of the 90th Infantry Regiment, a garrison artillery unit, a garrison engineer company, a garrison signal company, and a garrison motor transport company. The infantry regiment was the "A3" type as used elsewhere only by the 23rd Division, but with only two AT companies instead of the usual three. The garrison artillery unit was a small regiment, with two battalions with a total of five batteries.

The Allied drive through the central Pacific led to the creation of similar units for that theater. The 1st and 2nd South Seas Garrison Units were dispatched in April 1943 for the Gilbert Islands and Marcus, respectively. Each of these had a strength of 601 men organized into two regular rifle companies (each nine light MGs and nine grenade dischargers), a machine gun company (12 MGs), and a two-platoon anti-tank company (four 37mm AT guns), supported by 1 car and 10 trucks. The 1st SSGU was lost at sea to US submarine attacks and a second batch of units was formed and dispatched in June 1943: the 3rd South Seas Garrison Unit to Wake Island and the 4th South Seas Garrison Unit, originally planned for the Gilberts, diverted to Bougainville. The 3rd SSGU had 989 men in four rifle companies, an anti-tank unit, an artillery battery and a light tank company;

while the 4th SSGU was a regimental-size unit of 128 officers and 2,780 enlisted in three infantry battalions (each four rifle companies, a machine gun company, a 70mm platoon and a regimental gun platoon), three artillery batteries and a signal company. The rifle companies of the 4th SSGU had a unique organization, three rifle platoons each of 30 men (after deducting those detached for duties elsewhere in the battalion) built around three 9-man squads, two of riflemen (each squad with a light MG) and one of grenadiers (with four grenade dischargers)

The American reconquest of the Aleutians brought similar pressures to bear to the north and on 8 August 1943 IGHQ ordered the formation of the 1st Kuriles Garrison Unit for the northern Kuriles, followed a month later by the 2nd and 3rd Kuriles Garrison Units for Matsuwa and Uruppu/Etoforu Islands, respectively. The 1st Kuriles Garrison was the largest such unit fielded, consisting of three infantry "units" (each of three infantry battalions), nine field and mountain artillery batteries, and two heavy artillery batteries. The 2nd Kuriles Garrison, on the other hand, was one of the smallest, consisting of a single infantry battalion and a 15cm gun battery. The 3rd Kuriles Garrison was made up of two infantry battalions, a field artillery battery and two 105mm gun batteries. These infantry battalions consisted of four rifle companies (each 3 rifle platoons and an MG platoon) and an infantry gun company (2 each of regimental guns, battalion guns and 37mm AT guns).

Late 1943, however, represented the high-water mark of the Army's dedicated garrison units. The Kwantung Army's IGUs were largely drawn off by the constant need to reinforce other theaters. The 3rd and 4th IGUs were the first to go, being converted to the 1st Amphibious Brigade and the 2nd South Seas Detachment, respectively, in October 1943. In early 1944 the 2nd IGU HQ was used to form the 8th Expeditionary Unit HQ, the 6th IGU was reorganized into the 10th Expeditionary Unit, and the 8th IGU into the 9th Expeditionary Unit. Shortly thereafter, the 5th IGU was converted to the 11th Expeditionary Unit, the 7th IGU to the 14th Border Garrison Unit and the 9th IGU into the 108th Division. The Aershan Garrison Unit was used to form the nucleus for the 107th Division. This left only the 1st IGU, which was eventually redesignated the 101st IGU.

As the Allies advanced into the newly-expanded Japanese Empire the garrison units in occupied territories were faced with threats considerably more substantial than the internal security duties originally envisioned. Artillery battalions and engineer units were added and the units redesignated as independent mixed brigades. Thus, in September 1943 the two garrison units in Sumatra were reorganized into the 25th and 26th IMBs. These were followed shortly by the conversion of the 13th and 14th IGUs on Java to the 27th and 28th IMBs. On 16 November 1943 the three garrison units in the Philippines, 10th, 11th and 17th, were reorganized and redesignated the 30th, 31st and 32nd IMBs, respectively. The 12th and 18th IGUs in Malaya were used to form the 94th Division in 1944.

Of the South Seas Garrison Units, the 1st was lost at sea, the 2nd became the 1st Battalion/12th IMR and the 3rd became the 1st Battalion/13th IMR in the 31st Army reorganization of May 1944. The 4th South Seas Garrison Unit was diverted to Bougainville and remained there, depleted by combat losses, through to the end of the war.[56]

The Kuriles Garrison Units were also reorganized, the 1st Kuriles Garrison into the 91st Division on 12 April 1944, and the 2nd and 3rd Kuriles Garrisons into the 43rd IMR

56 The General Staff's *Unit Organization Tables* also lists a 5th and 6th South Seas Garrison Units among the elements no longer in existence at the end of the war, along with the 1st, 2nd and 3rd SSGUs. No further information is provided on these elements, however.

a few days later.[57] Thus, by mid-1944 only the 1st Independent Garrison Unit and the 4th South Seas Garrison Unit remained of the Army's garrison forces.

The need for garrison units for internal security, however, remained. Pressed on all sides, the Army could not afford to release the top-quality personnel that characterized the first nine Kwantung Army IGUs, nor the mediocre personnel used to form the second-generation garrison units. Instead, units had to be extemporized.

On 5 August 1944 IGHQ ordered the activation of the 101st, 102nd, 103rd and 104th Garrison Units to serve under the Kwantung Army. The hard core of these units, such as they were, was to be provided by seven battalion-size garrison units drawn from existing line of communications sector units. These were to be supplemented by nine "special garrison battalions" made up of elderly reservists and the medically infirm. These were of two types: the (A) battalion with five subordinate companies, and the 550-man (B) type with three.

101st GU:	LoC Garr Units 62, 69, 79
	Special Garr Bns 604 (B), 652 (A), 653 (A)
102nd GU:	LoC Garr Units 74, 77
	Special Garr Bns 601 (B), 654 (A)
103rd GU:	LoC Garr Unit 60
	Special Garr Bns 605 (B), 655 (A)
104th GU:	LoC Garr Unit 46
	Special Garr Bn 602 (B)

The final group of security units were raised by the China theater. The 1st through 7th Independent Guard Units were ordered activated on 1 February 1945 and raised on 6 March 1945, and the 9th through 14th Independent Guard Units ordered and apparently raised on 4 April.[58]

1st IGdU: Ind Guard Inf Bns 1, 2, 3, 4, 5, 6
2nd IGdU: Ind Guard Inf Bns 7, 8, 9, 10, 11, 12
3rd IGdU: Ind Guard Inf Bns 13, 14, 15, 16, 17, 18
4th IGdU: Ind Guard Inf Bns 19, 20, 21, 22, 23, 24
5th IGdU: Ind Guard Inf Bns 25, 26, 27, 28, 29, 30
6th IGdU: Ind Guard Inf Bns 31, 32, 33, 34, 35, 36
7th IGdU: Ind Guard Inf Bns 37, 38, 39, 40, 41, 42
9th IGdU: Ind Guard Inf Bns 43, 44, 45, 46, 47, 48
10th IGdU: Ind Guard Inf Bns 49, 50, 51, 52, 53, 54
11th IGdU: Ind Guard Inf Bns 55, 56, 57, 58, 59, 60
12th IGdU: Ind Guard Inf Bns 61, 62, 63, 64, 65, 66
13th IGdU: Ind Guard Inf Bns 67, 68, 69, 70, 71, 72

57 Some portions of the 1st Kuriles Garrison were not incorporated into the 91st Division, for the General Staff's *Unit Organization Tables* show a much reduced 1st Kuriles Garrison consisting of the 2nd, 4th and 29th Independent Garrison Infantry Battalions existing from 8 April 1944 to 16 July 1945. The *Unit Historical Tables* credit it with a strength of 126 officers and 4,128 enlisted starting 8 April 1944.

58 Some wartime intelligence and immediate post-war demobilization documents refer to these elements as garrison units, but their actual title was guard (*keibi*) rather than garrison (*shubitai*). The distinction between the guard and garrison units is important, as considerable confusion between units can arise.

14th IGdU: Ind Guard Inf Bns 73, 74, 75, 76, 77, 78

Each of these independent guard units consisted of the following elements:

	Off	WO	MSG	SGT	PVT	Total	Horses
Guard Unit Headquarters	27	3	9	26	195	260	22
Six Guard Infantry Battalions, each	35	7	9	130	1,204	1,385	21
Labor Unit	3	1	3	19	205	231	12
Total	240	46	66	825	7,624	8,801	160

A 1,385-man guard infantry battalion had five rifle companies (each six light MGs and six grenade dischargers) and a heavy weapons company (two medium MGs and two 70mm howitzers).

These independent guard units, almost completely immobile, served on guard duties in China through 1945 and were still in service at the surrender. The 8th Independent Guard Unit, although ordered activated on 1 February 1945 with the rest of the first group, was an anomalous formation. It consisted of the 33rd Guard Battalion and the 5th Battery of the Tsugaru Fortress Heavy Artillery Regiment for a total strength of 1,644 men for duty in Sapporo.

On 6 February 1945 lighter security units were formed for the homeland, the 1st, 2nd and 3rd Guard Brigades, and in early June they were assigned to the Tokyo Defense HQ. Each of these brigades had a strength of 4,078 men and consisted of an 82-man headquarters, six 639-man guard infantry battalions and a 162-man signal unit.

1st Guard Brigade (Guard Infantry Bns 1, 2, 3, 4, 5 & 6)
2nd Guard Brigade (Guard Infantry Bns 7, 8, 9, 10, 11 & 12)
3rd Guard Brigade (Guard Infantry Bns 13, 14, 15, 16, 17 & 18)

The guard infantry battalions were each made up of a small 30-man headquarters and three 203-man guard companies. The guard company, which consisted of three platoons, had no heavy weapons and was armed solely with 195 rifles. The brigade signal company consisted of two wire platoons and a radio platoon. In addition, Korea raised twenty guard battalions, the 141st through 160th.

Also for the homeland (broadly defined, to include Korea) were special guard battalions and companies. These were home guard units of civilians who could be called up in an emergency, together with a small Army cadre. In the home islands special guard battalions 51–69, 151–164, 202–25, and 251–257 were raised, along with special guard companies 13, 14, 21–34, 101–107, 110, 111, 201–208, 213–222, 225, 229, 326, 332 and 333. Each special guard battalion had a cadre of between 5 and 14 soldiers who supervised about 500 civilians. The exception was on the island of Kyushu, where invasion was regarded as imminent, where army strength increased to 218. The special guard companies had a cadre of about 4 soldiers (usually including one junior officer) to supervise 122 civilians. The civilians were armed for the most part with old shotguns, spears and swords and were issued armbands but not uniforms.

In addition, depot divisions raised their own special guard companies. On Kyushu the Fukuoka Regimental District raised 28 companies, Kagoshima 19, Kumamoto 17, Miyazaki 13, Nagasaki 15, Oita 17, and Saga 10. Each of these was made up of 6 soldiers and 288 civilians.

Korea also raised 16 special guard battalions (402, 403, 405, 409, 410, 413, 451–58, 461 and 463), but they differed from the homeland units in being made up of full-time soldiers, although in this case of very low quality, having been inducted in the final mobilization of July 1945 after previously being rejected or deferred.

The LoC Sector Units

The final type of garrison units were the lines of communications sector units. These were established to organize and supervise activities in army and area army rear areas. Under the 1941 mobilization plan an LoC sector unit consisted of a 203-man headquarters, a 1,035-man LoC garrison unit, and a 500-man duty company.

The LoC garrison unit was essentially a semi-mobile infantry battalion, consisting of a 129-man HQ, four 190-man rifle companies, an 84-man machine gun company and a 62-man mortar platoon. The rifle companies were of the older pattern, having three platoons, each of six squads – three with rifles only, two each including a light machine gun, and one with two grenade dischargers. The small machine gun company had two platoons, each with two squads, for a total of four medium MGs. The mortar platoon (nominally an infantry gun platoon) was equipped with two 70mm Type 11 mortars, weapons discarded by the line infantry with the advent of the 70mm howitzer. The LoC garrison unit was provided with only 13 rounds for each grenade discharger, but had the normal 6,480 rounds of medium MG ammunition per gun and 160 rounds per mortar. For transport the mortar platoon was provided with eleven pack horses to carry ammunition (the mortars themselves being manpacked), while the machine gun company had five Type 39 one-horse carts. The garrison unit HQ had seven riding horses and 32 pack horses as well as four cargo trucks for its combat and field trains.

The duty company, which was armed with only 99 Type 38 rifles, was divided into a 31-man HQ and three 160-man platoons. Each of these platoons, in turn, consisted of three 53-man sections, each of four 13-man squads. The duty company was essentially a manual labor unit, with its main functions being the loading and unloading of supplies in the forward area, construction of buildings and roads, and other similar duties as needed. For these roles, however, the duty company had no authorized vehicles or heavy equipment, although large numbers of man-pulled carts were often supplied locally. Unlike the sector HQs and garrison units, which adhered to the standard TO&Es throughout the war, the duty companies varied widely in strength among the various units and through time depending on the availability of manpower and the local requirements.

The first group of LoC sector units to be deployed were the 5th, 50th, 51st, 52nd, and 56th which were raised in July 1940 and sent to China in July to September. A second group was raised in July 1941 for operations to the south and was dispatched to Manchuria for intensive training in August of that year. These were the 41st–43rd, 46th–48th, 57th, 58th, 61st–64th, 72nd–74th, and 77th–80th. A month later the 41st, 42nd and 57th LoC Sector Units were reassigned from the Kwantung Army to the China theater. In November half the remainder were reassigned to the armies that were to drive to the south: the 61st and 63rd to the 14th Army for the Philippines; the 73rd to the 15th Army for Burma; the 43rd and 48th to the 16th Army for the East Indies; and the 47th, 58th and 78th to the

25th Army for Malaya.

A third major wave was raised in 1942/43. The 46th, 65th, 69th, 70th, 76th, 150th and 151st LoC Sector Units were assigned to the Kwantung Army to replace earlier transfers. The 44th, 54th, 55th and 56th LoC Sector Units were assigned to the south Pacific for western New Guinea, Hollandia, NW New Guinea, and Halmahera, respectively. The 53rd was sent to Burma, and the 45th and 59th to the East Indies.

The pressing need for front-line units took its toll on the LoC Sector Units. The 45th and 59th Units were disbanded to form the nucleus of the 26th IMB. The 57th Unit was transferred from the Kwantung Army to the Palau Sector Group in mid-1944 and was almost immediately disbanded to contribute to the 53rd IMB. The 70th and 150th Units were used to form the 2nd Amphibious Brigade. In these and many other cases the LoC HQ and duty companies were left untouched to continue their functions, but the garrison unit was disbanded to form badly needed combat units. Indeed, some of the LoC sector units formed during 1944/45 (excepting those destined for China) did not even include garrison units to start with.

Reinforcements for China included the 66th LoC Sector Unit in 1943, the 81st and 82nd in February 1944, the 86th and 87th in July 1944, and the 83rd and 84th in January 1945. Also in 1944 the 85th and 88th LoC Sector Units (HQ only) were raised for the Philippines; as were the 90th and 91st for Burma; the 49th (HQ only) for Okinawa; and the 60th, 68th and 69th for the Kwantung Army. Most of the LoC garrison units in Manchuria were withdrawn from their parent sector units in 1945 to form the new Independent Garrison Units.

The Fortress Units

A type of unit important in the pre-war force structure, but which actually saw little combat, was the "fortress." For the most part, the fortresses on the main islands of Japan consisted simply of coastal artillery elements, but the more outlying fortresses included infantry units as well, sometimes in substantial numbers as the war went on. The 1941 mobilization plan provided for 26 fortresses as part of the permanent force structure. Tokyo Bay, Shimonoseki, and Maizuru Fortresses were established on Honshu; Hoyo and Nagasaki on Kyushu; Soya, Muroran, Nemuro and Tsugaru on Hokkaido; Yura in the Inland Sea; Takao and Kiirun (Keelung) on Formosa; Rashin (Najin), Eiko Bay (Yongnung), and Reisui in Korea; Tsushima and Iki on islands in Tsushima Strait; Ryojun (Port Arthur) on the Kwantung Peninsula; Nakagusu Bay on Okinawa and Amami Oshima in the Ryukyus; Chichi Jima in the Bonins; Paramushio (northern Kuriles) at the northern end of the Kuriles; Karimata on Miyako Jima and Funauke on Inomoke in the Sakashima Group; Pescadores at Moko in that island group; and Chiukaiwan.

Generally, the fortresses early in the war consisted of a headquarters, a hospital detachment and a fortress heavy artillery regiment. The HQ had a variable strength ranging from about 50 to about 150 men, depending on the size and number of subordinate units. The hospital detachment had a strength of 25 men.

The fortress heavy artillery regiments were organized under one of five different TO&Es. The Tokyo Bay, Chichi Jima, Tsushima, Chiukaiwan, Shimonoseki, Iki, Amami Oshima, and Ryojun Fortress Heavy Artillery Regiments used the largest TO&E: a 50-man regiment HQ and two battalions each of a 25-man HQ and three 230-man batteries. The Yura, Tsugaru, Hoya, Nagasaki and Pescadores Fortress Heavy Artillery Regiments

Table 2.29: Personnel and Assigned Weapons of Fortress Heavy Artillery Regiments, from October 1940

	Personnel	Pistols	Rifles	Medium MGs	75mm Type 38 Gun	90mm Schneider gun	10cm Krupp Gun	10cm Type 38 gun	12cm Type 38 How	12cm Canet gun	12cm Krupp gun	12cm Schneider gun	15cm Armstrong gun	15cm Type 7 gun	15cm Type 45 gun	15cm Type 96 gun	24cm How	28cm How	30cm turret	7cm AA guns
Tokyo Wan	1,418	40	-	-	-	-	-	-	-	-	-	-	-	-	-	-	-	-	2	-
Yura	1,184	36	-	-	-	-	-	-	-	-	-	-	-	-	-	-	-	-	-	-
Shimonoseki	1,418	40	-	-	-	-	-	-	-	-	-	-	-	-	-	-	?	-	-	-
Tsushima	1,418	40	-	-	-	-	-	-	-	-	-	-	-	-	-	-	-	-	-	-
Nagasaki	475	13	116	12	12	-	-	-	-	-	2	-	-	-	-	-	?	-	-	-
Iki	1,418	40	-	-	-	-	-	-	-	-	-	-	-	-	-	-	-	4	-	-
Tsugaru	1,184	36	-	-	-	-	-	-	-	-	-	-	-	-	-	-	-	-	-	-
Rashin	710	14	174	-	-	4	-	4	8	-	-	-	-	-	4	-	-	2	-	-
Chinkai Wan	1,418	40	-	-	-	-	-	-	-	-	-	-	-	-	-	-	-	4	-	-
Ryojun	1,413	6	-	-	-	2	4	-	-	-	-	-	-	-	-	-	-	-	-	-
Kiirun	934	21	-	-	-	-	-	-	-	-	-	-	-	-	-	-	-	-	-	-
Pescadores	1,184	36	-	-	-	-	-	-	-	-	-	-	3	-	-	-	-	-	-	-
Nemuro	475	13	116	8	12	-	-	2	-	-	-	-	-	-	-	-	-	-	-	-
Soya	704	17	-	-	12	-	-	4	8	-	-	-	-	4	2	4	-	-	-	2
Muroran	475	13	-	-	-	-	-	-	-	-	-	-	-	-	-	-	-	-	-	8
Paramushio	704	17	172	16	16	-	-	8	-	-	-	-	2	-	2	-	-	-	-	-
Nakagusuku Wan	475	13	116	12	12	-	-	-	-	2	-	-	-	-	-	-	-	-	-	-
Karimata	475	13	116	8	8	-	-	-	-	-	2	-	-	-	-	-	-	-	-	-

Table 2.29: Personnel and Assigned Weapons of Fortress Heavy Artillery Regiments, from October 1940

	Personnel	Pistols	Rifles	Medium MGs	75mm Type 38 Gun	90mm Schneider gun	10cm Krupp Gun	10cm Type 38 gun	12cm Type 38 How	12cm Canet gun	12cm Krupp gun	12cm Schneider gun	15cm Armstrong gun	15cm Type 7 gun	15cm Type 45 gun	15cm Type 96 gun	24cm How	28cm How	30cm turret	7cm AA guns
Funauka	?	?	?	?	8	-	-	-	-	-	-	2	-	-	-	-	-	-	-	-
Reisui	475	13	16	-	8	-	-	-	-	-	-	-	-	-	-	-	-	-	-	2
Amami Oshima	1,418	40	-	-	-	?	?	?	?	?	?	?	?	?	?	?	?	?	?	-
Chichi Jima	1,418	40	-	-	-	?	?	?	?	?	?	?	?	?	?	?	?	?	?	-
Hoyo	1,184	36	-	-	-	?	?	?	?	?	?	?	?	?	?	?	?	?	?	-
Takao	934	21	-	-	-	?	?	?	?	?	?	?	?	?	?	?	?	?	?	-
Eiko Wan	475	13	-	-	-	?	?	?	?	?	?	?	?	?	?	?	?	?	?	-

were similar but lacked one battery. The Maizuru, Kiirun and Takao Fortress Heavy Artillery Regiments were each made up of a 50-man HQ and four 230-man batteries. The Soya, Paramushio, and Rashin Fortress Heavy Artillery Regiments were similar, but had only three batteries; while the Eiko Bay, Muroran, Nakagusuku Bay, Karimata, Funauke, Nemuro, and Reisui Regiments had only two batteries each.

The fortress heavy artillery regiments did not have any fixed complement of weapons, instead manning a combination of assigned weapons and whatever coastal defense artillery had been emplaced in the area. Some weapons were assigned by the mobilization plan, while others were not. The basis for the distinction is not clear, as both categories included both fixed and mobile artillery. Some of the weapons were clearly fit for little more than museum display, such as the old Schneider and Canet guns, and the 24cm howitzers. Others were more modern, including the 30cm turret guns in Tokyo Bay and the 15cm Type 96 guns.

As examples, the Soya Fortress Heavy Artillery Regiment actually manned eight 150mm guns, six 120mm howitzers, four 105mm guns, and four 75mm guns; while the Tsugaru Fortress Heavy Artillery Regiment manned two 305mm guns in a turret, eight 280mm howitzers, one 240mm howitzer, twelve 150mm guns, two 105mm guns, four 75mm guns, and four searchlights.[59] The Chichi Jima FHAR had four 24cm Type (Meiji) 45 howitzers, four 15cm Type (Meiji) 45 guns, four 12cm Type (Meiji) 38 howitzers, nineteen 75mm Type (Meiji) 38 field guns, and four 75mm Type (Meiji) 11 dual-purpose guns, spread out over three islands in the Bonins.

The assigned guns had both anti-ship and beach defense roles, as indicated by their assigned ammunition load. For the 12cm Type (Meiji) 38 howitzers the Rashin FHAR was allocated 3,200 rounds of AP and an equal quantity of shrapnel; while the Sōya FHAR also had 3,200 rounds of AP but only 1,280 rounds of shrapnel. The 15cm guns were provided solely with HE rounds, no APHE projectiles being held except in the case of the Rashin FHAR. The 28cm howitzers were provided with a mix of HE and AP ammunition.

Most of the artillery regiments were provided with a depression-type rangefinder and associated plotting gear. Being static organizations they were well provided with telephones but only rarely had radios. A few had 75mm AA guns and more had Type (Taisho) 3 medium machine guns on high-angle mounts but most had either three or six Type 93 150cm searchlights paired with Type 90 small sound locators, presumably for early warning purposes.

In addition to these basic elements, a fortress could also include a fortress infantry unit (type A or B), a fortress field artillery unit, a fortress mountain artillery unit, a fortress anti-aircraft unit, a fortress engineer unit, and a fortress duty company. The 340-man fortress duty company was a manual labor unit armed solely with 24 carbines. The 264-man engineer unit provided skilled labor, but little in the way of specialized equipment other than what was found within the fortress itself. Unlike the duty company, the engineer unit was provided with 251 rifles and could be used for security duties as well.

A fortress mountain artillery unit was a battery-sized element with 177 men and provided with four 75mm Type 41 (later Type 94) mountain guns. The unit was provided with 28 pack horses to carry the guns, twelve for general supplies and 41 to carry a total of 480 rounds of HE ammunition. Similarly, a field artillery unit with a strength of 120 men was a battery equivalent, equipped with four 75mm Type 38 field guns. This unit was

59 The Fortress Heavy Artillery Regiments are also discussed in the Artillery section.

equipped with six caissons (carrying 744 rounds of HE ammunition), a stores wagon, and 35 draft horses. Unlike the heavy artillery regiments the mountain and field artillery units, although smaller, were mobile units.

Such fortress AA units as were mobilized had varying organizations. The Chichi Jima and Keelung Fortress AA units each consisted of a HQ, an anti-aircraft battery and a searchlight battery. The latter operated six 150cm Type 93 searchlights, each associated with a Type 90 small sound detector. The 110-man AA battery was provided with four 75mm Type 88 guns, as well as two Type 90 light height finders, a Type 94 height finder, a Type 96 night height finder, and a Type 90 calculator. The AA unit did not have any early-warning equipment itself, but the fortress HQ on Chichi Jima did have a single Type B ultra-short wave radar mounted on four trucks.

The fortress infantry units were of two varieties: the battalion-size Type A, and the company-size Type B. The Type B infantry unit had a strength of 204 men and was armed with 195 Type 38 rifles. The Type A infantry unit consisted of a 34-man HQ and four Type B companies. The fortress infantry units, like the fortress heavy artillery regiments, primarily manned whatever heavy weapons were assigned to the fortress itself.

In September 1942 new tables for the fortress infantry units were issued. The Type B now had a strength of about 190 with 149 Type 38 rifles, 9 Type 96 light machine guns and 12 Type 89 grenade dischargers. The Type A infantry unit consisted of three rifle companies (similar to the Type B units) and a 112-man heavy weapons unit with eight Type 92 machine guns and two battalion guns. In addition, the 43rd Fortress Infantry Unit (B) and the B-Type units in the northern Kuriles received two 37mm anti-tank guns.

Fortress infantry units also manned whatever supplemental weapons were available at their location. By 1944 the norm called for each company to consist of three platoons (each with four light MG squads and a four-weapon grenadier squad) and a machine gun platoon of three squads. An examination of the Chichi Jima Fortress's infantry units in 1944 shows that all four of its Type B infantry units conformed to this organization, as did the companies of the 41st, 44th and 45th Type A Fortress Infantry Units. Notable here was that the medium machine guns had been decentralized to the infantry companies and the battalions (Type A units) homogenized as a variable number of identical companies.

Not all fortress infantry organization was so straightforward, however, and an analysis of one company's organization in the 61st Fortress Infantry Unit, also subordinate to the Chichi Jima Fortress, shows a very different picture. Like the other units, it consisted of a HQ, three rifle platoons and a machine gun platoon. Each rifle platoon, however, was made up of a platoon leader and six squads. Three of the squads in each platoon were made up of a corporal, seven riflemen and a grenadier. One squad in each of the 1st and 2nd platoons was made up of a sergeant and ten privates with two light MGs. The remaining two squads (three in the 3rd platoon) each had a corporal and eight privates, armed only with rifles. Thus, in these units the grenade discharger became the squad weapon, while the light machine guns became the platoon support weapons, a reversal of normal policy. The company's fourth (machine gun) platoon was an extremely small unit, consisting solely of the platoon leader and three 3-man medium MG squads. The company HQ was made up of the company commander, four NCOs and twelve privates (including two medics).

Signal assets also varied considerably. The "normal" 41st, 42nd, 44th and 45th FIUs (as well as the B-Type units) had no communications equipment whatsoever. The 54th, 56th, 57th, 61st and 66th FIUs, however, each had a Type 94 Mk 3A radio, two Mk 5 radios and two Mk 6 radios, along with three telephones.

The Paramushio Fortress received a Type A infantry unit bearing the fortress name on activation in August 1941. The activation of further fortresses in September–November 1941, however, led to the creation of a numbered series of fortress infantry units (FIU):

42nd FIU to Chichi Jima Fortress
43rd FIU to Chichi Jima Fortress
60th FIU to Amami Oshima Fortress
61st FIU to Nakagusuku Bay Fortress
65th FIU to Funauke Fortress

In addition, the Paramushio Fortress was assigned the 5th and 6th Fortress Mountain Artillery Units, while the Tsugaru Fortress received the 19th independent AA Battery, and the Takao Fortress the 25th independent AA Battery. Further, Paramushio and Chichi Jima Fortresses, at least, received fortress engineer units.

The Paramushio, or northern Kuriles, Fortress saw limited front-line service when the 2nd Company of the 1st Paramushio Fortress Infantry Unit was directed to proceed to and occupy Attu Island in October 1942, carrying "as many anti-aircraft weapons as possible." The survivors returned when the Army withdrew from the Aleutians. Shortly after that the Paramushio Fortress was disbanded to provide a nucleus for the North Kuriles Garrison force.

Few changes were made to fortress organization until the approach of Allied forces in early 1944. In March 1944 the Chichi Jima Fortress was reinforced by the 41st, 44th, and 45th FIUs (Type A), the 64th, 65th, 67th and 68th FIUs (Type B), the 7th Fortress Mountain Artillery Unit, and the 22nd Fortress Engineer Unit. A week later these were joined by the 54th, 56th, 57th, and 66th FIUs (Type A).

When thus reinforced the Chichi Jima Fortress became the largest fortress organization and was actually responsible for the defense of the entire Bonins chain. As can be seen from Table 2.30, however, the fortress infantry component, although numerically strong, was quite weak in supporting weapons, lacking mortars and anti-tank guns completely. To remedy this shortcoming, as well as smooth out the chain of command, the Chichi Jima Fortress was dissolved in May 1944 and its personnel and equipment reorganized into the unique 109th Division.

The shortcomings of the fortress organization caused the dissolution of other fortress units as well in 1944, including Amami Oshima, Funauke, Nakagusuku Bay, Nemura and Muroran.

In contrast, the two fortresses guarding Tsushima Strait were strengthened in March 1945, when the 1st to 6th Iki Fortress Infantry Battalions and the 1st to 3rd Tsushima Fortress Infantry Battalions were activated. A few months later the fortresses were further strengthened, and by the end of the war the Iki Fortress included a 369-man HQ, nine 826-man infantry battalions, a 126-man special garrison company and the 1,426-man fortress heavy artillery regiment. The Tsushima Fortress controlled a 347-man HQ, six 826-man infantry battalions, three 126-man special garrison companies and the 1,766-man fortress heavy artillery regiment. Each of the Iki and Tsushima fortress infantry battalions was composed of a headquarters, four rifle companies (each nine light MGs and nine grenade dischargers) and a small machine gun company (two platoons each of two medium MGs).

Table 2.30: Chichi Jima Fortress March/April 1944

	TO&E			Actual																					
	Officers	Enlisted	Other*	Officers	Enlisted	Other	Pistols	Rifles	Light MGs	Medium MGs	Heavy Gren Lnchrs	20mm AT Rifles	37mm Tank Guns	70mm Inf Howitzers	75mm Mountain Guns	75mm Field Guns	105mm Field Guns	120mm Howitzers	150mm Guns	240mm Howitzers	75mm AA Guns	Carts	Wagons	Cars	Trucks
Chichi Jima																									
Chichi Jima Fortress Headquarters	12	94	93	11	106	103	0	371	0	2	0	0	0	0	0	0	0	0	0	0	0	12	10	2	4
Chichi Jima Fort. Heavy Art. Regiment (-)	41	857	43	50	1003	46	0	240	0	8	0	0	0	0	0	6	4	2	2	4	0	0	0	1	0
Chichi Jima Fort. AA Defense Unit	11	309	13	11	364	13	0	100	0	4	0	0	0	0	0	0	0	0	0	0	4	0	0	0	2
42nd Fortress Infantry Unit (A)(-)	24	666	21	29	687	21	4	147	9	1	12	0	0	0	0	0	0	0	0	0	0	0	0	0	0
44th Fortress Infantry Unit (A)	24	807	19	26	917	19	24	927	46	12	46	0	0	0	0	0	0	0	0	0	0	0	0	0	0
54th Fortress Infantry Unit (A)(-)	24	807	19	24	779	17	0	167	3	3	10	0	0	0	0	0	0	0	0	0	0	0	0	0	0

Table 2.30: Chichi Jima Fortress March/April 1944

	TO&E			Actual																						
	Officers	Enlisted	Other*	Officers	Enlisted	Other	Pistols	Rifles	Light MGs	Medium MGs	Heavy Gren Lnchrs	20mm AT Rifles	37mm Tank Guns	70mm Inf Howitzers	75mm Mountain Guns	75mm Field Guns	105mm Field Guns	120mm Howitzers	150mm Guns	240mm Howitzers	75mm AA Guns	Carts	Wagons	Cars	Trucks
56th Fortress Infantry Unit (A)	24	807	19	23	775	18	0	734	10	14	41	0	0	0	0	0	0	0	0	0	0	0	0	0	0
57th Fortress Infantry Unit (A)(-)	24	807	19	23	776	18	0	159	2	3	9	0	0	0	0	0	0	0	0	0	0	0	0	0	0
22nd Fortress Engineer Unit (-)	6	252	6	6	252	6	0	222	0	0	0	0	0	0	0	0	0	0	0	0	0	0	0	0	0
17th Shipping Engineer Regiment** (-)	?	?	?	?	?	?	38	717	0	8	0	4	4	0	0	0	0	0	0	0	0	0	0	0	1
Chichi Jima Army Hospital	0	0	33	0	0	36	0	9	0	0	0	0	0	0	0	0	0	0	0	0	0	0	0	0	0
Haha Jima																									
Det, 42nd Fortress Infantry Unit (A)	***	***	***	***	***	***	16	359	18	15	24	0	0	2	0	0	0	0	0	0	0	0	0	0	3
45th Fortress Infantry Unit (A)	24	807	19	24	919	17	0	927	46	15	46	0	0	0	0	0	0	0	0	0	0	0	0	0	0

Table 2.30: Chichi Jima Fortress March/April 1944

	TO&E			Actual																					
	Officers	Enlisted	Other*	Officers	Enlisted	Other	Pistols	Rifles	Light MGs	Medium MGs	Heavy Gren Lnchrs	20mm AT Rifles	37mm Tank Guns	70mm Inf Howitzers	75mm Mountain Guns	75mm Field Guns	105mm Field Guns	120mm Howitzers	150mm Guns	240mm Howitzers	75mm AA Guns	Carts	Wagons	Cars	Trucks
Det., Chichi Jima Fort. Heavy Artillery Regiment	***	***	***	***	***	***	0	40	0	2	0	0	0	0	0	5	0	2	0	0	0	0	0	0	0
Det., 17th Shipping Engineer Regiment**	?	?	?	?	?	?	8	277	0	4	0	2	2	0	0	0	0	0	0	0	0	0	0	0	0
15th Fortress Engineer Unit (-)	5	249	6	6	236	6	0	53	0	0	0	0	0	0	0	0	0	0	0	0	0	0	0	0	0
Iwo Jima																									
41st Fortress Infantry Unit (A)	24	807	19	29	912	19	24	927	46	12	46	0	0	0	0	0	0	0	0	0	0	0	0	0	0
Det, 57th Fortress Infantry Unit (A)	***	***	***	***	***	***	0	575	14	9	32	0	0	0	0	0	0	0	0	0	0	0	0	0	0
61st Fortress Infantry Unit (A)	24	807	19	26	773	19	0	735	16	12	41	0	0	0	0	0	0	0	0	0	0	0	0	0	0

Table 2.30: Chichi Jima Fortress March/April 1944

	TO&E			Actual																						
	Officers	Enlisted	Other*	Officers	Enlisted	Other	Pistols	Rifles	Light MGs	Medium MGs	Heavy Gren Lnchrs	20mm AT Rifles	37mm Tank Guns	70mm Inf Howitzers	75mm Mountain Guns	75mm Field Guns	105mm Field Guns	120mm Howitzers	150mm Guns	240mm Howitzers	75mm AA Guns	Carts	Wagons	Cars	Trucks
64th Fortress Infantry Unit (B)	5	196	3	4	214	3	0	255	12	3	12	0	0	0	0	0	0	0	0	0	0	0	0	0	0
65th Fortress Infantry Unit (B)	5	196	3	4	215	3	0	255	12	3	12	0	0	0	0	0	0	0	0	0	0	0	0	0	0
66th Fortress Infantry Unit (A)	24	807	19	28	772	19	28	736	16	12	41	0	0	0	0	0	0	0	0	0	0	0	0	0	0
67th Fortress Infantry Unit (B)	5	196	3	4	214	3	0	255	12	3	12	0	0	0	0	0	0	0	0	0	0	0	0	0	0
68th Fortress Infantry Unit (B)	5	196	3	4	210	3	4	255	12	3	12	0	0	0	0	0	0	0	0	0	0	0	0	0	0
7th Fortress Mountain Artillery Unit	4	165	8	4	165	8	13	94	0	0	0	0	0	0	5	0	0	0	0	0	0	0	0	0	0
Det., Chichi Jima Fort. Heavy Artillery Regiment	***	***	***	***	***	***	0	80	0	4	0	0	0	0	0	8	0	0	0	0	0	0	0	0	0

Table 2.30: Chichi Jima Fortress March/April 1944

	TO&E			Actual																					
	Officers	Enlisted	Other*	Officers	Enlisted	Other	Pistols	Rifles	Light MGs	Medium MGs	Heavy Gren Lnchrs	20mm AT Rifles	37mm Tank Guns	70mm Inf Howitzers	75mm Mountain Guns	75mm Field Guns	105mm Field Guns	120mm Howitzers	150mm Guns	240mm Howitzers	75mm AA Guns	Carts	Wagons	Cars	Trucks
Det., 15th Fortress Engineer Unit	***	***	***	***	***	***	0	248	0	0	0	0	0	0	0	0	0	0	0	0	0	0	0	0	0
9th Ind Engineer Regiment (-) **	?	?	?	?	?	?	0	206	1	0	0	0	0	0	0	0	0	0	0	0	0	0	0	0	1
Muko Jima																									
Det., 54th Fortress Infantry Unit (A)	***	***	***	***	***	***	0	570	13	11	31	0	0	0	0	0	0	0	0	0	0	0	0	0	0
Det., 22nd Fortress Engineer Unit	***	***	***	***	***	***	0	29	0	0	0	0	0	0	0	0	0	0	0	0	0	0	0	0	0

* *includes non-combatant personnel and a small number of civilian officials*
** *attached unit*
*** *personnel figures included in parent unit shown above*

The Shimonoseki Fortress was also strengthened, but with lower quality troops. By the end of the war the fortress had not only its 108-man HQ and 1,025-man fortress heavy artillery regiment, but also an 853-man garrison battalion, four 550-man special garrison battalions, four 930-man special engineer garrison units, and a 145-man tractor company.

Another unit to be strengthened was the critical Tokyo Bay Fortress. By April 1945 it had expanded to fill corps functions. Organic fortress troops included the Tokyo Bay Fortress Heavy Artillery Regiment, 1st and 2nd T.B. Fortress Artillery Units (= battalions), 1st and 2nd T.B. Fortress Engineer Units (=companies), and the T.B. Fortress Signal Unit. In addition to these organic elements the Tokyo Bay Fortress HQ was also made the superior command for the 65th and 96th IMBs and the 6th Medium Artillery Battalion (A) in its role as command HQ for the Tokyo Bay Defense Group.

While some fortresses were thus being strengthened in 1945 the process of disbanding others did not stop. The Nagasaki Fortress Heavy Artillery Regiment (which essentially comprised all the troops in the fortress) was redesignated the 17th Heavy Artillery Regiment and provided the artillery component for the 122nd IMB, and the same process was applied to the Hoyo Fortress Heavy Artillery Regiment in favor of the 118th IMB.

The deactivation of fortresses in 1945 was not restricted to Japan proper. In the first three months of that year the Pescadores Fortress was converted into the 75th IMB, Keelung Fortress into the 76th IMB, and Takao Fortress into the 100th IMB, all in the Formosa area.

The Border Garrisons in Manchuria

Closely related conceptually to the fortress units were the border garrison units raised by the Kwantung Army to man fixed positions along the Soviet border in Manchuria. The 1st through 8th Border Garrisons (*Kokkyō shubitai*) were formed in March 1938, followed by the 9th through 13th Border Garrisons in April 1940, although they were not mobilized to war strength until 16 July 1941.[60] The 14th Border Garrison was activated in June 1944 from a nucleus of two battalions of the former 7th IGU; while the 15th, and final, Border Garrison was activated on 26 July 1945 from an infantry battalion and two artillery batteries of the 4th BG. In addition, the Arshaan Garrison Unit, located on the Mongolian border, served the functions of a border garrison unit as well as a regular garrison unit.

There were large variations in strength among the border garrison units. There were two basic types: the small BGU generally consisting of an infantry battalion, an artillery battalion of two batteries, and an engineer company for a total of 1,000 to 3,000 men; and a large BGU consisting of several "sector commands" each similar to the smaller BGUs. The larger units, the 1st (4 sectors) and later the 4th (3 sectors) and 8th (5 sectors) Border Garrisons, were also known as border garrison commands (*Kokkyō shubitai shireibu*) as a result.

Strengths for the border garrisons at mobilization were:

	Off	WO	MSG	SGT	PVT	Total	Horses
1st Border Garrison							
Headquarters	16	0	7	10	8	41	9
1st Sector	44	11	19	82	858	1,014	46

60 The 1st–4th and 9th–12th were on the eastern border, facing the Soviet Maritime Provinces, the 5th–7th, 13th and 14th were on the northern border, and the 8th on the western border.

2nd Sector	34	9	15	68	672	798	0
3rd Sector	44	11	19	82	858	1,014	46
4th Sector	59	17	21	119	1,338	1,554	61
2nd–4th Border Garrisons, each	74	22	24	167	1,821	2,108	47
5th Border Garrison	48	13	19	94	1,020	1,194	46
6th–10th Border Garrisons, each	114	30	30	241	2,480	2,895	94
11th–13th Border Garrisons, each	44	11	19	82	864	1,020	46
14th Border Garrison	49	14	11	112	1,125	1,311	84

An example of a large BG is the 1st Border Garrison Command, which consisted of four sector commands. The 1st and 3rd Sector Commands were each made up of an infantry unit (three companies), an artillery unit (two batteries), and an engineer company. The 2nd Sector Command differed in having only a single battery for its artillery unit and no engineer company; while the 4th Sector Command had a five-company infantry unit and a three-battery artillery unit as well as its engineer company. The 4th Border Garrison Command had three sector units and included a total of two 300mm howitzers, two 240mm howitzers, six 150mm guns, eight 105mm howitzers, eight 75mm Type 90 field guns, two 75mm Type 38 field guns, seventeen mountain guns, sixteen 70mm Type 92 infantry guns, eight medium trench mortars, eighteen anti-aircraft guns and ten AA machine guns. The other large BG was the 8th, expanded about 1943, which had five sector commands, each with an infantry unit and an artillery unit.

The smaller BGs were an equally disparate group. The 5th BG consisted of a 65-man HQ, a 740-man infantry unit, a 306-man artillery unit, and an 83-man engineer unit. The 6th–10th BGs each consisted of a five-company infantry unit, a three-battery artillery unit and a company-size engineer unit; while the 11–13th BGs had four infantry companies, one artillery battery and an engineer company.

The BGs remained on duty longer than most of their contemporaries in the Kwantung Army. The large 8th BG was deactivated in October 1944 to form the 119th Division. For the rest of the border garrisons, however, 1944 and early 1945 was simply a period of gradual diminution of strength as manpower levies were made against them by IGHQ to replace losses in more active theaters. Finally, the ax fell on the remaining BGs in July 1945, when the 1st, 2nd and 10th BGs were used to form the 132nd IMB; the 3rd and 4th BGUs to form the 135th Division; the 5th, 6th and 7th BGs to form the 135th and 136th IMBs; the 9th BG to form the 280th regiment of the 127th Division; and the 14th BGU to form the 134th Division. A few personnel from the 4th BGU were used as a cadre for a new 15th BGU, now the only such unit in the army. By TO&E this unit was to have a strength of 15 infantry companies and four artillery batteries, but in fact it only had four infantry companies and one battery at the time of the Soviet invasion.

The Army also formed a small number of infantry-type combat support units, including machine gun, mortar and anti-tank battalions. The fortunes of these units waxed and waned according to the IGHQ perception of requirements, but overall they were apparently not very popular.

Independent Machine Gun Battalions

Although not a part of the permanent army force structure, the 1st through 10th and 21st Independent Machine Gun Battalions were activated in 1938 for duty in China. Such a unit consisted of a headquarters and three 150-man line companies, the latter almost certainly very similar to those in the transitional organization of 1935 discussed earlier, each armed with eight medium machine guns, with two pack horses provided for each MG.

Although these MG battalions were useful from an economy of force perspective, combat in China, which emphasized infantry in the mobile, striking role against a lightly-armed foe, apparently did not provide much employment for these units. It can also be theorized that the defensive, less mobile, character of the machine gun units did not appeal to the infantryman's samurai spirit. In any event, the machine gun battalions were withdrawn to Japan in 1940 and disbanded, with the last two (2nd and 3rd) being ordered home for demobilization on 2 October 1940.

The nature of defensive fighting in the central Pacific, however, caused a fundamental rethinking of requirements. In June 1944 the Army made a desperate effort to activate a new series of independent machine gun battalions for service on threatened Saipan. The 1st through 4th Independent Machine Gun Battalions were quickly organized by drawing off machine gun companies from existing infantry regiments and depot units. This haste notwithstanding, Saipan fell before the battalions could be dispatched.

The new battalions apparently used a TO&E similar to that the earlier units had used in China. A machine gun squad consisted of nine men and manned a Type 92 medium MG. Two such squads, plus a 2-man HQ, made up the machine gun platoon. A machine gun company had four MG platoons, a company commander and a 22-man HQ section, for a total company strength of 103. The company had 40 rifles in addition to the eight Type 92s, but no grenade dischargers or any other supporting weapons. The battalion HQ, commanding three companies, was a tiny unit consisting of the battalion commander, adjutant, intendance officer, medical officer and twenty-one enlisted men for a total battalion strength of 17 officers, 3 warrant officers, 33 NCOs and 270 privates of the infantry branch, 3 men from the intendance branch, and 8 medical personnel, for a grand total of 334.

The US Tenth Army intelligence section noted that:

> The MG battalions were typically attached directly to a division, brigade, or regiment and ordered by those commands to "cooperate" with lower echelons in the defense of a particular line. Of course, as the campaign progressed, and communication deteriorated, this "cooperation" took on more and more the characteristics of attachment ... Ordinarily the MG Bn itself had no grenade dischargers or light machine guns, but an effort was always made to have an infantry unit disposed so as to be able to render supporting fire with such weapons.

It should be noted that centralized command of the battalion was rendered impossible in any event by the small size of the battalion HQ and the complete lack of signal facilities, rendering attachment of subordinate companies to the line infantry necessary.

These first four machine gun battalions were followed in July through September 1944 by the 6th, 12th–15th and 17th–26th Independent Machine Gun Battalions raised in Japan, the 7th and 8th battalions raised on Formosa, 10th battalion raised in China, and

A 90mm Type 94 trench mortar firing in China in 1938.

the 11th battalion raised in the Philippines. The 5th, 9th and 16th Battalions were ordered raised, but later cancelled.

As soon as organization was complete the bulk of the Japan-raised MG battalions were dispatched overseas. The 1st and 2nd Battalions were sent to Iwo Jima; the 3rd, 4th, 14th and 17th Battalions were sent to Okinawa; the 18th and 19th Battalions were sent elsewhere in the Ryukyus; the 20th and 22nd Battalions were sent to the Borneo Garrison Army; the 24th Battalion went to Formosa; and the 12th, 13th, 25th and 26th went to Luzon.[61]

The US Tenth Army's intelligence section's after-action report noted that "one of the most important lessons to be learned by the Japanese from the Okinawa campaign, in the opinion of most POWs interrogated, was the fact that the HMG battalions are relatively ineffective." The primary reason for this appears to have been the almost complete lack of heavy fire support for the machine guns. The long-range accuracy of tripod-mounted machine guns was negated by waiting until the enemy came close to open fire, but opening fire at long range gave away the Japanese positions to murderous US heavy weapons counter-fire, for which the Japanese, with no effective field artillery, had no answer. Unsupported by long-range weapons the medium machine guns quickly fell victim to US artillery and tanks.

No further machine gun battalions were formed, although three separate machine gun companies (35th and 36th of 138 men each and 37th of 150 men) were raised in Japan near the end of the war.

61 The 11th and 25th Battalions were attached to the 105th Division, and the 26th Battalion to the 103rd Division.

The Mortar Battalions

Unlike the machine gun battalions, the mortar battalions were an established part of the Army's permanent force structure. They were built around two weapons developed in the early 1930s, the 90mm trench mortar (also known as a light trench mortar) and the 150mm medium trench mortar.

Originally the 90mm mortar was designed as a chemical weapon and to this end a structure known as the trench mortar regiment (*hakugeki rentai*) was formulated. This consisted of a 130-man headquarters, a 650-man mortar battalion and a 410-man gas battalion. The mortar battalion was organized with three 210-man companies each with twelve 90mm mortars, initially Type 94, later Type 97. The gas battalion had six Type 97 tankettes serving as prime movers and 12 small tracked trailers, half for gas spraying the other half for decontamination, and ten trucks, one of which mounted a steam cleaning plant for clothes. The 1st and 3rd Trench Mortar Regiments remained in Japan until 1945, when they were apparently disbanded. The 2nd Trench Mortar Regiment was stationed in Manchuria under the Kwantung Army until 1945 when it, too, was apparently disbanded.

In action trucks carrying 200 rounds brought ammunition up to about 500 meters behind the mortar positions, from where it was carried by hand. The crew consisted of a mortar commander, a gunner, a loader, two ammunition handlers, and four ammunition carriers. They set up 600 to 1,000 meters behind the front lines and usually fired 1,000 to 1,500 meters beyond the front line. Although they held and fired stocks of HE ammunition, chemical warfare was regarded as their primary mission and they practice-fired DM, DC and mustard rounds during training.

The 1st to 5th Trench Mortar Battalions (*hakugeki daitai*)(also known as Mortar Battalions and Infantry Mortar Battalions) were activated in July 1937 and, along with the slightly later locally-raised 20th and 21st Trench Mortar Battalions, served in China during 1938–41 primarily as HE-throwing units. As originally configured under the 1937 organization, a battalion had a strength of 963 men and 429 horses and was divided into three companies (each with twelve 90mm mortars) and an ammunition trains.[62] The 2nd and 20th Trench Mortar Battalions were demobilized in December 1940, but the other five remained on duty in China.

Table 2.31: Infantry Branch Mortar Battalions, from October 1940

	Personnel	Pistols	Carbines	Rifles*	Light Machine Guns	90mm Mortars	150mm Mortars	Riding Horses	Draft Horses	Carts	Cars	Trucks	Lt Repair Vehs
Trench Mortar Battalion													
Battalion Headquarters	113	10	53	180	0	0	0	0	0	0	2	23	0
Three Mortar Companies	204	21	44	0	1	12	0	0	0	0	2	20	0
Ammunition Trains	104	9	49	0	0	0	0	0	0	0	2	21	1

62 A decrement of 80 privates was applied to these battalions in September 1940.

Medium Trench Mortar Battalion													
Battalion Headquarters	130	11	2	150	0	0	0	10	30	30	0	0	0
Three Mortar Companies, each	140	2	0	0	0	0	4	1	56	54	0	0	0
Ammunition Trains	42	2	31	0	0	0	0	0	0	0	1	15	0
* *rifles are undistributed weapons for battalion self-defense*													

A new trench mortar battalion organization was promulgated in October 1940 by the 1941 mobilization plan, which called for smaller, entirely motorized formations. The new trench mortar battalions were made up of a HQ, three mortar companies, and an ammunition trains. A mortar company consisted of three platoons, each with four squads. The company was provided with 20 trucks that carried not only the company's twelve mortars, but also a hundred rounds of ammunition for each. The battalion ammunition trains carried a further 4,500 rounds of 90mm ammunition and provided a small vehicle maintenance section. A set of "Type 94 light infantry mortar observation equipment" was provided to each company HQ and to the battalion HQ, giving four FO teams for the battalion.

This new organization was not applied retroactively to the trench mortar battalions in China, which remained on their old horsed organization, and indeed appears to have been implemented with only a single battalion, the 10th Trench Mortar Battalion, activated in July 1941 by the Kwantung Army for use on the flat plains of Manchuria.

With the decision to invade southward, some realignment of forces was obviously required. The 3rd and 5th Trench Mortar Battalions were dispatched from China to participate in the invasion of Malaya, with the 3rd then moving on to participate in the siege of Bataan.

At the conclusion of these campaigns the 3rd Trench Mortar Battalion moved further south being disbanded on Bougainville in June 1944. The 5th Trench Mortar Battalion was assigned to the defense of the Dutch East Indies and garrisoned Timor to the end of the war. The 21st Battalion served in the 17th, and then the 18th Armies, finishing the war in New Guinea. The 1st and 4th Trench Mortar Battalions remained in China to the end of the war.

A new series of trench mortar battalions were activated by the Kwantung Army in July 1942 to take advantage of the stockpiles of 90mm mortars available. About 15 battalion sets had been produced during 1934–40, but only about a third of them were actually issued. These units, the 11th–14th Battalions, reverted to a horsed organization, as did the 15th and 16th Trench Mortar Battalions raised in China in February 1944, and the 17th–20th and 22nd Trench Mortar Battalions also raised in China in February 1945. A second 21st Battalion served in the Philippines.[63] These were considerably smaller than the preceding battalion, and each of these battalions were organized into a headquarters, three mortar companies and an ammunition trains. These elements gave a strength of a field-grade commander, 22 company-grade officers (including 2 medical and 2 veterinary), four warrant officers, six master sergeants (2 each of administrative, medical and veterinary branches), 58 line sergeants, and 771 privates (including 15 medical) for a total of 866 men with 486 horses. Armament comprised 256 rifles, 18 light machine guns and eighteen 90mm mortars.

63 One company of this battalion, plus some reinforcements, were rearmed with 24 20cm rocket launchers on Luzon.

By early 1943 the heavy mortar was becoming much more attractive. Where the IJA had not been particularly fond of mortars before, they now adopted them with a fervent enthusiasm. They were cheap and (more importantly) simple to build, light and easy to move and hide, and could fire from a deep defilade that protected them from much of the effects of American artillery superiority. Production of mortars (other than those for the infantry regiments), which had totaled a mere 60 units in fiscal year 1941 and 57 units in FY-1942, soared to 704 in FY-1943, and then to 1,053 in FY-1944, before dropping somewhat to 199 in April–July 1945. Production was a mixture of the older 90mm weapons and the new 120mm Type 2 mortar. In March 1945 IGHQ activated the 23rd and 25th Trench Mortar Battalions, followed in May by the 24th and 26th–40th Trench Mortar Battalions, all in Japan for the defense of the homeland. They retained the general strength and organizational characteristics of the preceding series of trench mortar battalions. Their organization was:

	Men	Horses
Battalion Headquarters	145	86
Three Mortar Companies, each	182	89
Trains	177	134

Each of the mortar companies was armed with six 120mm mortars, to give a battalion total of 18, a somewhat low figure by international standards for the manpower involved.

Although the IJA had never successfully practiced massing of fire before, they apparently decided that the time had come to give it a genuine try. The result was an order of 23 May 1945 directing the activation of no fewer than 32 large mortar battalions for the defense of the homeland. In light of their large size these units were given a new and rather confusing designation, the 1st to 32nd Trench Mortar Artillery Battalions (*Hakugekihō Daitai*).[64] These were raised with two strength authorizations depending on the availability of officers:

	Off	WO	MSG	SGT	PVT	Total	Horses
Type 1	15	20	19	100	1,260	1,414	471
Type 2	34	6	13	100	1,254	1,407	471

These were similar to the earlier trench mortar battalions, but had six mortar companies instead of only three, thus thirty-six 120mm mortars. They were also similar to the trench mortar artillery regiments of the 200-series divisions, but lacked the intermediate battalion HQs. Apparently delays in weapons production meant that these units were not actually activated until June (Battalions 4, 5, 12, 13, 16, 17, 24 and 25), July (Battalions 6, 7, 18–20 and 26–29) and August (1, 2, 14, 15, 21, 22, 30 and 31). The remaining battalions were never actually formed. Those that were formed may have still suffered from shortages of weapons, possibly averaging about eighteen weapons apiece.

Related to the trench mortar battalions were the medium trench mortar battalions (*Chūhakugeki Daitai*) formed to use the 150mm mortar in its various configurations (Types 94, 96 and 97). The October 1940 mobilization plan provided for this type of unit as a part of the army's permanent force structure, but only one such unit had been formed by

64 Not to be confused with the series of artillery mortar battalions (*kyūhō daitai*) raised with the same weapon at about the same time. These are discussed in the artillery section.

mid-1944. The 1st Medium Trench Mortar Battalion apparently remained in Japan until mid-1943 when it was shipped to the Southern Army. In November it was deactivated to provide personnel for the artillery units of the 24th and 26th IMBs.

The 2nd, 3rd and 4th Medium Trench Mortar Battalions were activated in June 1944, and the 5th through 7th Battalions a month later. The 2nd and 3rd Battalions were sent to Iwo Jima, the 4th Battalion went to Luzon, the 5th and 6th Battalions to the 32nd Army in the Ryukyus (but thence on to the Philippines), while the 7th Battalion went to China and then to Luzon.

All of the medium trench mortar battalions appear to have conformed to the organization set out in the October 1940 mobilization plan. It seems likely, however, that the units destined for Iwo Jima would have been stripped of their transport assets, as was common with most units sent to the Central Pacific theater.

The medium trench mortar battalion consisted of three mortar companies, each with four 150mm mortars and 54 two-wheeled one-horse carts. The battalion trains held all the motor vehicles, a passenger car and fifteen cargo trucks. Communications assets are unknown, but are unlikely to have been extensive.

In addition to the mortar battalions, the infantry raised independent trench mortar companies. Each of these had a strength of 4 officers, a warrant officer, 17 NCOs and 126 privates of the line branch, together with two men from the technical and intendance branches and four medics, for a total strength of 154. Each company was assigned one riding horse and eighteen draft horses with Type 39 carts. Armament consisted of 39 rifles, one light MG and twelve 81mm Type 94 or Navy Type 3 mortars.

The 1st through 10th Independent Trench Mortar Companies were activated on 25 August 1944 in Japan. The 1st Company was sent to Iwo Jima and the 2nd Company to Shanghai, while the remainder were assigned to the 32nd Army and sent to Okinawa.

Once there it was agreed that centralized control of these mortars would work best, and the companies were consolidated into two light mortar battalions in February 1945. The 42nd Trench Mortar Battalion comprised the 3rd–6th companies, while the 43rd Trench Mortar Battalion was made up of the 7th–10th companies.[65] Personnel for the new battalion HQs were found by drawing off men from the companies, resulting in new company strengths of 3 officers, a warrant officer, and 134 NCOs and privates. The battalion HQ had 6 officers and 70 enlisted. This reduction in strength was compensated for by the addition of 25 local Japanese conscripts and 20 *Boei Tai* Okinawans to each company (the conscripts serving as line troops and the *Boei Tai* for ammunition and rations carriers) shortly after the reorganization. In addition, an ammunition trains was formed for each battalion using another hundred *Boei Tai* each.

Independent Anti-Tank Units

The lessons of Nomonhan notwithstanding, it was not until July of 1940 that the Army began forming separate anti-tank units to support to infantry in the field. In that month the 1st, 2nd and 5th Independent Anti-Tank Battalions were formed, each consisting of a headquarters, three AT gun companies, and a trains group with a total strength of 20 officers and 471 enlisted.

A further expansion of the anti-tank force took place a year later, in July 1941, when

65 For reasons unclear US intelligence referred to them as the 1st and 2nd Light Trench Mortar Battalions, even after the campaign.

four more battalions (3rd, 4th, 6th and 7th) and twelve independent AT companies (1st–12th) were raised. These battalions were smaller, with only two line companies for a total of 15 officers and 338 enlisted. The components of the two types of battalions were almost identical.

The AT battalion HQ consisted of a lieutenant colonel (battalion commander), three captains (two staff and one medical), a lieutenant (adjutant), one transport and four medical sergeants, a corporal and 36 privates.

A gun squad normally consisted of a corporal, a gunner, a loader, two spotters, and six ammunition bearers. These were supplemented by two drivers with a Nissan cab-over-engine truck that carried the gun portee-fashion in the rear, along with the gun crew. A gun platoon consisted of the lieutenant platoon leader and a runner to command two gun squads. These two men rode in the cabs of the gun trucks, having no transport of their own. In some cases, a portion of the ammunition bearers in the gun squads were concentrated under the platoon leader to form an ammunition squad.

The AT company was made up of three such platoons plus the headquarters and an ammunition platoon, the latter consisting of a senior NCO and two squads, each of ten men with a truck. The company HQ consisted of the company commander, four sergeants and 15 privates (including 4 signalers and 2 AA lookouts). The company was authorized a total of 264 rounds of AP and 132 rounds of HE for each of the six 37mm guns.

Table 2.32: Independent Anti-Tank Units

	Personnel	Pistols	Carbines	Rifles*	Anti-Tank Guns	Field Cars	Trucks	Light Tractors	Lt Repair Vehicles
Independent Anti-Tank Battalion, 1940 Pattern									
Battalion Headquarters	53	1	10	100	0	3	12	0	0
Three Anti-Tank Companies, each	126	0	30	0	6	3	10	0	0
Trains Group	60	0	20	0	0	1	9	0	1
Independent Anti-Tank Battalion, 1941 Pattern									
Battalion Headquarters	47	1	29	100	0	2	11	0	0
Two Anti-Tank Companies, each	125	0	23	0	6	2	10	0	0
Trains Group	56	0	21	0	0	1	8	0	1
Independent Anti-Tank Battalion, 1944 Pattern									
Battalion Headquarters	23	0	0	109	0	2	3	0	0
Three Anti-Tank Companies, each	110	0	0	29	4	2	8	5	0
Trains Group	50	1	0	46	0	1	15	0	1
Independent Anti-Tank Company, 1944 Pattern	140	0	0	59	8	2	8	8	0

** rifles in battalion HQs are undistributed weapons for battalion self-defense*

Some variations in organization and equipment invariably occurred. The companies of the 2nd and 6th Independent AT battalions at Guadalcanal, for instance, split their load evenly between AP and HE ammunition, the latter unit carrying 660 rounds of 37mm ammunition per gun. Of these, 48 rounds per gun were carried with the gun squads, and the remainder with the ammunition platoon. In addition, infantry weapons were more liberally distributed in the 6th than the original mobilization tables allowed, with eight men in each gun squad being allocated Type 38 rifles, along with fourteen more rifles in the ammunition platoon and twelve in the company HQ, for a total of 76 weapons, plus a light MG in each of the two company ammunition squads for local defense.

The battalion trains group had a strength of 50 men and was provided with one field car and eight trucks, carrying fuel and a further 120 AP and 30 HE rounds for each gun in the battalion. The trains also included a small maintenance section with a light repair vehicle.

The independent AT company was larger than its battalion-based brethren in that it had four gun platoons for a total of eight guns. The gun squad was similar, but the 1½-ton trucks used towed the guns and carried 150 rounds of ammunition each, but not the crewmen, who walked alongside and behind the trucks. In this case the platoon included an ammunition section of 8 men in addition to the ammunition handlers of the gun squads, along with a liaison corporal, to give a platoon strength of 37 men with two guns.[66]

At least two companies, the 6th and 9th, were formed as a pack units, with 8 officers and 242 enlisted men and 77 horses. These had four gun platoons, each of two sections, each with a 37mm gun and six horses. The company's four ammunition sections, each of 12 men with 6 horses, were centralized under an ammunition platoon.[67]

With the decision finally made in the autumn of 1941 to move south against the Western colonial holdings instead of northwest against the Soviets, anti-tank units assumed a lower priority. No further separate units were raised until the summer of 1944. Those that had been formed, in the meantime, found themselves busy. The 1st, 2nd, and 5th Battalions were sent to south China in July 1940, and in early 1942 they were reassigned southward, with the 1st Battalion going to Burma and the other two to the East Indies and Bougainville, with the 2nd then moving to Guadalcanal with the 2nd Division. Other AT battalions of the early levies, the 3rd and 4th Battalions, were assigned to the Kwantung Army in July 1941, joining the 6th and 7th Battalions that were raised there. The 6th Battalion left Manchuria for Guadalcanal in September 1942, while the other three battalions remained in Manchuria until June 1944, when the 3rd and 7th Battalions were transferred to Okinawa and the 4th Battalion went to Formosa. As the newer 47mm Type 1 guns became available they were gradually introduced to the force, so that the 1st Battalion in Burma in mid-1944, for instance, was authorized three 37mm and three 47mm in each of its companies.

Faced with increasing armor threats from the counterattacking Allies, the Army ordered the mobilization of a third wave of AT units, much more numerous than the previous two. Carried out in June to September 1944, the Army raised twenty-one new AT battalions (numbered 8–28)[68] using a different organization from the earlier versions.

66 One company, the 8th, was distinctive for using 37mm Rheinmetall AT guns captured in China. It took these guns to Rabaul and Guadalcanal.

67 The 9th Company gradually transitioned to a motorized 47mm gun unit, with the first platoon-set arriving in mid-1942.

68 These represented, in part, an expansion of existing assets. The 13th Independent AT Battalion, for

The new battalions were triangular in configuration, which undoubtedly increased their flexibility as compared with their predecessors, but which also substantially increased their manpower overhead. The 1944 battalion consisted of sixteen officers (including one technical and one medical), three warrant officers, three master sergeants, 33 line sergeants, and 348 privates divided into a HQ, an ammunition trains and three 4-gun companies. The new, smaller, AT company was made up of a company HQ, an ammunition platoon, and two gun platoons. The gun platoon consisted of a small HQ, two 13-man gun squads (each with a light tractor towing a 47mm Type 1 AT gun and an ammunition truck), and an ammunition squad of 14 men with a truck. The company ammunition platoon was provided with two more ammunition trucks and a reserve tractor. The company HQ included two field cars, a personnel truck, and a Type 93 light field rangefinder. The AT company carried with it 192 AP and 48 HE rounds per gun, which were supplemented by the 224 AP and 56 HE rounds per gun carried in the battalion ammunition trains. The 47mm guns authorized for the 1944-pattern battalions were considerably more effective than the 37mm weapons previously used, but some of the AT battalions did not receive the 47mm guns until 1945, relying in the meantime on the old 37mm Type 94s.

Since the line companies acted primarily in direct support of infantry units signal facilities within the battalion were not particularly strong. Each company included six optical/telephone signalmen, while the battalion HQ had a small signal section with 12 radiomen and 10 optical/telephone signalers. Combat service support for the battalion was found in the trains group, which included an 18-man maintenance section and an 18-man ammunition section. The trains were provided with nine cross-country trucks (one for personnel, three ammunition, two equipment and three reserve vehicles) and six road-bound trucks (four for ammunition and two equipment).

Since the AT battalions were oftentimes split up to support dispersed infantry units, some adopted a modified organization in which the battalion trains and some of the HQ elements were disbanded and the personnel assigned directly to the subordinate AT companies. In addition, a small number of light machine guns and grenade dischargers were sometimes issued to the company HQs for unit self defense.

Further independent AT companies were also formed, the 13th–32nd in June 1944, the 37th–40th in November 1944 and the 41st–45th in February 1945. The first group, each of 145 men with six truck-towed 47mm, was raised in Manchuria then dispatched to Taiwan and the Pacific. The second group of companies, raised in Burma, was slightly larger, at 179 men, while the third group, each 118 men with six guns and 8 trucks, were raised in China.

A final group of anti-tank battalions, the 28th to 33rd, were activated in the first half of 1945. The first five, most raised in February, were destined for the Kwantung Army and had strengths of 496 men each, presumably organized along the same lines as the 1944 battalions. The 33rd Battalion, raised in and for Hokkaido, used the same TO&E as that adopted for the new divisional AT battalions.

Reserve and Replacement Units

Early in the war a small number of Second Reserve (*kobi*) infantry battalions were formed. Each had a strength of about 945 men divided into a headquarters, four rifle companies

instance, was formed in Burma using the 6th Independent AT Company as a nucleus, while other guns and personnel came from infantry regiment AT units, and the trucks and drivers from motor transport units.

(each of three platoons armed with rifles only) and a machine gun company of two 2-gun platoons. The battalion had trains personnel but no transport. To control these units the 1st and 2nd Second Reserve Infantry Groups were formed under the Kwantung Army and served until about 1942.

Much more significant were the organizations set up in the divisional districts to train new soldiers and hold them until needed at the front. During peacetime this function was carried out by the divisions themselves, but once the divisions mobilized and departed their administrative and training functions were taken over by a "depot division" (*Rusu Shidan*) bearing the numerical designation of either the regular army division resident in that district or the initial "B" division mobilized there. Thus, the Western District Army contained, during 1943–44 the 5th Depot Division, the 6th Depot Division, the 55th Depot Division (in the 11th Division district) and the 56th Depot Division (in the 12th Division district).

Each of these depot divisions was, in essence, a "shadow" division to the active duty unit, consisting of three infantry depot regiments, an artillery depot regiment, a cavalry or reconnaissance depot regiment, an engineer depot regiment, etc. Each of these units trained new soldiers in their respective specialties and held them until they were ordered dispatched to the front to form new units or replace casualties in units already supported by that district.

A major reorganization of this system went into effect on 9 April 1945. The purpose was apparently to provide more organizational flexibility to meet new requirements and to more closely match training capacity with actual population size in each district. The new depot divisions, largely continuations of the earlier divisions, bore the names of their districts, rather than numbers, and each consisted of a variable number of infantry replacement units (or regiments), an artillery replacement unit, an engineer replacement unit, a transport replacement unit, and assorted other service replacement elements. Notable was the absence of the former cavalry and reconnaissance replacement units, these having been distributed to the infantry replacement units as mounted companies.

The new depot divisions (with the number of infantry replacement units in parentheses) were as follows: in the North East District Army were the Hirosaki (2) and Sendai (3) Depot Divisions; in the Eastern District Army were the Tokyo (5), Utsonomiya (3) and Nagano (3) Depot Divisions; in the Tokai District Army were the Nagoya (4) and Kanazawa (2) Depot Divisions; in the Central District Army were the Kyoto (2), Osaka (4) and Hiroshima (6) Depot Divisions; in the Western District Army were the Zentsuji (3), Kurume (3) and Kumamoto (3) Depot Divisions; in the Northern District Army was the Ashigawa (3) Depot Division; and in the Korea District Army were the Seoul (3), Pyongyang (2), Rana (2), Taikyuzan (2) and Koshu (2) Depot Divisions. The Formosa Army contained no depot divisions.

Originally designed as training and holding units, these depot divisions were allotted combat roles in the plans for the defense of the homeland. In common with the training and replacement units in all other armies they held a higher proportion of infantry than the field units they supported because of the higher casualty rates normally sustained by the infantry.

The components that formed the basis of these divisions were generally standardized as follows:

	Men	Horses
Division Headquarters	108	13
Signal Replacement Unit	345	38
Each Infantry Replacement Unit	3,202	338
Artillery Replacement Unit	760	343
Engineer Replacement Unit	705	29
Transport Replacement Unit	659	178

The infantry replacement unit was usually a regiment-size formation with the following composition:

	Men	Horses
Regiment Headquarters	39	4
Signal Company	162	11
Mounted Company	163	163
Two Infantry Battalions, each		
Battalion Headquarters	7	2
Four Rifle Companies, each	239	0
Machine Gun Company	249	11
Infantry Gun Company	209	11
Pioneer Company	205	0

The infantry replacement rifle company held 220 rifles and was divided into three platoons each with two light MGs and two grenade dischargers. The machine gun company had three 2-gun machine gun platoons and a 2-gun 70mm howitzer platoon. The infantry gun company was armed with two Type 41 mountain guns and two Type 94 anti-tank guns. The mounted company was provided with four light MGs and four grenade dischargers.

The divisional artillery replacement unit was a battalion-size unit, usually consisting of four batteries, one with three 75mm Type 95 field guns, one with three 105mm Type 91 field howitzers, one with three 75mm Type 94 mountain guns, and one with four 120mm mortars. Designed for training crewmen on a wide variety of weapons in service, concentration of fire by such a disparate formation would have proven all but impossible.

The engineer replacement unit was built around four 169-man engineer companies, while the signal replacement unit had only a single 331-man company. The transport replacement unit had one horsed company of 359 men with 173 horses, and one motor transport company of 267 men with one car, 35 trucks and a light repair vehicle.

There were also a few specialist infantry replacement units, to include the 1st to 3rd Mortar Regiment Depot Units, which could have been used *in extremis*.

The theaters also set up their own "field replacement units" to act has holding pools and refresher training centers for personnel who were temporarily without a permanent unit. Such soldiers included those released from hospitals after substantial periods and some new arrivals and transfers. These field replacement units had no fixed strength and their composition varied considerably over time. The 12th Field Replacement Unit, for instance, served China and was made up of a variable number of subordinate companies scattered about for training and garrison duties. Such a company usually approximated a rifle company in strength, organization and equipment.

3

Cavalry

Unlike the armies of most other nations, the Japanese Army had no tradition of a cavalry arm. Where others entered the 1930s with a large cavalry branch that had to be either mechanized or reduced drastically in size, the Japanese Army simply retained its organization of four non-divisional cavalry brigades.

The peacetime structure of the army called for:

1st Cavalry Brigade (Guards, 13th and 14th Cavalry Regiments)
2nd Cavalry Brigade (1st, 15th and 16th Regiments)
3rd Cavalry Brigade (8th, 23rd and 24th Regiments)
4th Cavalry Brigade (3rd, 25th and 26th Regiments)

For support of these cavalry elements GHQ maintained a single battalion-sized Horse Artillery Regiment. The Guards, 1st, 3rd and 8th Cavalry Regiments were withdrawn from the brigades about 1936, leaving each brigade a two-regiment formation. In partial compensation three more horse artillery battalions were activated, permitting the assignment of one to each brigade, although these did not always deploy with the brigades before the war.

Three of these brigades were shipped to Manchuria in 1936/37 for duty with the Kwantung Army. Two, the 1st and 4th Cavalry Brigades, were formed into an extemporized Cavalry Group in August 1937, while the 3rd Cavalry Brigade operated independently. The 2nd Cavalry Brigade, which had not left Japan, was demobilized in 1939/40.[1]

The 4th Cavalry Brigade moved to central China under the 2nd Army in early 1938, being replaced in the cavalry group by the 3rd Brigade. In mid-1939 it moved north, becoming subordinate to the China Occupation Army, where it was joined by the HQ of the Cavalry Group and the 1st Cavalry Brigade, thus reconstituting the Group, but now in an active theater. In late 1942 the Cavalry Group HQ and the 1st Cavalry Brigade were demobilized and the personnel used in forming the 3rd Armored Division. The 3rd Cavalry Brigade was returned to Japan and disbanded on 16 January 1945. Thus, by mid-1945 the Japanese non-divisional cavalry consisted of the 4th Cavalry Brigade (25th and 26th Cavalry Regiments, field artillery unit and transport unit) in northern China and the 171st Cavalry Regiment in Manchuria.[2]

1 The subordinate 15th and 16th Cavalry Regiments continued to serve in Japan until 1942 when they were combined into a new 73rd Cavalry Regiment that was finally disbanded about a year later. The 71st and 72nd Cavalry Regiments were formed in late 1939 first for north China, then the Mongolia Garrison Army but were transferred to the Mechanized Army in mid-year, the latter presumably being broken up for personnel. The 71st Regiment served almost to the end of the war in Manchuria.

2 The 171st Cavalry Regiment was a 529-man divisional cavalry regiment that had been split off from its parent 71st Division in February 1945 and assigned to 3rd Area Army HQ under the Kwantung Army. By war's end the cavalry branch also still included the 3rd, 9th, 11th, 28th, 47th, 75th, 77th, 79th and 93rd divisional cavalry regiments, along with the remnants of the 49th and 55th in Burma.

A member of a scout/rifle squad poses in 1938.

The peacetime organization of a cavalry regiment called for a 33-man HQ, four 153-man rifle squadrons each of four platoons, and a 143-man machine gun squadron. The rifle squadrons were armed solely with rifles, all automatic weapons being concentrated in the MG squadron, which included a light MG platoon (six weapons), a medium MG platoon (six weapons), and an ammunition platoon.[3] When deployed to Manchuria, however, the regiments were enlarged to 1,260 men with each rifle squadron receiving six light MGs and the MG squadron rearmed with eight heavy MGs.[4]

A cavalry brigade in Manchuria consisted of two cavalry regiments, a nine-man HQ, a 143-man battery from the Horse Artillery Regiment with four 75mm guns, an 85-man machine gun squadron (with eight medium MGs), a 40-man engineer platoon, a 60-man tankette company with nine tankettes or cavalry combat cars, and a 470-man transport battalion with two companies totaling 288 pack horses.

By the time the 1941 mobilization plan was published in October 1940, a cavalry brigade was built around two cavalry regiments (the 13th and 14th in the 1st Brigade, the 23rd and 24th in the 3rd Brigade, and the 25th and 26th in the 4th Brigade) and was supported by a two-battery horse artillery "regiment," a transport unit, and smaller support elements.

3 In practice the MG squadron appears to have been often missing in peacetime, being formed for deployment.

4 The 1938 cavalry regulations show each squadron made up of four line platoons, each of 2 or 3 rifle sections (each a light MG) and 1 grenadier section, indicating that the Type 89 grenade discharger had made its way to the cavalry as well as the infantry on a semi-official basis.

The pack horse for the Type 11 LMG carried the machine gun on one side and two ammunition chests on the other.

The basic building block of the cavalry regiment was the mounted platoon, which consisted of two scout (or rifle) squads, two light MG squads, and a grenadier squad. The scout squad consisted of seven men, each with a Type 44 carbine, a cavalry sword, and a riding horse. The light MG squad had eight men each with a riding horse plus one pack horse carrying a light MG. The grenadier squad consisted of five men on riding horses plus one pack horse carrying a Type 89 heavy grenade discharger and 48 rounds of HE ammunition. The total mounted platoon thus had a strength of 40 men and a principal armament of two Type 11 light machine guns and one 50mm Type 89 grenade discharger. A rifle squadron consisted of four such platoons along with its 26-man headquarters and trains group.

The cavalry regiment consisted of a headquarters, four rifle squadrons, and a machine gun squadron. The machine gun squadron was made up of a HQ, two machine gun platoons, an AT rifle platoon, and an ammunition platoon. Each 40-man MG platoon was divided into two sections, each of which consisted of two gun squads. One of the four sections in the squadron used draft transport, utilizing two gun carts, an ammunition cart and a tools cart, each pulled by two horses. The other three sections used pack horses. The small AT rifle platoon consisted of two mounted squads, each with one pack horse carrying a 20mm AT rifle on a pack harness and a second horse carrying 69 AP and 35 HE rounds. The MG squadron had 26 pack horses carrying ammunition for a total ammunition load of 8,000 rounds per gun.

The regiment HQ carried rations and fodder in 49 Type 96 two-horse wagons, but did

Table 3.1: Cavalry Brigade, from October 1940

	Personnel	Pistols	Carbines	Rifles	Light Machine Guns	Medium Machine Guns	Heavy Gren Lnchrs	20mm AT Rifles	20mm Auto Cannon	37mm AT Guns	75mm Field Guns	Light Tanks	Field Cars	Trucks	Lt Repair Vehs	Riding Horses	Pack Horses	Draft Horses	Carts	Wagons
Brigade Headquarters	119	8	35	0	2	0	0	0	0	0	0	0	1	0	0	62	6	18	0	9
Signal Unit	176	3	132	0	0	0	0	0	0	0	0	0	0	0	0	0	19	24	0	12
Water Purification Section	67	0	19	0	0	0	0	0	0	0	0	0	2	7	0	0	0	0	0	0
Medical Unit	290	24	84	0	0	0	0	0	0	0	0	0	7	25	0	0	0	0	0	0
Veterinary Unit	84	0	15	0	0	0	0	0	0	0	0	0	1	4	0	20	0	14	0	6
Anti-Tank Company	95	11	28	25	0	0	0	0	0	4	0	0	2	12	0	0	0	0	0	0
Machine Cannon Company	117	12	34	30	0	0	0	0	4	0	0	0	3	14	0	0	0	0	0	0
Tank Unit	125	25	92	0	0	0	0	0	0	0	0	12	5	14	1	0	0	0	0	0
Two Cavalry Regiments, each																				
Regiment HQ	214	8	31	0	0	0	0	0	0	0	0	0	0	0	0	78	0	100	0	49
Four Rifle Squadrons, each	192	3	146	0	8	0	4	0	0	0	0	0	0	0	0	187	20	0	0	0
Machine Gun Squadron	174	3	98	0	0	8	0	2	0	0	0	0	0	0	0	165	42	8	0	4
Horse Artillery Regiment																				
Regiment Headquarters*	154	10	155	0	0	0	0	0	0	0	0	0	0	0	0	73	0	69	0	32
Two Batteries, each	138	11	10	0	0	0	0	0	0	0	4	0	0	0	0	102	3	39	0	7
Ammunition Trains	96	5	0	0	0	0	0	0	0	0	0	0	0	0	0	70	0	30	0	9
Transport Unit																				
Unit Headquarters*	48	3	5	100	0	0	0	0	0	0	0	0	3	0	0	10	0	12	0	6

Table 3.1: Cavalry Brigade, from October 1940

	Personnel	Pistols	Carbines	Rifles	Light Machine Guns	Medium Machine Guns	Heavy Gren Lnchrs	20mm AT Rifles	20mm Auto Cannon	37mm AT Guns	75mm Field Guns	Light Tanks	Field Cars	Trucks	Lt Repair Vehs	Riding Horses	Pack Horses	Draft Horses	Carts	Wagons
Two Horse Companies, each	208	3	22	0	0	0	0	0	0	0	0	0	0	0	0	28	0	176	0	85
Two Motor Companies, each	98	10	87	0	0	0	0	0	0	0	0	0	4	30	1	0	0	0	0	0

* *includes undistributed rifles and carbines for unit self-defense*

A cavalry lineman with the pole that allowed him to run wire above ground in trees.

not carry additional ammunition for the line elements. It also provided the usual range of trains personnel, including blacksmiths, cobblers, etc.

Fire support within the brigade was primarily provided by a horse artillery battalion (nominally a "regiment") of two batteries. Each battery was provided with four 75mm Type 41 cavalry guns (not to be confused with the Type 41 mountain gun), six caissons, a stores wagon, an observation wagon and a total of 520 rounds of HE ammunition. No smoke or other special ammunition was carried. The battalion ammunition trains carried 608 rounds of HE ammunition in eight caissons, while the battalion HQ had 32 Type 96 transport wagons for rations and fodder, and an observation wagon.

There were three fully motorized combat units in the brigade, the light tank company, the anti-tank company and the machine cannon company. The anti-tank company provided four truck-drawn 37mm Type 94 AT guns and was provided with 360 rounds of AP and 90 rounds of HE ammunition per gun. The machine cannon company manned four 20mm Type 98 automatic cannon, each provided with 1,000 rounds of HE and 500 rounds of AP ammunition. The light tank company provided 12 Type 95 light tanks and carried 136 HE and 136 AP rounds of 37mm ammunition for each tank, along with 9,000 rounds of MG ammunition per tank.[5]

The brigade's transport unit consisted of four companies, two with horse-drawn wagons and two with trucks. One of each type was used to carry ammunition and gas masks,

5 The Type 2 combat cars originally used had been replaced by the start of the war.

while the other two carried rations, fodder, equipment, etc. The ammunition companies carried a total of 320 rounds of grenade discharger HE ammunition, 217 HE and 863 AP rounds of 37mm tank/AT gun ammunition, 1,873 rounds of 75mm HE ammunition, and 2,640 rounds of 20mm AA ammunition, along with 85,320 rounds of ammunition for the medium MGs.

The brigade medical unit was built around two stretcher-bearer platoons each of 83 men and one treatment platoon of 19 men. The unit HQ operated the trucks, which sometimes served as ambulances.

The cavalry brigade was lightly armed, especially in terms of HE weapons and AT guns. The organization of the unit, however, does not appear to have changed through the war. This was probably due to the fact that of the two brigades, one was facing only Chinese forces while the other belonged to the Kwantung Army, which was starved of equipment in order to maintain the war on the active fronts. Neither force could be said to have contributed much to Japan's war effort, although the 4th Brigade in China did see some combat.

Further, it seems unlikely that the cavalry brigades ever had their full complement of supporting units at all times. Although the 3rd Brigade had received its anti-tank company in August 1940 and its tank unit in June 1942, by late 1944 it had apparently lost not only those but the 3rd Horse Artillery Regiment as well, and was reporting the following TO&E strengths, partially reflecting the 1941 Mobilization Table:

Brigade HQ	170 men	93 horses
Signal Unit	176 men	193 horses
23rd Cavalry Regiment	1156 men	1253 horses
24th Cavalry Regiment	1156 men	1253 horses
Transport Unit	685 men	439 horses
Medical Unit	290 men	?
Veterinary Unit	84 men	38 horses

The 3rd Cavalry Brigade was dissolved in February 1945 in Manchuria to form the 77th Independent Mixed Brigade.

It is unclear whether the 4th Cavalry Brigade ever received its allotment of mechanized support units. At the end of the war it consisted simply of the headquarters, two cavalry regiments, the horse artillery regiment, and a transport unit. Most of these were still on the authorized strengths adopted on mobilization in December 1939 as follows:

	Off	WO	MSG	SGT	PVT	Total	Horses
Brigade Headquarters	12	0	4	8	0	24	21
Two Cavalry Regiments, each	49	11	17	92	940	1,109	1,226
Horse Artillery Regiment	25	6	16	47	430	524	414
Transport Unit	21	5	20	40	404	490	197
Total	352	33	74	279	2,714	3,452	3,084
Change of June 1942:							
Horse Artillery Regiment	-3	-2	-	-12	-92	-109	-
Change of March 1945:							
Brigade Headquarters	-	-	+2	-	+18	+19	-

4

Armor

The Army clearly discerned the importance of tanks from the reports of the First World War and launched a real, if modest, effort to incorporate them into their force. They bought about two dozen tanks in the early 1920s, mostly Renault FTs, with a few British Mk A Whippets as well. With these, two tank companies were activated in 1925, one at the 12th Division depot at Kurume, and the other at the Chiba Infantry School.

In 1931 one platoon of Renault tanks was sent to Manchuria and there it was mated with the Armored Car Platoon of the Kwantung Army to form a provisional armored company. Additional Renaults and personnel were sent in 1932 and a full tank company was formed. The tank company participated in the Jehol campaign, performing well.

These early operations were regarded as highly successful and the Army pressed ahead with plans to expand the tank arm. The numbers of available and projected tanks was sufficient by 1933 to allow a major increase in the armored arm, with the 1st Tank Company (Kurume) being reorganized as the 1st Tank Battalion, the 2nd Tank Company (Chiba) becoming the 2nd Tank Battalion, and a 3rd Tank Battalion raised *de novo* in Manchuria.

The Army Staff continued to regard tanks as infantry support devices, but the Kwantung Army, facing the wide-open spaces of Manchuria, envisioned early a more independent role. In 1934 they formed an experimental combined arms force, known as the Kungchuling Mechanized Mixed Brigade.

The Kungchuling Brigade consisted of the new 4th Tank Battalion (a light tank formation), a motorized infantry regiment, a field artillery battalion and an engineer company, with small service support elements. The 4th Tank Battalion was built around three tank companies, each with fifteen light tanks, along with a headquarters with five more light tanks (plus two cars and five trucks) and a depot. The depot held ten more light tanks as an equipment reserve, and also supplemented the supply efforts of the ten trucks in each tank company with 30 more of its own. The depot also included a small maintenance section with four repair trucks.

The motorized infantry regiment consisted of a tankette company and three infantry battalions. Each battalion was made up of a headquarters, three rifle companies (each 150 men with fifteen trucks), and a battalion gun company (ten trucks and four 37mm infantry guns). The artillery battalion was a motorized formation with three batteries, each with four 75mm Type 90 field guns. The engineer company, with its twelve trucks, was reinforced by a flamethrower platoon with five engineer assault tanks.

Co-located with the brigade at Kungchuling but not an organic part of it, was the 3rd Tank Battalion, which consisted of two companies of Type 89 medium tanks and a depot. Elements of this battalion were routinely attached to the brigade for trials purposes.

The mechanized mixed brigade did not prove an unqualified success. The wheeled vehicles of the brigade were not able to keep up with the tanks in the trackless steppes, particularly in winter. The equipment was not particularly well suited to the extreme colds often encountered in Manchuria and the engineer unit, configured for the infantry assault role, did not prove helpful. Elements of the brigade and the 3rd Tank Battalion participated in combat in north China in 1937 with undistinguished results.

The 1st and 2nd Battalions, each a mix of Type 89 mediums and Type 94 tankettes in three companies, were sent to China and participated in operations in north China from September 1937 into mid-1938. The 5th Battalion was activated in July 1937 for use in Shanghai.

The staffs were found to be overwhelmed in combat and in 1938 additional staff officers and clerks were added. With this minor change the battalions were redesignated as regiments, the 1st and 2nd Battalions as the 7th and 8th Regiments in 1938, and the 3rd and 4th Battalions as the 3rd and 4th Regiments in 1939, with the 5th Tank Battalion being brought home and dissolved.

In addition, new regiments were raised. The Kwantung Army raised the first, the 5th Tank Regiment, using new Type 97 medium tanks, at Mutanchiang in 1938. Combat operations drove more significant expansion and the 9th, 12th, 13th and 14th followed in China in 1939, although this was not really an accretion to strength since the last three were simply consolidations of tankette companies using their original equipment until actual tanks became available. The 13th Tank Regiment, for instance, was formed by consolidating and reorganizing the 2nd, 6th, 7th and 9th Tankette Companies into a regiment of three light companies, each of three platoons.

It would appear that the Kwantung Army had no early intention of using tanks at Nomonhan, but faced with the increasing Soviet threat, mechanized units were ordered west to supplement the efforts of the infantry. The 3rd Tank Regiment arrived at Nomonhan with a strength of 376 officers and men divided as follows:

HQ Company	two Type 89 mediums, two tankettes
2 Medium Companies, each	two Type 97 and ten Type 89 mediums, five tankettes
Supply Company	four Type 89 mediums, three tankettes

Each of the line companies consisted of three medium platoons (three Type 89 tanks each), a 4th (light) platoon with Type 94 tankettes, and a company HQ with one Type 89 medium and two Type 97 mediums. In addition, the regiment HQ held two Type 89 mediums and these, together with supply company holdings, gave the regiment a total of four Type 97 and 26 Type 89 medium tanks, and five Type 97 and ten Type 94 tankettes.

The 4th Tank Regiment showed up with a strength of 565 as follows:

HQ Company	three Type 95 lights
3 Light Companies, each	nine Type 95 lights
Medium Company	eight Type 89 mediums, 2 tankettes
Supply Company	five Type 95 lights

The light and medium companies were each built around two platoons. The regiment had a total of 35 Type 95 light tanks, eight Type 89 medium tanks, and two (some sources say three) Type 94 tankettes.

The two regiments were formed into the "Yasuoka Detachment," which was also to consist of motorized infantry, artillery and engineers. In fact, however, synchronization of the various elements proved impossible and the tank regiments wound up fighting naked and alone. Both took extremely heavy losses and accomplished little.

The leadership of the tank regiments came in for almost immediate criticism for failure to properly coordinate with the other branches, but in fact there does not appear to have been much the tankers could have done about it, since the infantry and artillery proved totally incapable of keeping up with the tanks in even the shortest of moves. Possibly as a result of this criticism the Mechanized Brigade was disbanded in 1940 and in its place two tank groups activated: the 1st Tank Group at Tungning and the 2nd at Tungan. These groups were charged with the mission of infantry support and each consisted of three tank regiments.

The Army mobilization plan for 1941 attempted to standardize armor organization by providing set TO&Es for the tank group HQ, group maintenance company, group ammunition trains, and regular and light tank regiments.

Table 4.1: Armor Units, from October 1940

	Personnel	Pistols	Carbines	Light Machine Guns	Light Tanks	Medium Tanks	Field Cars	Trucks	Heavy Repair Vehs	Light Repair Vehs
Tank Group Headquarters	60	15	37	0	2	2	3	4	0	0
Tank Group Maintenance Company	162	30	115	0	4	6	3	11	1	3
Tank Group Ammunition Trains	176	18	151	2	0	0	3	55	0	0
Tank Regiment (except 4th Regiment)										
Regiment HQ	67	11	45	0	1	2	3	3	0	0
Light Tank Company	73	15	51	0	10	0	2	6	0	0
Three Medium Tank Companies, each	96	18	71	0	2	10	2	8	0	0
Trains Company	221	23	184	0	3	5	4	39	0	3
4th Tank Regiment										
Regiment Headquarters	52	9	33	0	2	0	3	3	0	0
Three Light Tank Companies, each	72	15	51	0	10	0	2	6	0	0
Trains Company	189	21	154	0	7	0	4	35	0	2

Under this organization the medium tank company consisted of three platoons, each with three medium tanks; and a company headquarters with one medium and two light tanks. In addition, the company included a small field car, a motorcycle and a truck in the 26-man company HQ and eight trucks in the company's trains elements. The company's basic ammunition load revealed its infantry support role, 107 HE and 53 AP rounds of 57mm for each medium tank and 80 HE and 80 AP rounds of 37mm for each light tank.[1]

1 In contrast to tanks, which were rigidly apportioned, local commanders apparently found it quite easy to acquire additional trucks locally, thus expanding their company trains units as time went by. A fairly

The company had four radios, one in the medium tank in the company HQ, and one for each platoon leader's tank.

The tank regiment consisted of three medium tank companies and a light tank company, together with a regiment HQ and a trains group. The regiment's light tank company consisted simply of three light tank platoons (each three light tanks), a company HQ with a light tank and two small field cars, and a trains element with six trucks. The ammunition basic load was the same as for the light tanks in the medium company.

The regimental trains carried the second echelon load of ammunition, consisting of 148 HE and 148 AP rounds of 37mm for each of the 20 light tanks and 200 HE and 99 AP rounds of 57mm for each of the 37 medium tanks, along with other supplies. The trains also included a small maintenance detachment with three light repair trucks, as well as three light and five medium tanks as reserve replacement vehicles.

Under the mobilization plan the 4th Tank Regiment remained a light tank unit, consisting of three light tank companies, identical to those in the normal tank regiment. The one significant change was in the basic ammunition load. Without the medium tanks, with their short-barrel low velocity 57mm guns to provide HE firepower, the 4th Regiment was forced to use the 37mm guns of the light tanks in this role, and the light tank company's ammunition load was altered to 107 HE and 53 AP 37mm rounds per tank, while the regimental ammunition trains carried 214 HE and 106 AP rounds per tank.

These figures, however, represented IJA plans rather than reality, and it seemed that every time enough tanks were produced to bring existing units up to full strength, additional tank regiments were raised. The 9th and 11th Tank Regiments had been mobilized in July 1941; the 1st, 6th and 23rd in September; and the 24th in November. Nevertheless, by the end of 1941 it would appear that most of the tank regiments were in pretty good shape.

The 1st, 3rd and 11th Tank Regiments were fully equipped with Type 97 medium and Type 95 light tanks. The 5th and 6th Regiments also used Type 97 mediums, but appear to have been slightly below strength in numbers. The 7th, 8th and 9th Tank Regiments were up to full strength, but with the older Type 89 mediums. The 4th Regiment was, as noted, a light tank unit, as was the 14th Regiment, which was formed with a strength of 33 officers and 469 enlisted in three companies of Type 95 light tanks pending arrival of medium tanks (which finally showed up in 1943). The 13th Regiment, nominally a medium unit, was also still equipped largely with light tanks. Probably still typical was the 3rd Company of the 13th Regiment in April 1941, which had twelve men, two Type 95 light tanks and one Type 97 tankette in each of its three platoons, 23 men in the company HQ (with one Type 95 and two Type 97s) and 48 men in the trains. These trains provided three more tankettes as reserve equipment, along with ten trucks for supply duties and two repair trucks.

The two tank groups controlled six tank regiments in Manchuria: the 1st Tank Group with the 3rd, 5th and 9th Tank Regiments; and the 2nd Tank Group with the 2nd, 4th and 11th Regiments. Although the group HQs had been formed in early 1941, it was not until 1 August 1941 that IGHQ sanctioned the mobilization of the group maintenance companies and ammunition trains.

The group headquarters were small affairs that would be hard-pressed to control

---

common inventory during 1942 being 3 trucks for the 13-man forward section, 4 trucks for the 9-man ammunition section, 4 more for the 10-man fuel section, and a parts truck and a repair truck with trailer for the 15-man repair section. Such a unit carried 8,400 liters of diesel fuel for the tanks, said to be good for seven days of operation.

A platoon of Type 89 mediums in decidedly tank-unfriendly terrain in China.

tactical operations. In fact, the only radios in the HQ were those in the four tanks. The maintenance company had four repair vehicles and eleven trucks, seven of which carried spare parts, along with replacement tanks. The ammunition trains used 55 trucks to carry 5,400 rounds of HE and 3,700 rounds of AP for the 57mm guns but, surprisingly, none for the 37mm weapons.

With the decision made to invade southward a major reshuffling of armor assets was ordered. In November 1941 the 3rd Tank Group HQ was formed, along with maintenance and trains elements, and the 1st Tank Regiment (from the homeland), and the 6th and 14th Tank Regiments (from China) were pulled together under this new HQ and made subordinate to the 25th Army that was to invade Malaya. At the same time the 4th Tank Regiment was taken from the Kwantung Army and the 7th Tank Regiment from China for the 14th Army (for the invasion of the Philippines).

For its new relatively independent role the 3rd Tank Group was also assigned an organic engineer unit with a motorised bridging company (with six Type 99 pontoon bridge sets) and two regular engineer companies, substantially increasing its capabilities compared with the tank groups in Manchuria.

These moves left the 2nd Tank Regiment on Formosa; the 3rd, 5th, 8th, 9th, 10th, and 11th Tank Regiments in the Kwantung Army (the 8th and 9th having been moved up from China to compensate the Kwantung Army for its losses); and the 12th and 13th Tank Regiments in China. In addition, the 23rd and 24th Tank Regiments were raised on the standard organization with Type 97 mediums and Type 95 lights, and these were assigned to the Kwantung Army.

The tank elements in the southern invasion forces performed well and contributed to the spectacular initial advances. The exception noted in an after-action report on armor in the Malaya operation was that the Type 95 light tank was severely hampered in operations by the one-man turret crew, which required the tank (and sometimes unit) commander to also serve as gunner and loader.

Once Singapore had fallen and the Fil-American forces retreated into the jungled hills of Bataan, the tanks involved were free for reassignment. HQ of the 3rd Tank Group and the 6th Tank Regiment were sent back north to the Kwantung Army. The 1st and 14th Tank Regiments were assigned to the 15th Army in Burma in March 1942, where the 14th Regiment served out the rest of the war (the 1st Regiment had the good fortune to be reassigned to the Kwantung Army a mere four months later). The 4th Tank Regiment was shifted from the Philippines to the 16th Army for the invasion of Java in January 1942, while the 7th Tank Regiment remained in the Philippines until May when it was reassigned to the Kwantung Army. The 4th Tank Regiment remained in the East Indies for the remainder of the war, serving on garrison duty in Timor.

Thus, by mid-1942 only the two light tank regiments, the 4th on garrison duty in the East Indies and the 14th fighting in Burma, were left in Japan's new colonial conquests. The rest of the units had been pulled back and concentrated primarily under the Kwantung Army.

The Resurgence of Combined Arms

Even while the Army staff in Tokyo directed its lightning drives to the south the Kwantung Army kept its eyes firmly on their presumed opponent, the Soviet Union. The success of the German panzer formations in the first year of the war would not have escaped anyone's notice and the forces in Manchuria were not about to be left behind.

On 4 July 1942 the Mechanized Army was officially formed by IGHQ order. The units assigned to it were the 1st and 2nd Tank Groups (with their subordinate tank regiments); the 6th and 7th Tank Regiments; the 1st, 2nd, 4th, 7th and 12th Independent Anti-Tank Companies; the 1st Independent Field Artillery Regiment (minus one battery); the 27th, 28th and 29th Independent Field AA Companies; the 16th and 18th Field Machine Cannon Companies; the 5th Independent Engineer Regiment (minus one company); the engineer unit of the 3rd Tank Group (minus one company), and the 3rd Heavy River Crossing Engineer Company.

The same IGHQ order assigned the 341-man 15th Tank Regiment to the 4th Army and the 276-man 16th Tank Regiment to the 6th Army, both under the Kwantung Army, in the infantry support role. These were not new units, but simply redesignations of the former 1st and 23rd Division Tank Units respectively a month earlier. In the case of the 16th Tank Regiment this subordination did not last long and within a few months it had been shipped out to the central Pacific to join the Wake Island garrison, where it remained for the rest of the war.

The IGHQ plan for the Mechanized Army, however, had an even shorter existence, as the tank groups and other units mentioned in the July order had been disbanded a month earlier in favor of two new armored divisions by the Kwantung Army, which had its own plans.

By using the six tank regiments in the two tank groups and the two independent tank regiments as well as the other units assigned to the Mechanized Army, the Kwantung Army was able to form two armored divisions, each with four tank regiments. From 24 June 1942 the 1st Armored Division consisted of the 1st Tank Brigade (1st and 5th Tank Regiments) and the 2nd Tank Brigade (3rd and 9th Tank Regiments); while the 2nd Armored Division had the 3rd Tank Brigade (6th and 7th Tank Regiments) and the 4th Tank Brigade (10th and 11th Tank Regiments).

The tank regiments were reorganized on 24 June 1942 to dramatically increase their size. Regiment strength rose from 649 men (including six maintenance NCOs and a 15-man medical detachment) to 1,017 (including 8 NCOs and 30 privates in the technical branch for maintenance and a seven-man medical detachment). Because the medical detachment included no privates an additional 24 medical and 30 general-duty privates were authorized for all regiments, bringing strength up to 1,071. The detailed strength of the two organizations were:

	Off	WO	MSG	SGT	PVT	Total
1941-pattern	26	6	10	103	484	649
From June 1942	49	24	17	273	708	1,071

Each of these tank regiments was now to consist of a headquarters, a light tank company, three medium tank companies, a gun tank company and a maintenance company. A light tank company had a headquarters (with a light tank), three tank platoons (each three light tanks) and a maintenance platoon. A medium tank company contained three Type 95 light tanks and eleven Type 89 or Type 97 medium tanks, divided into a headquarters platoon (with three light and two medium tanks, and a field car), three tank platoons (each three medium tanks), and a maintenance platoon with six trucks. The gun tank company was a new component, brought about by the rearming of the Type 97 mediums with new high velocity 47mm guns in place of the earlier low velocity 57mm guns. The gun tank company was identical to the medium tank company but featured these improved Type 97s. The regiment HQ had four light tanks for liaison and two medium tanks for command duties. The maintenance company also carried out supply duties with 30 trucks and operated a reserve pool of 10 additional tanks. This new organization of the tank regiments increased the firepower of the regiment , but also reduced the logistical support somewhat, apparently in recognition of the increased support available from a divisional structure.

It should be noted, however, that the organization given above represented Kwantung Army plans that were only slowly implemented. The scale of issue of the improved Type 97s also increased as production progressed, so that within two years most of the regiments were in fact made up of one light, three improved Type 97 and one basic Type 97 companies. In this configuration roles were reversed, and the company with the old 57mm-equipped tanks was designated the Gun Tank Company.

The armored division also included a motorized infantry regiment, a reconnaissance battalion, an anti-tank battalion, a field artillery regiment, an anti-aircraft battalion, an engineer battalion, a maintenance battalion and a transport regiment.

These elements gave the division the following strengths:

	Off	WO	MSG	SGT	PVT	Total
Division Headquarters	48	5	34	47	123	257
Reconnaissance Battalion	n/a	n/a	n/a	n/a	n/a	551
Two Brigade HQs, each	5	0	0	6	11	22
Four Tank Regiments, each	49	24	17	273	708	1,071
Mechanized Infantry Regiment	97	28	27	340	2,526	3,029
Mechanized Artillery Regiment	74	20	21	179	1,212	1,506
Anti-Tank Battalion	25	7	8	56	348	444
Anti-Aircraft Unit	n/a	n/a	n/a	n/a	n/a	1,014
Engineer Unit	46	14	16	155	918	1,149
Transport Unit	32	9	10	99	615	765
Maintenance Unit	26	6	48	50	648	778
Total	635	213	255	2,289	10,425	13,821

The divisional organization emphasized anti-tank protection, with each of its nine rifle companies including 47mm anti-tank guns, the anti-tank battalion three 6-gun companies of the same weapons, and 37mm Type 94 AT guns sprinkled liberally throughout the unit.

A distinctive feature of the Japanese armored division was that in addition to the tank regiments, the engineer unit was also an armored component. The engineer unit was made up of six companies, each of which consisted of two armored engineer platoons carried in APCs and one assault platoon with four SS armored engineer vehicles. In addition to the Type 96s, the engineers also had 62 APCs, 44 non-armored half-track trucks, 29 trucks, 3 workshop trucks and one passenger car.

The divisional reconnaissance unit was made up of 1st and 2nd light tank companies, 3rd rifle company, 4th gun tank company and 5th maintenance company. The light tank companies were identical to those in a tank regiment, while the gun tank company had a HQ (two gun tanks and two medium tanks) and two platoons (each three gun tanks).

The motorized infantry regiment consisted of three battalions plus infantry gun, maintenance and medical companies. A rifle company consisted of three platoons, each of four sections. One section in each platoon was an anti-tank unit with a 47mm Type 1 AT gun and 70 rounds of ammunition. The other three sections were rifle units, either with a light machine gun, a grenade discharger, and 11 rifles each, or two sections with light machine guns and one with grenade dischargers, depending on the company commander's preference. It may have been intended to equip the regiments with full-track Type 1 APCs, but in fact, they never became available in those numbers and instead the infantry relied on 6x4 Type 94 trucks. As with the rest of the division, the rifles were, surprisingly, old 6.5mm Type 38s.

The artillery regiment had three 12-gun battalions, one with 75mm Type 90 field guns and the other two with 105mm Type 91 howitzers, all towed by Type 98 4-ton tractors. The anti-aircraft regiment was made up of two gun batteries (each four 75mm Type 88) and four autocannon batteries (each six 20mm Type 98).

The maintenance unit consisted of four maintenance companies and a parts company. Each maintenance company had three maintenance platoons and a recovery platoon. A maintenance platoon had one heavy repair section, two light repair sections, an ordnance/radio repair section and a parts section. The parts company had three platoons, one with fifteen half-track trucks, one with fifteen 6x4 trucks, and one as a trains with five fuel trucks, five ammunition trucks, and five rations trucks. The transport unit had six motor transport companies and a maintenance company.

Shortly thereafter, the 3rd Armored Division was formed in China. This was built around the 5th Tank Brigade (8th and 12th Tank Regiments), the 6th Tank Brigade (13th and 17th Tank Regiments), and the usual infantry, artillery and support units mostly found through conversion of elements of the 3rd Cavalry Brigade.

The raising of these three armored divisions, it will be noted, did not involve the creation of any additional tank regiments, and indeed no new tank regiments were formed between mid-1942 and early 1944.[2] Part of the reason for this was that Japanese tank production had peaked in early 1942 and shortly thereafter went into a precipitous decline that lasted until the end of the war. By 1943 the priority being given to shipping and aircraft production had taken its toll and finally, Japan managed to produce only 127 tanks in the eight months of 1945 before the surrender.

Such tanks as were being produced, mainly the modified Type 97 and a similar tank called the Type 1, were being used to replace older models in existing tank regiments. Any tanks left over from this effort, mainly light models, were used to form the organic tank elements mandated for the amphibious brigades and island defense divisions.

In November 1943 the Mechanized Army was disbanded and the 1st Armored Division reassigned to the 3rd Army and the 2nd Armored Division to the 5th Army. In February 1944 the organization of the armored division was changed. One of the tank regiments and one of the brigade HQs were deleted from the divisional structure to create a triangular unit. Thus, the 9th Tank Regiment was removed from the 1st Armored Division and the 11th Tank Regiment from the 2nd Armored Division. The 3rd Armored Division, which had already lost its 8th Tank Regiment to duty in Rabaul, was not effected. At the same time, and for unknown reasons, the divisional AA units were also detached from the divisions, although they continued to retain their designations (i.e., 2nd Armored Division Anti-Aircraft Unit) to the end of the war. By early 1944 most of the divisional tank regiments had been substantially re-equipped, with each divisional tank regiment being provided with a total of 31 light tanks and 50 medium tanks. The internal organization of the regiments differed from division to division, depending on the commander's preference and equipment available.

In addition, the 1st Armored Division lost its 3rd Tank Regiment to duty in China in April 1944, leaving it with only two tank regiments. Amazingly, the division's constituent units retained, at least on paper, their 1942 organization right to the end of the war, even after moving to the homeland in March 1945. In fact, there was some redistribution of personnel within the division as can be seen from this comparison:

	Authorized	Actual
Division Headquarters	221	477
1st Tank Regiment	1,071	887
5th Tank Regiment	1,071	886
1st Mechanized Infantry Regiment	3,029	2,844
1st Mechanized Field Artillery Regiment	1,506	1,674
1st Armored Division Anti-Tank Unit	444	481

2 Nor were the armored divisions made part of the permanent army force structure by the Army Staff in Tokyo.

	Authorized	Actual
1st Armored Division Engineer Unit	1,149	1,204
1st Armored Division Transport Unit	765	726
1st Armored Division Maintenance Unit	778	854
Total	10,034	10,033

In addition, a large number of trucks were withdrawn from the organization, apparently on the theory that mobile operations of the kind envisioned for Manchuria would not be possible in the crowded homeland. Each tank regiment was left with only three cars and eight cargo trucks, while the mechanized infantry regiment was reduced to a "leg" infantry formation with only 39 APCs, 8 cars and 21 trucks. The transport unit was left with only 61 trucks.

Each of the two tank regiments had, at the time of surrender, 30 Type 95 light tanks, 35 Type 97 medium tanks and 12 Type 97 improved medium tanks, along with three full-tracked and nine half-tracked personnel carriers. The mechanized infantry regiment held 39 full-tracked APCs and 15 tankettes.

The 2nd Armored Division, on the other hand, was sent into combat in the Philippines. To conserve shipping space, always a high priority in IJA deployments, the division was drastically reduced in strength on 21 July 1944. This time, IGHQ did not specify detailed strengths for each component, but only a single overall strength figure for each unit, mandating that the division HQ was to be reduced to 100 men, each tank regiment was to be reduced to 800, the infantry regiment to 1,500, the artillery regiment to 1,200, the anti-tank unit to 200, the engineer unit to 800, the transport unit to 400, and the maintenance unit to 500.

Given the rather general organizational guidance received from Tokyo it is not surprising that the division's organization was rather flexible. Each medium tank company consisted of three platoons, each with three Type 97 improved medium tanks. The company HQ section included one or two light tanks for liaison and one or two medium tanks for command duties. As was common, the platoon leaders' tanks and the company command tanks were equipped with radios, but the others were usually not. The company HQ also usually included a three-man fuel team and two medics. The company also included a 32-man trains section that included six reserve tank crew, and fifteen skilled maintenance personnel (two radio repairmen, six mechanics, two electricians and five metalworkers) under a warrant officer. A gun tank company was similar, but used the old Type 97 tanks with the 57mm short-barrel gun, presumably retained for their HE capability. The maintenance company was divided into a recovery platoon (with five medium tanks), a repair platoon (12 trucks), an ammunition platoon (two trucks) and a fuel platoon (three trucks), and an HQ with kitchen and generator trucks. The 6th Tank Regiment, roughly similar to the other two regiments, had the following nominal strength:

	Off	WO	NCO	PVT	Lt Tks	Mdm Tks (47mm)	Mdm Tks (57mm)	Cars	Trucks
Regiment HQ	9	4	29	36	2	5	0	1	2
Light Tank Company	5	2	27	54	12	0	0	0	2

	Off	WO	NCO	PVT	Lt Tks	Mdm Tks (47mm)	Mdm Tks (57mm)	Cars	Trucks
3 Medium Companies, each	5	2	37	57	2	11	0	0	2
Gun Tank Company	5	2	37	57	2	0	11	0	2
Maintenance Company	6	3	19	165	0	0	5	0	22
Total	40	17	223	483	22	38	16	1	34

The 7th Tank Regiment differed only in detail, with each medium/gun tank company having an extra four enlisted and the maintenance company an extra 14, with no additional vehicles. In the case of the 10th Regiment the gun tank company had lost its vehicles to submarine attack *en route* to the Philippines, so tanks from the other companies were redistributed to it.

The infantry regiment left most of their vehicles, including their few APCs, behind and deployed as normal leg infantry, albeit more heavily-armed.

The mobile artillery regiment had one battalion of 75mm Type 90 field guns (one battery of which used Type 1 SP weapons) and two of 105mm Type 91 field howitzers, each battalion consisting of a headquarters and three batteries. A battery had a nominal strength of 181 men, but each left 50–60 behind when they moved to the Philippines, apparently to conserve shipping space, and further drains for details for higher HQs and other units generally brought strength down to about half that. Each battery had four main guns, but other equipment varied. In the SP battery the firing unit had four Type 1 SP guns and four APCs for ammunition, while the battery HQ had 2 armored observation vehicles, two cars and a truck for observation equipment and personnel. The battery's maintenance section had four APCs for ammunition, fuel and parts. A towed battery should have included a Type 98 4-ton tractor for each gun, along with two armored observation vehicles for the battery HQ. In fact, one battery was noted as having five Type 98 4-ton tractors, a tracked cargo vehicle, a Type 92 light observation car and four trucks. Another had 4 tractors, 2 half-tracks and 3 trucks.

The Division's 75-man brigade HQ was provided with five medium tanks (including one for the brigade commander), two light tanks, two passenger cars, a radio truck, one truck each for fuel, ammunition, and medical supplies, and two for miscellaneous supplies.

The division's anti-tank battalion had three companies, each with three 2-gun platoons of 47mm Type 1 guns. In February 1945, however, the guns were abandoned during a retreat and the unit issued machine guns instead. The engineer regiment took its engineer assault tanks to the Philippines, but no other armored vehicles. The transport regiment was mobilized with 603 men in four companies, each of 114 men with 33 trucks. For the movement to the Philippines strength was reduced to 400 men, but the all the trucks were taken.

The 2nd Armored Division was shipped from Pusan in mid-September 1944, but broken up on Luzon and its elements assigned to different commands. It was destroyed without ever having effected the battle.

The 3rd Armored Division served the entire war in China as that theater's mobile reserve, although it rarely moved far. Its distance from Tokyo apparently enabled it to customize its organization to a considerable degree. In May 1944 it had the following organization:

	Off	WO	NCO	PVT	Lt Tk	Mdm Tk	APC	Car	Truck	Tractor
Division HQ	45	0	93	365	2	7	0	16	41	0
Recon Unit	23	6	96	404	34	10	0	8	45	0
12th Tank Regiment	35	15	243	646	31	50	0	11	68	0
13th Tank Regiment	38	14	229	585	31	50	0	11	70	0
17th Tank Regiment	43	6	225	675	31	50	0	12	72	0
Mechanized Infantry Regiment	86	29	276	2477	9	0	14	48	239	0
Mechanized Artillery Regiment	77	8	145	985	0	0	0	39	124	47
Anti-Tank Unit	24	3	92	331	0	0	0	12	39	16
Engineer Unit	36	14	152	828	0	0	*12	19	122	0
Transport Unit	23	3	90	562	0	0	0	31	238	0
Maintenance Unit	26	3	85	500	3	3	0	17	84	0
Medical Unit	9	2	27	250	0	0	0	3	40	0
Field Hospital	15	1	27	178	0	0	0	2	25	0

* *includes 6 armored engineer vehicles*

The three tank regiments of the division had slightly different organizations, although similar in overall structure:

	12th Regt	13th Regt	17th Regt
Regiment Headquarters	89	87	90
Light Tank Company	104	102	111
Four Medium Tank Companies, each	143	129	143
Trains Company	175	161	176
Total	940	866	949

In each case the light tank company had twelve light tanks, while the medium companies each had four light tanks and eleven medium tanks. In each case a company consisted of three platoons. The regiment HQ had a further three light and four mediums, while the trains company held two mediums as reserve. In light of the limited armored threat in China, all the medium tanks of the 3rd Division were nominally the basic Type 97s, although by May 1944 thirty of the mediums in each regiment were the improved Type 97 versions. Other elements of the division were not so lucky. The anti-tank unit was supposed to have sixteen 47mm AT guns, but in fact only had six, while in the mechanized infantry regiment the rifle companies had only two 47mm guns instead of three. The artillery regiment had its full quota of sixteen 105mm howitzers, but only nine of the

High-speed Type 91 howitzers were produced, but apparently not supplied to the 3rd Tank Division in China. Here, one of their howitzers on a Type 98 artillery trailer behind its tractor at war's end.

twelve 75mm Type 90 guns authorized.

The 4th Armored Division had been ordered raised in July 1944, but it was not until the arrival of the 30th Tank Regiment to supplement the earlier 28th and 29th Tank Regiments that it began to take shape. Nevertheless, the 4th Armored Division never approached the status of the first three, and by the end of the war still resembled a large tank brigade rather than an armored division.

The new tank regiments were raised on a new TO&E of 966 men when activated on 16 July 1944. On 6 April 1945 they were reorganized onto the new standard 1,198-man TO&E, supplemented a month later by two additional technical personnel to bring strength up to 1,200 men, although it would seem that this later addition was not always implemented. These compared in strength to the nominal regiments of the other three armored divisions as follows:

	Off	WO	MSG	SGT	PVT	Total
In 1st, 2nd & 3rd Armored Divisions (nominal)	49	24	17	273	708	1,071
4th Armored Division, from 16 July 1944	50	17	18	194	687	966
4th Armored Division, from 6 April 1945	50	18	10	197	923	1,198
4th Armored Division, from 23 May 1945	51	18	11	197	923	1,200

The tank regiments of the 4th Division were a completely new kind of organization that included a large infantry/pioneer company in the battalion, presumably in lieu of a divisional infantry regiment. The lack of transport assets that had plagued Japanese tank units, however, was not remedied. The vehicles held by one of these regiments was:

	Lt Tks	Mdm Tks	APCs	SP Guns	Cars	Trucks
Regiment HQ	1	3	0	0	1	0
Two Medium Tank Companies, each	2	10	0	0	0	0
Two Gun Tank Companies, each	2	10	0	0	0	0
SP Gun Company	0	0	4	6	0	0
Infantry/Pioneer Company	0	0	8	0	0	0
Maintenance Company	1	2	0	0	0	11

A medium tank company had a strength of 114 men divided into a company HQ (one light and one medium tank) and three platoons (each three medium tanks), and was equipped with 47mm-gunned tanks. A gun tank company was identical except for an increase in strength of five men and was equipped with Type 3 medium or gun tanks with 75mm guns. The SP artillery battery had 152 men to operate and support six SP guns, either 75mm Type 1 guns or 105mm Ho-Ni II howitzers. The battery was also provided with four half-track APCs. The infantry/pioneer company was a large formation of three rifle platoons and a pioneer platoon with 368 men and was provided with 32 pistols, 292 rifles, 10 light MGs, 10 grenade dischargers, one light tank and eight tracked armored personnel carriers. The maintenance company had 129 men and was provided with one light and two medium tanks as reserve and eleven trucks.

Impressed though the Japanese may have been by German successes with their panzer arm, they either misread the message and failed to comprehend the magnitude of logistical problems attendant on an armored force, or they simply gave up and restricted their tanks to the infantry support role for the defense of the homeland. The tank regiments of the 4th Armored Division had a nominal "maintenance" company, but it was provided with only a single light repair truck, along with ten cargo trucks that had to serve the logistical needs of the whole regiment. The regiments of the 1st Armored Division had three light repair trucks, but only eight cargo trucks.

Clearly, ten cargo trucks (plus the solitary light repair truck) would not be able to carry even the 752 men of the battalion who were not tank crews or could not fit into the 15-man APCs. It is doubtful that they could even meet the regiment's needs for fuel, ammunition and repair parts. The division's transport unit with four companies each with forty trucks could help, but only at the expense of general carriage for the division. In this they were actually much more fortunate than the 1st Armored Division, whose transport unit had a total of only 61 trucks to support tank regiments with only eight trucks apiece.

In addition to the three tank regiments, the division included a 135-man HQ, a 686-man machine cannon unit, a 189-man signal company, a 524-man maintenance unit, and a 766-man transport unit. The machine cannon unit had five batteries, each with four 20mm autocannon, a passenger car and seven trucks, while the headquarters had a car, two trucks and a light repair truck. The divisional maintenance unit had two companies, each of which held six medium tanks as reserve vehicles, along with three tracked tank recovery vehicles, four light and two heavy repair trucks and ten cargo trucks.

Table 4.2: Homeland Armored Division, August 1945

1st Armored Division	
Division HQ (221) [6 Lt Kts, 13 Mdm Tks]	
1st Tank Regiment (1,071)	
Regiment HQ	[distribution unknown: total regiment
Light Tank Company	30 Lt Tks, 47 Mdm Tks]
3 Medium Companies	
Gun Tank Company	
5th Tank Regiment (1,071)	
as above	
1st Motorized Infantry Regiment (3,029)	
Regiment HQ	
Three Infantry Battalions, each	
Battalion HQ	
3 Rifle Companies, each [9 LMG, 9 GD, 3 47mm AT]	
Machine Gun Company [16 MG]	
Infantry Gun Company [4 75mm regtal guns]	
Maintenance Company	
Medical Company	
1st Motorized Artillery Regiment (1,506)	
Regiment HQ	
Two Field Gun Battalions, each	
Battalion HQ [1 MG, 2 37mm AT]	

4th Armored Division
Division HQ (135)
Signal Company (189)
28th Tank Regiment (1,200)
Regiment Headquarters (83) [1 Lt Tk, 3 Mdm Tks]
2 Medium Companies, each (114) [2 Lt Tks, 10 Mdm Tks]
2 Gun Tank Companies, each (119) [2 Lt Tks, 10 Gun Tks]
SP Company (152) [6 105mm SP how]
Inf/Pioneer Company (368) [10 LMG, 10 GD]
Maintenance Company (129) [1 Lt Tk, 2 Mdm Tks]
29th Tank Regiment (1,200)
as above
30th Tank Regiment (1,200)
as above
Machine Cannon Battalion (686)
Battalion HQ (26)
5 Batteries, each (132) [4 20mm AA]
Motor Transport Unit (766)
Unit HQ (26)
4 Transport Companies, each (163)
Maintenance Company (89)
Maintenance Unit (524)

Table 4.2: Homeland Armored Division, August 1945

1st Armored Division	4th Armored Division
3 Batteries, each [1 GD, 3 75mm guns]	Unit HQ (22)
Two Field Howitzer Battalions, each	2 Maintenance Companies, each (251) [6 Mdm Tks]
Battalion HQ [2 MG, 2 37mm AT]	
3 Batteries, each [2 GD, 4 105mm How]	
Anti-Tank Battalion (444)	
Battalion HQ	
3 AT Companies, each [6 47mm AT]	
Engineer Battalion (1,149)	
Battalion HQ	
3 Engineer Companies, each [2 MG, 4 GD, 2 37mm AT]	
Motor Transport Unit (765)	
Battalion HQ [3 MG]	
4 Trans Companies, each [2 LMG, 1 MG, 2 GD, 1 37mm AT]	
Maintenance Unit (854) [1 Mdm Tk]	

Separate Tank Regiments

The consolidation of tank forces into the two armored divisions in June 1942 left the field armies without close tank support. Thus were formed at the same time five more tank regiments on the smaller 1941 pattern, the 15th and 16th in Manchuria, the 17th in north China, and the 18th and 19th in Japan, to take over the infantry support role.

At the same time the Mechanized Army HQ was dissolved three more tank regiments were added to the Army's order of battle. Unfortunately, these did not represent any accretion to the Army's strength, but instead were redesignations and reorganizations of existing forces. Thus were formed the 25th Tank Regiment (from elements in China) and the 26th and 27th Tank Regiments (from the reconnaissance units of the Kwantung Army's two armored divisions).

For their new role one light tank company in each reconnaissance battalion was re-equipped with medium tanks, giving these two unique tank regiments an organization of one light and two medium tank companies, a motorized infantry company, and a maintenance company. The regiment HQ had a strength of 60 men and was provided with four improved Type 97 medium tanks. The light tank company had 100 men with 13 Type 95 light tanks, while each of the medium companies had 120 men with one light and eleven medium tanks. The infantry company had 200 men and was designed to protect the tanks close in and at night, and was also used for OPs and reconnaissance. The maintenance company's 100 men brought total regiment strength to 700 men, with 26 medium and 15 light tanks.

In April 1944 the two ex-reconnaissance regiments were reorganized, with the tank company personnel strengths reduced to partially offset the addition to each regiment of an artillery company and an engineer company. The artillery company had four 75mm Type 90 field guns drawn by 6-ton tractors, while the engineer company was specially trained in mines and demolitions and road maintenance and repair. This essentially transformed the two tank regiments into all-arms combat teams, although the patent inadequacies of the medium tanks' guns forced the artillery company to be trained and employed as an anti-tank unit rather than a field artillery unit.[3] These gave the two regiments the following organization:

Regiment HQ (40) [1 Lt Tk, 3 Mdm Tks]
Light Tank Company (69) [11 Lt Tks]
Two Medium Tank Companies, each (100) [1 Lt Tk, 11 Mdm Tks]
Rifle Company (150) [6 LMG, 4 MG, 4 GD]
Artillery Battery (100) [4 75mm Guns]
Engineer Platoon (80)
Maintenance Company (70)

The light tank company had a 10-man headquarters, three 13-man platoons (each three Type 95 light tanks) and a 20-man trains section. The medium companies each had a ten-man HQ (with one light and one medium tank), three 20-man platoons and a 30-man trains section. The rifle company was built around two rifle platoons (each three rifle and one grenadier squad) and a machine gun platoon with Type 92 machine guns. The engineer

3 The medium tanks of the 27th Regiment, at least, were armed exclusively with the low-velocity 57mm guns.

platoon had four squads. The maintenance company had a repair platoon with four repair trucks and a supply platoon with ten cargo trucks.

Table 4.3: 9th Tank Regiment Strength, 15 May 1944

			Officers	Warrant Officers	Sergeants	Privates	Total	Light Tanks	Medium Tanks	cars	light trucks	trucks
Saipan	Headquarters	Line	8	0	23	49	80	3	3	2	1	5
		attached	2	0	5	2	9					
	3rd Company	Line	5	2	41	57	105	3	11	0	1	6
		attached	0	0	0	3	3					
	4th Company	Line	6	3	40	55	104	3	11	0	1	6
		attached	0	0	0	4	4					
	5th Company	Line	7	2	34	54	97	3	11	0	1	6
		attached	0	0	0	4	4					
	Maint Company (-)	Line	4	1	20	54	79	0	0	0	2	24
		attached	0	1	3	35	37					
Guam	1st Company	Line	5	2	30	45	82	15	0	0	1	6
		attached	0	0	0	4	4					
	2nd Company	Line	5	2	41	52	100	3	11	0	1	6
		attached	0	0	0	4	4					
	Det/Maint Company	Line	1	0	6	23	30	0	0	0	0	11
		attached	0	0	2	16	18					
	unassigned	attached	1	0	3	0	5	0	0	0	0	0

In June 1944 the 26th Tank Regiment was embarked on ships for transport to Saipan but was rerouted to Iwo Jima on hearing of Saipan's fall. A month later the 27th Tank Regiment was transferred to Okinawa (less one medium company that was sent to Miyako Jima). Both regiments were destroyed in the island fighting without seriously disrupting the Allied effort.

At the other end of the spectrum were the 15th and 16th Tank Regiments, formed from the 1st and 23rd Division Tank Units in August 1942. Each consisted solely of a headquarters and two small light tank companies. Each of the companies had a strength of 81 men and manned eleven Type 95 light tanks, along with four tracked carriers (replaced by trucks in the 16th Regiment). The regiment HQ had a further five light tanks, while the maintenance company had three more as reserve.

In January 1944 the 15th Tank Regiment was sent to the Nicobars, ending the war there, while the 16th Regiment was sent to Marcus and Wake Islands, being absorbed by the 12th and 13th Independent Mixed Regiments in the 31st Army reorganization of May 1944.

In May 1944 the 1st and 2nd armored divisions each shed one tank regiment, releasing the 9th and 11th Tank Regiments for other duties. The 9th Tank Regiment nominally remained its basic 1,071-man TO&E, but in fact it was stripped down to save shipping before leaving for the Pacific. As deployed, it had an actual strength of 677 organic (line) personnel and 91 attached personnel (medical in the tank companies, technical in the maintenance company) divided into a HQ, a light tank company, four medium tank companies, and a maintenance company.

The basic combat unit was the medium tank platoon, which had a strength of one officer, five sergeants and nine privates with three Type 97 tanks. There were three such platoons in a company, along with a HQ platoon and a maintenance platoon. The former nominally included three officers and two warrant officers and 31 enlisted and was provided with three light tanks for liaison, two medium tanks for command, and a passenger car. The maintenance platoon was composed of one officer and 26 enlisted and was provided with six trucks. Of the eleven medium tanks in the company, nine were basic-model Type 97s, while the remaining two were improved models.

Two of the tank companies were sent to Guam, while the rest of the regiment went to Saipan. All elements were destroyed in the savage fighting that followed.

The 14th Tank Regiment was the army's original light tank unit and fought in Burma from the beginning of that campaign to the end. Originally, it had three companies of Type 95 light tanks, but in 1943 some Type 97 mediums were added. Two more companies were added at the same time, a 4th Company with captured M3 light tanks and a 5th Company with Type 95 lights. The regiment was reorganized in early 1944 to consist of three similar companies, each with one platoon of light tanks (five or six Type 95s), one platoon of captured light tanks (six or seven M3s) and one platoon of medium tanks (two or three Type 97s). By early 1945 actual strength had fallen to seven medium and 11 light tanks, along with a single Type 1 SP 75mm gun possibly sent for operational trials.

A new generation of tank regiments started raising in the late spring of 1944 with the 28th and 29th Tank Regiments from training elements at the Chiba Tank School, the 30th and 33rd Tank Regiments in Manchuria, and the 34th and 35th in China. Two provisional units, the 31st and 32nd Tank Battalions, were also formed at this time.

The ex-Chiba School regiments were initially formed with one company of Type 95 lights, three companies of improved Type 97 mediums, and one company with six Type 1 105mm SP howitzers each. They would be used to form the basis for the new 4th Armored Division.

The regiments raised in Manchuria each consisted of one light tank company and three medium tank companies, with a total of 12 Type 95 light tanks, 20 basic Type 97 medium tanks and 25 improved Type 97 mediums. In 1945 the 30th Regiment was transferred to Chiba and on arrival adopted the 1945 organization of the two regiments already there.

The two regiments raised in China had little need for modern tanks and consisted of one light tank company and three medium tank companies, with a regiment total of 14 Type 95 lights and 36 basic Type 97 mediums. In 1945 one medium company in each regiment was apparently re-equipped with improved Type 97s. In August 1945 these two regiments were moved north to Manchuria where they were destroyed by the advancing Soviet Army.

The two provisional battalions generally followed the old regiment organizational structure with a light tank company (ten Type 95s), three medium tank companies (each two Type 95 lights and 10 Type 97 mediums), and a 5th (trains) company with a total strength of 29 officers, six warrant officers, 13 master sergeants, 114 sergeants and 471 privates.

The final large wave of tank unit mobilization came in April 1945, when the 36th to 48th Tank Regiments were formed in Japan, along with the 49th Battalion in Malaya. The bulk of the homeland regiments were used to form the new independent tank brigades, but two regiments remained separate GHQ units, the 44th on a 556-man TO&E described later and the 46th on the 1,200-man TO&E used by the 4th Armored Division.

In April 1945 four light tank regiments (45th and 47th–49th) were raised using new-model light tanks. Each of these had a strength of only 390 men and was made up of a regiment HQ (three Type 95 light tanks), three tank companies (each ten Type 95 or Type 4 light tanks), and a trains/maintenance company (three light tanks and eight full-tracked APCs). These regiments were retained as GHQ units and assigned to the field armies as needed.

The Tank Brigades

The first nominal armored brigade to be formed was the Armored Operational Training Brigade (*Kyōdō Sensha Ryodan*) raised by the Kwantung Army on 14 June 1942, apparently to develop tactics and train leaders for the new Mechanized Army. With a strength of 198 officers and 4,077 enlisted it was composed of the 23rd and 24th Tank Regiments, and brigade artillery, engineer, signal and maintenance units. It was deactivated on 28 April 1945 and the personnel brought back to Japan as the basis for the 8th Tank Brigade. Although not a fully operational unit, it served as a precursor for similar units later.

The Kwantung Army initiated a new type of combined-arms armored unit with the activation of the 1st Independent Tank Brigade on 11 October 1944. This unit had the following strengths initially authorized:

	Off	WO	MSG	SGT	PVT	Total
Brigade Headquarters	6	0	2	15	43	66
Two Tank Regiments, each	26	6	10	103	504	643
Infantry Unit	23	0	25	53	459	560
Artillery Unit	24	4	9	47	444	528
Engineer Unit	17	4	10	48	270	349
Maintenance Unit	9	2	15	19	244	289
Total	131	22	81	388	2,468	3,090

Each of the two tank regiments had a light tank company and three medium tank companies for a total of fourteen Type 95 light tanks and thirty-six Type 97 medium tanks. In July 1945 each regiment received an additional intendance master sergeant and twenty privates of the technical branch to assist in maintenance. Presumably about the same time one company of the 34th Regiment (and possibly the 35th as well) traded in their old Type 97 mediums for improved Type 97s. Simultaneously the infantry unit had twelve technical personnel added, while the artillery unit and the engineer unit each had ten, again presumably to aid with maintenance.

Following the lead of the Kwantung Army the IJA in the homeland began raising similar independent tank brigades in April 1945. The 2nd through 7th Independent Tank Brigades were activated in Japan on 6 April 1945, followed by the 8th Independent Tank Brigade 22 days later. Aside from the two tank regiments, the only combat element in the brigade was the machine cannon unit, which consisted of three companies, each with only seven trucks and four 20mm Type 98 AA guns.

Table 4.4: Homeland Independent Tank Brigade, 1945

	Off	WO	NCO	PVT	Pistols	Rifles	20mm AA Guns	Light Tanks	Medium Tanks	APCs	Cars	Trucks	Repair Vehs
Brigade HQ	9	0	18	43	15	42	0	2	3	2	2	2	0
Signal Company	3	1	18	117	14	115	0	3	3	0	1	5	0
Machine Cannon Unit	22	6	59	354	16	86	12	0	0	0	4	21	0
Maintenance Unit	7	1	19	175	10	125	0	0	0	0	1	10	5
Transport Unit	17	4	54	302	12	307	0	0	0	0	3	72	3
Two Tank Regiments	composition varied												

There were, inevitably, a few deviations from this pattern. The 4th Tank Brigade apparently used a 551-man anti-tank battalion in lieu of the machine cannon battalion. The 8th Tank Brigade, under the 13th Area Army, had a larger brigade base, consisting of a 25-man HQ, a 205-man signal unit, a 960-man infantry battalion (HQ, 3 rifle companies, MG company and AT company), a 640-man artillery battalion (HQ and 4 batteries), a 469-man engineer unit (HQ and 3 companies), a 279-man maintenance unit and a 247-man transport unit. The 8th Tank Brigade was thus a considerably more balanced formation than the others.

The assignment of tank regiments to the brigades was as follows:

1st Tank Brigade:	34th and 35th Tank Regiments
2nd Tank Brigade:	2nd and 41st Tank Regiments
3rd Tank Brigade:	33rd and 36th Tank Regiments
4th Tank Brigade:	19th and 42nd Tank Regiments
5th Tank Brigade:	18th and 43rd Tank Regiments
6th Tank Brigade:	37th and 40th Tank Regiments
7th Tank Brigade:	38th and 39th Tank Regiments
8th Tank Brigade:	23rd and 24th Tank Regiments
9th Tank Brigade:	51st and 52nd Tank Regiments

For the brigade role the existing 2nd, 18th and 19th Tank Regiments were reorganized onto the TO&E described for those of the 4th Armored Division with two medium companies, two gun companies, an SP company, an infantry/pioneer company, and a maintenance company. The 36th to 40th Tank Regiments were activated with this organization, as was the independent 46th Tank Regiment.

For the first time Japanese tank regiments were equipped with tanks capable of knocking out opposing tanks. This gain, however, was largely offset by the reduced tactical utility of these formations due to decreased mobility. Trucks were scarce, and the tank regiments were not spared in the search for "excess" vehicles. Only the ten trucks in the maintenance company were left, reducing these new type regiments to the quasi-static infantry support role.

Extending this larger organization to all the tank regiments, or even all the new regiments, proved beyond Japan's industrial capabilities, however. The 41st to 43rd Tank Regiments (for brigades) and the independent 44th Tank Regiment were raised on a more modest version of this TO&E, calling only for the HQ, two medium tank companies, a single gun tank company, and the maintenance company, yielding a total strength of 556 men.

The final example of this series, the 9th Tank Brigade, was formed in Manchuria on 10 July 1945. Unlike the others this was simply a hasty grouping of whatever tanks remained in the Kwantung Army's inventory manned by the new 51st and 52nd Tank Regiments formed the same date. Each of the 876-man tank regiments was to be equipped with 18 light tanks and 27 medium tanks, although each actually had 12 Type 95 lights and 30 Type 97 mediums, the 51st Regiment having 720 men and the 52nd 645. The only other unit in the brigade was the 66-man headquarters. It existed only about a month before being destroyed.

Other Armor Units

For operations where a full regiment was not required, the army formed independent tank companies. To strengthen forces on Guadalcanal the 4th Company of the 2nd Tank Regiment was detached and redesignated the 1st Independent Tank Company in early October 1942. Interestingly, the company was not reinforced for its independent operations and retained its earlier organization. It was destroyed on that island.

The 13th Independent Tank Company was raised in November 1943, staying in Japan to the end of the war. This was followed by the 1st to 4th and 7th to 12th Companies in February–March 1944. Since these units were organized out of forces available there was no consistent TO&E but strengths varied only slightly from 110 to 130 men. The 1st and 2nd Companies in the Philippines were the former light tank companies of the 10th and 6th Tank Regiments, respectively, and each had ten Type 95 light tanks. The 13th was also a light unit with nine Type 95s. Others were equipped with left-over obsolete mediums, such as the 7th Independent Tank Company, formed on Leyte in June 1944, and the 8th to 11th Companies formed in Japan in mid-1944 and shipped to Luzon in August which had a 30-man HQ (with a truck and two medium tanks), three 18-man platoons (each three tanks) and a 26-man trains platoon (3 repair and 3 cargo trucks) and held 11 old Type 89 tanks.[4] The 12th Company, also on Luzon, was also identical, but the model of tank is not known. Generally, an independent tank company had three medium tanks in each of its three platoons, one or two more tanks in the company HQ, and a car, two light repair trucks and up to 12 cargo trucks in the trains platoon. Quite apart from their obsolescence, by 1944 the mechanical reliability of the old Type 89s was poor and they contributed little

4 The 10th company was lost at sea *en route*.

Tankettes crossing the Xinjiang River in Jiangxi Province in August 1941.

to any battle they were committed to.

The 5th and 6th Independent Tank Companies were formed in July 1945 for defense of Japan.

Thirteen independent tankette companies were also formed for duty in China in 1937–38. Details of organization varied, but they usually consisted of about 118 men divided into a HQ, four platoons (each four Type 94 tankettes), and a trains platoon with seven trucks. A platoon consisted of the platoon leader, three tankette commanders and four drivers, along with four relief drivers and two helpers, for a total of 14 men. Because these tankette units were to operate independently long distances from other armored units they needed large trains elements and these were often divided into four trains sections (each of about five men) and a maintenance section (of about ten men). A company normally had seventeen tankettes (all but the company commander's vehicle fitted with trailers, although these were often not used), ten Model 94 trucks (including two fitted as repair vehicles) and seven field cars. Additionally, a few tankette companies were split off for general duties when divisional reconnaissance regiments were disbanded in China in 1941–42.

About half of these were disbanded during the war to provide personnel for new tank regiments, and by the end of the war the remaining tankettes in north China had been consolidated into a tankette group operating directly under the North China Area Army. This had 380 men and 66 tankettes in total, organized into an 80-man HQ and six 50-man companies. Each company had ten tankettes in a HQ section, two tankette platoons and a maintenance platoon, with the remaining six tankettes in the group HQ. All the tankettes were still the Type 94 model, no Type 97s having been received, and serviceability suffered due to parts shortages.

5

Non-Divisional Artillery

As in other armies, the Japanese formed non-divisional artillery to support maneuver units and undertake specialized tasks. These included separate field artillery, mountain artillery, medium artillery (also known as heavy field artillery), heavy artillery and artillery mortar regiments, battalions and batteries.

Field Artillery

With a modest manpower pool available the Army initially opted not to form any independent field artillery units, these being duplicative of the divisional artillery already extant. One independent field artillery regiment, the 11th, was raised for China in September 1937, intended to be a temporary expedient. It had 55 officers and 1,642 enlisted, probably organized into two 3-battery battalions of horse-drawn 75mm guns and served on-and-off as part of the 11th IMB until May 1943 when it was ordered incorporated into the 26th Division then on occupation duty in Mongolia.

While the11th was not incorporated into the regular army the appearance of the 75mm Type 90 field gun, with its unique capabilities, brought about a limited change of heart. The Type 90 not only provided long range at low weight, but could also serve as a powerful anti-tank gun. A 1st Independent Field Artillery Battalion with three batteries of these weapons served as part of the mechanized brigade in Manchuria and in March 1938 the 1st Independent Field Artillery Regiment was activated and made part of the permanent force structure.

This did not represent an epiphany for the Army staff, however, for this one nominal regiment was still only a battalion in size. Under the 1941 mobilization plan of October 1940 the unit was built around field guns towed by Type 94 four-ton tractors. Four of the tractors in each battery towed guns and caissons, while the remaining eight each pulled a caisson alone, with each caisson carrying 47 rounds of ammunition. The battalion ("regimental") service battery included 12 ammunition trucks, each with 100 further rounds. The 75mm Type 90 gun was useful, but the tractor requirements proved a drain on Japanese resources, and no further independent units were formed with them. The Regiment served its entire existence with the Kwantung Army (most of it with a fourth battery not envisioned by the mobilization plan) until July 1942, when it was assigned to the Mechanized Army and broken up to provide cadres for the artillery elements of the 1st and 2nd Armored Divisions.

Five more independent field artillery regiments were subsequently formed but these followed the larger pattern of divisional field artillery regiments, rather than the battalion structure of the 1st Regiment, and were "special" not regular formations. The 2nd Independent Field Artillery Regiment was raised in May 1943 and assigned to the China theater where divisional artillery was scarce, where it remained through to the end of the war. This unit had a strength of 1,697 men and 1,212 vehicles and was made up of three battalions, each with two 75mm gun batteries and one 105mm howitzer battery, each with

Table 5.1: Non-Divisional Artillery, from October 1940

	Personnel	Pistols	Carbines	Rifles	Light MGs	Medium MGs	Artillery Weapons	Riding Horses	Pack Horses	Draft Horses	Carts	Wagons	Cars	Trucks	Tractors	Lt Repair Vehs
Artillery Headquarters																
Artillery Headquarters	119	13	31	20	0	0		0	0	0	0	0	6	8	0	0
Ind Field Artillery Regiment																
Regiment Headquarters	156	16	53	160	2	0		0	0	0	0	0	3	18	0	0
Three Field Batteries, each	114	11	29	0	0	0	4 75mm Type 90 Guns	0	0	0	0	0	2	1	12	0
Service Battery	89	7	31	0	0	0		0	0	0	0	0	0	13	0	1
Ind Mountain Artillery Regiment																
Regiment Headquarters	176	12	2	700	1	0		35	59	0	0	0	0	0	0	0
Three Mountain Battalions, each																
Battalion HQ	310	12	9	0	1	0		36	165	0	0	0	0	0	0	0
Three Batteries, each	197	13	0	0	0	0	4 75mm Type 94 Mtn Guns	17	88	0	0	0	0	0	0	0
Service Battery	156	8	0	0	0	0		13	79	0	0	0	0	0	0	0
Regimental Ammunition Trains	387	13	4	0	1	0		26	248	0	0	0	0	0	0	0
Medium Artillery Regiment (A1)																
Regiment Headquarters	134	15	0	400	2	0		0	0	0	0	0	3	8	0	0
Two Medium Battalions, each																
Battalion HQ	147	17	49	0	1	0		0	0	0	0	0	2	20	0	0
Three Batteries, each	148	12	0	0	0	0	4 15cm Type 96 Howitzers	0	0	0	0	0	1	3	11	0

Table 5.1: Non-Divisional Artillery, from October 1940

	Personnel	Pistols	Carbines	Rifles	Light MGs	Medium MGs	Artillery Weapons	Riding Horses	Pack Horses	Draft Horses	Carts	Wagons	Cars	Trucks	Tractors	Lt Repair Vehs
Service Battery	110	9	45	0	0	0		0	0	0	0	0	2	21	0	0
Regimental Ammunition Trains	237	11	100	0	1	0		0	0	0	0	0	3	34	6	3
Medium Artillery Regiment (A2)																
Regiment Headquarters	182	13	1	400	0	0		46	11	23	17	2	0	0	0	0
Two Medium Battalions, each																
Battalion HQ	209	10	5	0	0	0		35	0	90	81	2	0	0	0	0
Three Batteries, each	196	14	0	0	1	0	4 15cm Type 4 Howitzers	43	0	60	0	11	0	0	0	0
Service Battery	154	10	0	0	0	0		32	0	56	0	18	0	0	0	0
Regimental Ammunition Trains	202	15	77	0	0	0		0	0	0	0	0	3	35	0	0
Medium Artillery Regiment (A3)																
Regiment Headquarters	411	13	10	400	0	0		47	7	212	198	2	0	0	0	0
Three Batteries, each	196	14	0	0	1	0	4 15cm Type 4 Howitzers	43	0	60	0	11	0	0	0	0
Service Battery	468	22	0	0	0	0		86	0	185	0	56	0	0	0	0
Medium Artillery Regiment (B)																
Regiment Headquarters	119	11	20	200	0	0		0	0	0	0	0	3	8	0	0
Two Medium Battalions, each																
Battalion HQ	100	10	14	0	0	0		0	0	0	0	0	3	8	0	0
Two Batteries, each	154	13	30	0	1	0	4 10cm Type 92 Guns	0	0	0	0	0	1	4	10	0

Table 5.1: Non-Divisional Artillery, from October 1940

	Personnel	Pistols	Carbines	Rifles	Light MGs	Medium MGs	Artillery Weapons	Riding Horses	Pack Horses	Draft Horses	Carts	Wagons	Cars	Trucks	Tractors	Lt Repair Vehs
Service Battery	75	7	32	0	0	0		0	0	0	0	0	2	15	0	0
Regimental Ammunition Trains	167	9	65	0	1	0		0	0	0	0	0	1	24	0	0
Heavy Artillery Regiment (A)																
Regiment Headquarters	154	2	45	250	0	0		0	0	0	0	0	3	19	0	1
Two Heavy Battalions, each																
Battalion HQ	141	2	47	0	0	0		0	0	0	0	0	2	22	0	0
Two Heavy Batteries, each	171	2	13	0	0	1	2 24cm Type 45 Howitzers	0	0	0	0	0	1	6	17	0
Regimental Ammunition Trains	88	2	0	0	0	0		0	0	0	0	0	0	0	0	0
Heavy Artillery Battalion (C)																
Battalion Headquarters	118	2	23	170	0	0		0	0	0	0	0	3	8	0	0
Three Batteries, each	143	2	5	0	0	1	2 28cm Howitzers	0	0	0	0	0	0	3	4	0
Service Battery	53	2	1	0	0	0		0	0	0	0	0	1	0	0	0
Heavy Artillery Battalion (D1)																
Battalion Headquarters	240	1	36	150	0	0		0	0	0	0	0	2	48	0	0
Two Batteries, each	269	1	83	0	0	1	2 30cm Type 7 Short Howitzers	0	0	0	0	0	1	17	37	0
Service Battery	53	1	35	0	0	0		0	0	0	0	0	2	5	2	0
Heavy Artillery Battalion (D2)																
Battalion Headquarters	170	2	43	150	0	0		0	0	0	0	0	3	18	0	0

Table 5.1: Non-Divisional Artillery, from October 1940

	Personnel	Pistols	Carbines	Rifles	Light MGs	Medium MGs	Artillery Weapons	Riding Horses	Pack Horses	Draft Horses	Carts	Wagons	Cars	Trucks	Tractors	Lt Repair Vehs
Two Batteries, each	223	2	15	0	0	1	2 30cm Type 7 Long Howitzers	0	0	0	0	0	1	7	14	0
Heavy Artillery Battalion (E)																
Battalion Headquarters	164	1	71	70	0	0		0	0	0	0	0	3	28	0	0
Two Batteries, each	179	1	53	0	0	1	4 15cm Type 89 Guns	0	0	0	0	0	2	17	8	0
Service Battery	98	1	41	0	0	0		0	0	0	0	0	1	5	10	2
Artillery Mortar Battalion																
Battalion Headquarters	35	2	6	190	0	0		0	0	0	0	0	1	6	0	0
Two Batteries, each	296	2	0	0	0	0	4 25cm Type 98 Mortars	0	0	0	0	0	0	0	0	0
Service Battery	21	2	10	0	0	0	(8 reserve mortars)	0	0	0	0	0	0	4	0	0

Note: all rifles are undistributed weapons for self-defense.

four weapons.[1]

For the defense of the homeland the 3rd–7th Independent Field Artillery Regiments were formed in February 1945. Each of these had the following organization:

	Men	Horses
Regiment Headquarters	123	60
Three Field Artillery Battalions, each		
Battalion Headquarters	146	93
Three Batteries, each	137	111
Service Battery	167	47
Regiment Trains	158	103
Total	2,135	1,612

Each of the artillery batteries was provided with four pieces, 75mm guns in two batteries per battalion and 105mm howitzers in one. The 6th Independent Field Artillery Regiment was assigned to the 58th Army in June, but the other regiments of this group were quickly used to form divisional artillery components for 200-series divisions. The 8th–10th Independent Field Artillery Regiments were activated in the homeland to take their place in June, although they had not completed forming by war's end. These had the same structure as the earlier block of regiments, with each battalion consisting of two 75mm batteries and one 105mm howitzer battery, but with a slightly reduced allocation of manpower and horses:

	Men	Horses
Regiment Headquarters	120	60
Three Field Artillery Battalions, each	530	290
Regiment Trains	152	100
Total	1,862	1,030

Also outside the mobilization plan structure were a small number of independent field artillery battalions. The 1st Independent Field Artillery Battalion was raised in late 1943 by the Kwantung Army and dispatched south to the Nicobar islands. It had an authorized strength of 23 officers, 108 NCOs and 368 privates. The 5th Battalion was raised in Japan with a strength of 528 men in June 1944 and shipped to the Formosa Army. The 6th Battalion was raised on Luzon in mid-1944. Somewhat earlier, a block of field artillery battalions were raised by the Kwantung Army for service in China. These were the 7th, 8th, 9th and 10th Battalions, supplemented by the 11th Battalion locally raised by the North China Area Army. HQ, China Expeditionary Forces added another battalion, the 29th, in March 1945. Two of these (10th and 11th) were transferred back to the Kwantung Army in May 1945 to help form new divisions. Two others (7th and 8th) were also absorbed into larger units, this time in China. By the surrender only two independent field artillery battalions, the 9th and 29th, remained in China. These had authorized strengths of 617 men with 476 horses, 205 rifles and twelve 105mm howitzers.

Two independent field artillery battalions, 13th and 14th, were raised in Manchuria

1 During the Hunan operations of May–June 1944 the regiment temporarily operated as a mountain artillery unit, fielding two 75mm mountain guns per battery.

by 3rd Area Army. The 13th Battalion was formed by splitting off one battalion from the 23rd Division's 13th Field Artillery Regiment, the only motorized divisional artillery regiment, on 5 October 1944. The battalion was streamlined, in apparent anticipation of a move overseas, consisting of a headquarters, three batteries and a service battery with a total strength of 24 officers and only about 275 men. Each battery manned four 75mm Type 90 field guns pulled by tractors, and two trucks. The Battalion was shipped to the Philippines in November 1944 and was destroyed there in January 1945 by US artillery fire. The 14th was absorbed in the creation of a new division in Manchuria later in 1945. One more was raised in Singapore, the 25th in April 1945.

A number of battalions were also activated for the defense of the homeland. The 12th Independent Field Artillery Battalion had 415 men in three batteries to man twelve 75mm Type 90 guns. The 28th Battalion had 639 men with a car and 36 trucks to move its twelve pieces, while the 31st–34th Battalions each had 522 men. Of these, only the 31st was a horse-drawn unit with 260 horses for twelve 105mm howitzers, while the 34th had truck-drawn 105mm howitzers and the others truck-drawn 75mm field guns. Two more battalions, the 36th (820 men with 590 horses) and the 37th (650 men with 450 horses) were activated for the 5th Area Army on 16 July.

Mountain Artillery

If the Army was cool towards non-divisional field artillery that was not so as regards mountain, or pack, artillery. These added capability and flexibility by providing fire support in marginal terrain using pack horses exclusively for transport. Under the October 1940 mobilization plan each battery was provided with 30 pack horses to carry the 75mm Type 94 guns and equipment, and 48 more each carrying 10 rounds of HE ammunition. The battalion service battery included 72 more ammunition-carrying pack horses, and the regimental ammunition trains a further 192.

The Army's plans called for ten such regiments, but only the 1st to 3rd Independent Mountain Artillery Regiments were actually activated in August–October 1937, followed by the 10th in September 1938 for service in China during 1938–40; while the 4th and 12th Regiments served under the Kwantung Army. Of these, the 1st Regiment was ordered home and deactivated in September 1940. Its place in the army force structure was taken by the two-battalion 4th Regiment, which was formed in March 1939 then mobilized and transferred in July 1941 to the Kwantung Army. Two of the independent mountain artillery regiments participated in the subsequent drive to the south. The 3rd Regiment was transferred to the 25th Army (invasion of Malaya) in November 1941, and then to the 14th Army (Philippines) in January 1942 before being shipped to the Kwantung Army in May. The 10th Regiment was assigned to the 17th Army based at Rabaul in September 1942, and was itself stationed on Bougainville until it was absorbed by other units in mid-1944.

To replace the lost 10th Regiment the China theater raised the 5th Independent Mountain Artillery Regiment in June 1943, while the Kwantung Army, lost its 4th Regiment to the south Pacific in March 1944. Horses were useless in the jungles of New Guinea so they, along with their associated personnel, were eliminated from the 4th Regiment before it departed Manchuria. This reduced battery strength to 108 men and eliminated the battalion service batteries and the regimental ammunition trains. A few trucks were added for those rare instances where roads existed, but for the most part the battery personnel had to hand-carry their Type 94 guns, along with 60 rounds per gun, through the jungle undergrowth. Their deployment to the relatively small (but doomed) Noemfoor Island may have either contributed to the decision to leave the horses behind, or resulted from it. Signals equipment consisted simply of eight telephones and 12km of wire in each battery, along with one radio in each battalion HQ and the regimental HQ.

The 3rd Regiment, the most widely-travelled of the independent mountain artillery regiments, was reduced to a single battalion and transferred to the 31st Army as part of the 1st Expeditionary Unit in 1944. The regiment then had the misfortune of being shipped to Saipan, where its 1st Battalion formed the 47th IMB's artillery unit and the remainder of the regiment, including crews for a dozen 15cm Type 4 howitzers, served as a central artillery reserve.

Additional regiments were formed late in the war for the defense of the homeland. The 6th Independent Mountain Artillery Regiment was activated on 6 February 1945 with a nominal strength of almost 3,800 men, but with a reduction of almost 900 men to yield an authorized strength of slightly under 3,000 men as follows:

	Off	WO	MSG	SGT	PVT	Horses
TO&E strength	94	13	38	187	3,459	2,178
Reduction	-11	-1	-5	-18	-849	-1,420

The manpower reduction was restored on 23 May, bringing the unit up to full TO&E strength, but only 518 of the horses were reinstated, giving a total of 1,276. The regiment was divided into three battalions and was armed with twenty-four 75mm mountain guns and twelve 105mm mountain howitzers.

Shortly thereafter the 7th–11th Independent Mountain Artillery Regiments were activated on a more modest basis with two battalions, each with three 4-gun batteries of 75mm Type 94 mountain guns:

	Men	Horses
Regiment Headquarters	175	104
Two Mountain Artillery Battalions, each	1,005	565
Regiment Trains	311	208
Total	2,496	1,442

On 2 April, however, all but the 8th Regiment were redesignated and reassigned as divisional artillery units for the new 200-series divisions, reducing the GHQ pool in the homeland to two regiments.

As soon as weapons and equipment came available a new series of independent mountain artillery regiments were activated in the homeland, the 13th, 18th and 20th in June; the 12th, 14th, 15th, 16th and 19th in July; and the 17th in August. These reverted back to the 3-battalion structure of the 6th Regiment but with a leaner allocation of manpower:

	Men	Horses
Regiment Headquarters	136	83
Three Mountain Artillery Battalions, each	850	340
Regiment Trains	295	200
Total	2,981	1,303

The regiments were notable for their mixed complement of weapons. Eight of the nine batteries in each regiment were normal mountain units each with four Type 94 mountain guns, but one battery in each regiment was provided with three 105mm mountain howitzers instead. They were in varying states of readiness and equipment at the war's end.

In addition to these regiments, the various theaters formed independent mountain artillery battalions not envisioned by the mobilization plans. The 20th and 21st Independent Mountain Artillery Battalions were formed in China in 1939. The 20th, at least, was initially formed with mortars as its armament. The 21st was demobilized in December 1940, but the 20th Battalion re-equipped with captured Chinese Bofors mountain guns for the operation against Hong Kong in December 1941, then apparently re-equipped again with Type 94s in May 1942, when it was transferred to the 17th Army and shipped to Bougainville, where it ended the war.

These were followed by the 51st and 52nd Independent Mountain Artillery Battalions, formed in Japan for service in China in September 1940. The bulk of these two battalions were subsequently transferred to the 16th Army occupying the Dutch East Indies and were officially deactivated in November 1943 to provide the artillery units for the 27th and 28th IMBs, respectively. A portion of the personnel remained behind, however, and they were reactivated by the China theater in November 1944, although on a reduced scale. The China theater raised two more mountain artillery battalions, the 1st Battalion in February 1944 and the 24th Battalion in February 1945.

At the time of the surrender there were still four independent mountain artillery battalions serving in China. The 1st and 24th Battalions, formed in early 1944, each had a strength of 991 men with 558 horses and twelve 75mm mountain guns. The 51st and 52nd Battalions were much smaller units, composed of only two gun batteries and no service battery, consisting of 315 men with 169 horses and six 75mm mountain guns.

Only two independent mountain artillery battalions were formed in the homeland for late-war defense, the 22nd and 23rd. These unusual units had a strength of 415 men in three batteries one each with four 75mm Type 94 mountain guns, four 105mm Type 99 mountain howitzers, and four 105mm Type 91 field howitzers.

Medium Artillery

Heavy fire support at the corps level was to be provided by the medium artillery regiments.[2] What should have been a fairly potent force in fact contributed little. This was due to poor tactical communications, unfavorable terrain in the active theaters, and massive US and Soviet superiority in firepower and air support.

These regiments came in two main types, a Type A with 150mm howitzers, either horse-drawn or tractor-drawn, and a Type B with tractor-drawn 105mm guns. The permanent force through most of the 1930s consisted of two regiments of 105mm tractor-drawn guns and six of 150mm howitzers, initially all horse-drawn, but slowly converting to the motorized configuration as Type 96 howitzers came available.

The gun (Type B) regiment consisted of a 62-man headquarters, a combat trains of about 150 men, and two artillery battalions. A battalion was made up of a headquarters, a combat trains, a field trains and two batteries. The battalion HQ included a signal platoon, a platoon leader, a radio section of 7 men and a wire/optical section of 13 men, along with

2 In Japanese, field heavy artillery regiments (*Yasen juhōhei rentai*).

a 4-man team for placing airplane signal panels. It also included a reconnaissance officer, a liaison officer, and 13 instrument operators, 3 of them sergeants, to operate the liaison and observation post. The combat trains had three spare tractors and nine cargo trucks each carrying 80 rounds of 105mm ammunition. The field trains probably had about 46 men with six cargo trucks for fuel, rations and baggage.

Table 5.2: Medium Artillery Battalion (105mm Gun), 1935

	Officers	NCOs	Other Ranks	Pistols	Rifles	105mm Guns	Sidecar Motorcycles	Passenger cars	Reconnaissance cars	Cargo Trucks	Tractors
Battalion Headquarters	8	17	41	18	0	0	6	2	2	0	0
Combat Trains*	2	6	73	8	16	0	1	0	0	10	3
Field Trains	composition unknown										
Two Artillery Batteries, each											
Battery Headquarters	2	4	23	7	**16	0	4	1	1	0	0
Two Firing Platoons, each	1	2	24	3	0	2	0	0	0	0	2
Ammunition Platoon	0	1	18	1	0	0	0	0	0	0	2
Combat Trains*	1	1	43	2	0	0	1	0	0	6	0
** compositon approximate*											
*** for distribution within battery*											

The gun battery was commanded through a headquarters that included a battery detail to operate forward in a liaison/observation post, comprised of the reconnaissance officer, a nine-man signal section, five instrument operators, and drivers for three motorcycles and a recon car. The battery detail was provided with a large 6x4 truck that carried three aiming circles, two battery commander's scopes, a one-meter rangefinder, a transit, a tripod-mounted telescope tables, binoculars, telephones, and 22 reels with 11km of wire.

Each firing platoon had a platoon leader and two gun sections, each of 13 men with a gun with limber and a tractor. The firing platoon carried no ammunition, the forward carriage of which was entrusted to the ammunition platoon with two tractors towing six caissons with a total of 144 rounds. The combat trains carried a further 480 rounds in six cargo trucks. In the late 1930s an old Type (Taisho) 11 light machine gun team was added to each battery for local and air defense of the command post, two tractors were added to reduce the number of caissons per tractor to two, and two more tractors added as spares.

The 105mm gun regiment was motorized in the sense that all horses had been replaced by motor vehicles, but the tractors were slow-moving and many of the troops had to walk, reducing the speed of the unit to that of a walking man unless additional transport was provided, usually for administrative moves.

The other medium artillery unit was the 150mm howitzer regiment, all horse-drawn until the introduction of the Type 96 howitzer in 1937. A gun section consisted of a

A 105mm gun battery preparing to move out in 1940. At the left, a tractor with two caissons for ammunition, to the right a tractor with gun and limber, to the right of that the 7-passenger car for the commander, and to the right of that the battery detail truck.

sergeant, a gunner, six drivers and ten cannoneers with their Type (Taisho) 4 howitzer and two limbers, necessary since the howitzer broke down into two parts, each with six draft horses, for transport. Each limber carried four rounds of ready ammunition. Two sections made up a platoon, and two platoons formed the heart of a howitzer battery. The battery detail provided the forward observation/liaison post and the communications between it and the battery firing position. This included instrument and signal sergeants, five mounted telephone operators, four mounted instrument operators, and 20 dismounted signal and instrument men, and it was provided with a Type 90 light observation two-wheel cart and associated limber. The cart carried 21km of wire in 42 spools, the battery commander's periscope, two aiming circles and a battery-operated morse-code signal lamp. The limber carried the one-meter optical rangefinder (graduated, optimistically, out to 10km), a tripod-mounted telescope for triangulation, telephones, map boards, a ladder, etc. The battery also included an ammunition platoon with four squads, each with two caissons, each carrying 12 rounds. The battery combat trains had eight more caissons for ammunition plus two store carts with limbers for pioneer tools and camouflage equipment, one for the battery detail and one for the firing position. In the late 1930s a light machine gun team was added, as were some medical personnel.

Table 5.3: 150mm Howitzer Battalion, Horse-Drawn, 1936

	Officers	NCOs	Other Ranks	Rifles	150mm Howitzers	Riding Horses	Pack Horses	Draft Horses	Limbers	Caissons	Carts & Wagons
Battalion Headquarters											
Command Group	5	6	5	0	0	16	0	0	0	0	0
Forward Group	4	6	65	16	0	20	2	12	0	0	2
Combat Trains											
Trains HQ	1	0	1	0	0	2	0	0	0	0	0
3 Ammunition Platoons, each	1	2	64	0	0	11	0	48	0	16	0
Reserve Section	0	1	22	0	0	10	0	24	0	0	2
Field Trains											
Trains HQ	0	1	1	0	0	2	0	0	0	0	0
7 Sections, each	0	0	16	1	0	1	0	10	0	0	10
Three Batteries, each											
Battery HQ	1	2	3	40*	0	5	0	0	0	0	0
Battery Detail	1	2	33	0	0	13	0	6	0	0	1
2 Gun Platoons, each	1	2	34	0	2	5	0	24	4	0	0
Ammunition Platoon	0	1	32	0	0	5	0	24	0	8	0
Combat Trains	1	2	48	0	0	12	0	40	0	8	2

**for distribution within battery*

The 150mm howitzer battalion provided a combat trains with 48 more caissons, each with 12 rounds of ammunition,[3] and a field trains with 70 store carts with limbers for baggage, rations, etc. The battalion also had a forward group, analogous to the battery detail, that operated an observation/liaison post, but this time including a 5-man radio team and a 9-man section for placing aerial observation panels. The horse-drawn 150mm regiment had a headquarters and a combat train with 48 caissons for ammunition and 20 transport carts for rations, baggage, etc., along with the two howitzer battalions.

With the introduction of the 150mm Type 96 tractor-drawn howitzer a portion of the howitzer regiments were gradually converted to that weapon and adopted an organization approximating that of the 105mm gun regiments, except that they retained their structure of three batteries per battalion, with attendant increases in the size of the battalion and regimental trains. Whether horse-drawn or tractor-drawn, however, the medium regiments tended to be rather plodding units, rarely moving much faster than a man could walk. If required, the motorized battery could move 150km in one 14-hour day, a slight improvement. Nor would they have been terribly responsive, the fire from each battery being controlled by that battery's forward detail, with communications between the forward and firing elements provided entirely by telephone.

If the plan had been to replace most of the older medium pieces with the new family of

3 Over time the combat trains apparently shrank; in August 1944 that of the I Battalion, 14th Medium Artillery had 130 men with 12 six-horse wagons, each holding 24 rounds.

A horsed 150mm howitzer unit. Top left, shifting the trails; top right, a stores cart with limber; lower left the 1-meter rangefinder; lower right the Type 90 observation cart with limber.

weapons that was thwarted. The campaigning in China forced an expansion of the arm that resulted in the new weapons supplementing, rather than supplanting, the older ones. As a result by the time the 1941 mobilization plan was issued, in October 1940, the regular force had expanded to no fewer than 20 nominal regiments, still in the two basic types, armed with 150mm howitzers (Type A) or 105mm guns (Type B). There were further variations within the types based on means of transport and unit configuration. The mobilization plan allotted the organizations as follows:

Type A1: 1st, 3rd, 4th, 17th & 20th Regiments
Type A2: 2nd, 5th, 6th, 9th, 10th, 11th, 12th, 13th & 14th Regiments
Type A3: 15th Regiment
Type B: 7th, 8th, 16th, 18th & 19th Regiments

Even within these neat groupings, however, there were variations caused by the initial retention of older weapons. The 11th and 12th (A2) Regiments used the 150mm Type (Meiji) 38 Howitzer through to the end of the war, and the 13th and 14th Regiments (A2) used the 120mm Type (Meiji) 38 Howitzer in lieu of the intended 150mm Type 4 howitzers. Similarly, the 16th and 19th (B) Regiments used 105mm Type (Taisho) 14 guns instead of the called-for Type 92 guns. These changes in equipment, however, did not alter the manpower or other authorizations of the units.

In the motorized 150mm howitzer regiments (type A1) each battery was provided with four Type 96 150mm howitzers, twelve caissons (each carrying 12 rounds of HE ammunition) and eleven Type 98 six-ton tractors. The battalion service battery carried a further 504 rounds of HE ammunition, while the regimental ammunition trains carried 532 more. For observation posts, each battery included a light truck fitted as an observation vehicle, while the battalion and regimental HQs each had two more.

The firing platoon of the horsed (A2) regiment had two Type (Taisho) 4 howitzers and five squads (two gun squads per gun to provide 24-hour manning, and one ammunition/ spares squad). Each howitzer was provided with a two limbers each carrying 4 rounds of ammunition and was towed by six horses. The platoon's fifth squad was provided with a stores wagon to carry spare parts and two more caissons for extra ammunition (this in lieu of the ammunition platoon formerly provided). The battery combat trains had a further four caissons, for a battery total of eight. The total authorized basic load for the battery was 192 rounds, but it would seem that not all of this could be carried in one lift. The battalion service battery was provided with a further 17 caissons, although again these could not have carried the full authorized basic load of 408 rounds. The regimental ammunition trains, now motorized, carried 792 rounds of ammunition in 35 trucks. The command and control network for the regiment was provided by a Type 90 light observation wagon in each firing battery, and a light and a heavy observation wagon (also horse-drawn) in each battalion/regiment HQ.

The A3 type medium artillery regiment was a derivative of the A2 horse-drawn type but was, in fact, simply a separate battalion. The battalion service battery/ammunition trains was authorized 1,320 rounds of ammunition as its basic load, along with 55 caissons (sufficient to carry half that). The only unit to organize on these lines was the 15th Regiment.

The B-type medium artillery regiments used 105mm guns, Type 92 (18,200 meter range) or Type 14 (15,300 meter range), primarily for long-range interdiction and counter-battery roles. All the B-type regiments were motorized formations, using the Type 92 five-ton tractor as a prime mover.

A four-gun battery was provided with ten tractors and 12 caissons and had a basic load of 144 Type 91 "long" HE rounds and 144 Type 14 HE rounds. These holdings were supplemented by a further 180 rounds of each type held by the battalion service battery and 360 of each by the regimental ammunition trains. Not being intended for massed fire, a B-type medium artillery regiment was organized as only two battalions each of two batteries.

The 1st through 8th Medium Artillery Regiments were part of the permanent peace-time force structure of the army although, like the rest of the army, they were usually maintained at cadre strength only. The remaining twelve regiments specified by the mobilization plans were initially represented by little more than equipment stocks held in depots in anticipation of mobilization.

In July 1937 the 2nd, 3rd, 5th and 6th Medium Artillery Regiments were mobilized from their depot status, followed in August by the 10th and 14th, then the 12th in September. All were 150mm howitzer units, one a motorized formation with Type 96s, the others using horse-drawn Type 4s. In August 1938 the small 15th Regiment was mobilized, then the 1st and 7th in June 1939, and the 17th and 18th Regiments in November 1939. These five were all motorized units, the 7th and 18th with 105mm Type 92 guns, the others with 15cm howitzers.

The medium regiments participated to a great extent in the early campaigns in China, serving under medium artillery brigades. The 1st Medium Artillery Brigade (organized 1922 and mobilized August 1936) commanded the 2nd and 3rd Regiments, the 2nd

Brigade (organized 1922 and mobilized March 1937) the 5th and 6th Regiments, the 5th Brigade (1936) the 11th and 12th Regiments, and the 6th Brigade (September 1937) the 13th and 14th Regiments. In addition, the 9th Regiment served independently in China.

Surprisingly, the situation was the complete opposite in Manchuria, where the potential foe was the well-equipped Soviet Army. Through mid-1939 there were few medium artillery regiments in Manchuria, only the 3rd Medium Artillery Brigade with two regular regiments rotating in and out. With the failure of the first Japanese attacks in the Nomonhan area, however, a call went out for artillery support and the 3rd Medium Artillery Brigade with two regiments received its full mobilization orders on 24 June 1939.

The 3rd Medium Artillery Brigade was mobilized with the two medium artillery regiments generally considered the best in the army: the 1st (150mm motorized) and 7th (105mm motorized). The brigade hurried over, arriving on the battlefield on 17–18 July, but in spite of the elite personnel of these units, their performance did not match expectations. This was due in part to chronic ammunition shortages (possibly reflecting lower efficiency against targets than initial plans foresaw) and open, treeless terrain that left them exposed, but also to the poor infantry–artillery liaison that was to plague Japanese forces throughout the war.

In 1940 the policy of rotating regular army units between Manchuria and the homeland ceased and instead the 1st, 4th, 7th, 9th, 20th and 22nd Medium Artillery Regiments became a permanent part of the Kwantung Army. The 20th and 22nd were new units, formed in Manchuria in 1939–40.

The medium artillery in China, on the other hand, was apparently regarded as superfluous in light of the ill-armed opposition and in late 1939 the 2nd, 3rd, 5th, 8th, 10th, 11th, 12th, 13th and 14th Regiments were brought home and returned to their parent commands. There they were demobilized. The 14th Regiment remained in China through the war.

This reduction in strength, however, was not to last long. With war appearing imminent, IGHQ ordered on 11 July 1941 that "to improve security against the USSR" the 5th, 8th, 17th and 18th Medium Artillery Regiments were to be dispatched to the Kwantung Army, mobilizing those not already mobilized. The next month the 2nd, 4th, 9th, 10th, 12th, and 20th Medium Artillery Regiments, along with a new 22nd (150mm) Medium Artillery Regiment were sent to Manchuria as well. In addition, the diminutive 15th Medium Artillery Regiment was activated on 15 July staying in Korea through the war. The 13th Regiment was also reactivated in September, but with only one battalion and never left the homeland.

With the new direction to the war the medium artillery regiments moved to the south with the Japanese invasion forces. The 1st and 8th Regiments went to the Philippines, the 3rd and 18th Regiments to Malaya, while the 17th Regiment participated in the Java campaign. The regiments were not tapped for garrison duty once the objectives had been seized. The 1st and 17th Regiments were returned to Manchuria, the 3rd and 18th Regiments were sent on to Burma, and the 8th, and 16th Regiments were returned to Japan for demobilization.

Except for the regiments in Burma and China the medium artillery remained quiet until September 1942. At that time the Kwantung Army was directed to transfer the 4th and 7th Regiments to the 17th Army on New Britain, where both regiments finished the war. To partially compensate the Kwantung Army the 23rd Regiment (an A-1 style unit with 150mm Type 96 howitzers) was activated there in March 1942.

Thus, during the pivotal year of 1943 the Kwantung Army held about half the medium artillery regiments in the army: the 1st, 2nd, 5th, 9th, 10th, 12th, 20th and 23rd, all 150mm units. The bulk of these were horse-drawn units, although on 15 July 1943 the 10th Regiment was converted to the A-1 type organization. The China theater held two more: the 6th and 14th, both horsed 150mm units.[4] The 3rd and 18th Regiments were fighting in Burma, but inexorably losing strength in the protracted combat, while the second battalion of the 3rd Regiment was the one medium artillery unit to perform garrison duty, in the Dutch East Indies. The 5th Regiment arrived in Burma in October 1944, having left its horses in Manchuria and picking up some trucks on arrival. The 4th and 7th Regiments were around Rabaul awaiting an invasion that never came. The 7th Regiment had an unusual organization in the south Pacific, consisting of 800 men manning five 105mm guns and seven 150mm howitzers, undoubtedly a product of extemporization.

As the Allied drives picked up steam the Japanese threw a few of their medium artillery regiments into the fray. In March 1944 the I Battalion 9th Medium Artillery (with 12 15cm Type (Taisho) 4 howitzers) was shipped from Pusan to Saipan where it was absorbed by the 1st Expeditionary Unit. In June the 1st Regiment was transferred to the 32nd Army, followed a month later by the 23rd Regiment, each with two battalions of 150mm Type 96 howitzers. The 1st Regiment sent its 1st Battalion on to Miyako Jima but the remainder of the two regiments were destroyed on Okinawa. Neither regiment landed with its full complement of gun tractors, the I/1 battalion having only seven 6-ton Type 98 tractors, while the 23rd Regiment had a total of twenty-six 6-ton Type 96 tractors. As with other types of artillery, the medium artillery proved ineffective on Okinawa. Fire was almost never massed, almost all firing being done by individual batteries, and communications proved so poor that timely shifting of fire proved almost impossible. The 12th Regiment moved from Manchuria to Luzon in September 1944, its two battalions still equipped with 15cm Type (Meiji) 38 howitzers, fired on US forces early in 1945 and then abandoned its pieces in February when their horses were killed and served as infantry in the hills. The 22nd Regiment was transferred to Luzon and the 5th Regiment to Burma at about the same time, again contributing little.

These failed to stem the tide and IGHQ began preparing for an invasion of the homeland. The 2nd and 10th Medium Artillery Regiments were returned to Japan from Manchuria in March 1945.

At the same time the 8th (Type B), 11th (A-2) and 19th (B) Medium Artillery Regiments were reactivated in Japan in their original configurations. These were supplemented by the 26th, 27th, 28th and 29th Regiments activated in April 1945 and the 30th, 52nd, 53rd and 54th Regiments activated in July, all in the homeland. One, the 30th, was a Type B unit with 105mm guns, two, the 28th and 29th, were horse-drawn A-2 units with old Type 4 howitzers, the others were motorized A-1 with Type 96 weapons.

In any event the homeland regiments were understrength at the end of the war, at least some of the shortages had to be made up by impressing civilians into duty for the less-martial roles.

Manchuria re-activated the 17th Regiment (A-1 150mm howitzer) in early 1945, followed shortly by the new 30th Regiment (105mm gun) on 31 July.

The army also formed a small number of independent medium artillery battalions.

4 The 14th Regiment had re-equipped from its old 12cm howitzers to 15cm Type (Taisho) 4 models before moving to China in early 1942.

The most active was the 21st Battalion, raised in July 1940, which served in China during 1941 before being transferred to the 25th Army for the invasion of Malaya. In March 1942 the battalion (less one battery) was sent to Burma. The detached battery was sent to the 17th Army in mid-1942, but returned to its parent unit in March 1943. This was apparently the only separate medium artillery battalion until March 1943 when the 5th Battalion was formed in Rabaul by redesignation of the artillery unit of the 21st IMB. The battalion consisted of one 3-gun battery of 105mm Type 92 guns, one battery with four 150mm Type 96 howitzers and one transport battery with ten trucks. The battalion was dispersed following losses at sea shortly thereafter. In July 1944 the 6th and 7th Battalions were activated for duty in China. The 6th had a strength of 620 men in a HQ, two firing batteries and a service battery. It was a motorized unit with 78 trucks, tractors and cars, and was armed with 288 rifles and eight 150mm howitzers. The 7th was a horse-drawn unit with 695 men had 765 horses.

The 9th and 11th Medium Artillery Battalions were activated in Burma in February 1945, out of the remnants of the 18th and 3rd Medium Artillery Regiments, respectively. The 12th Medium Artillery Battalion was activated on Timor at the same time. Four more independent medium artillery battalions, the 16th–20th, were raised in Japan in 1945 for defense of the homeland. Each consisted of 20 officers, 6 warrant officers, 7 master sergeants, 54 sergeants and 613 privates in a headquarters, three batteries, and a service battery, and each was provided with twelve 150mm tractor-drawn howitzers, Type 96 in the 16th and 17th Battalions, Type (Meiji) 38 in the 18th Battalion, and Type (Taisho) 4 in the 19th and 20th Battalions.

Coastal and Heavy Artillery

During the period up to the Russo-Japanese War the coastal artillery had occupied a very important and prestigious niche in the IJA. During that war their weapons proved extremely useful in the siege role as well and led to the coast artillery being divided into two components: the fortress (or seacoast) artillery with its weapons emplaced in permanent fortifications guarding harbor entrances and other strategic water sites, and the heavy artillery with similar weapons mounted on (barely) mobile carriages and sited in improved, but not permanent, firing positions, also along the coast.

During the period up to and through the First World War a new generation of weapons was produced for the coastal artillery, including the 24cm Type 45 (1912) howitzer, the 30cm Type 7 (1918) howitzer in two models, the 10cm Type 7 (1918) gun, and the 15cm Type 45 (1912) gun. At this point, however, stagnation set in. A new generation of weapons was planned for the 1920s, but was abandoned due to the availability of weapons from vessels scrapped or stopped under the Washington Naval Treaty. Thus only one new coastal gun, the 150mm Type 96 (1936) was developed and placed into production during the inter-war period. Similarly, no new fire control systems were developed, nor was there any major expansion of the branch.

The peacetime structure of the army prior to WWI provided for 13 coast artillery regiments, comprising six in Japan, two each in Formosa and Korea, and four in Manchuria. Aside from those in Manchuria, each of these consisted of one fortress (coastal) artillery battalion and two heavy artillery battalions. The heavy and fortress artillery were separated and in 1920 the heavy artillery was defined as five regiments and six battalions, although two of the regiments were downgraded to battalions in 1922. Thus, through the 1930s the

Troops of the 1st Heavy Artillery Regiment in the Philippines 1942.

heavy artillery consisted of three regiments and eight (later nine) separate battalions, with a total of 34 batteries.

The dominant characteristic of the heavy artillery units, in addition to the size of weapons, was their relative immobility. Only the 150mm Type 89 (1929) gun was a mobile unit using a conventional split trail design, but even this had to be disassembled into two loads for transport and took a long time to bring into action compared to similar weapons in service elsewhere. The newer Type 96 (1936) field gun added 8,000 meters to the Type 89's range of 18,100 meters, but was a non-mobile pedestal-mounted weapon. The other main weapons were the 240mm Type (Meiji) 45 howitzer (16,000 meters), an antique 280mm howitzer that entered service in 1887 (7,900 meters), and the 305mm Type (Taisho) 7 howitzer in short-barrel (11,700 meters) and long-barrel (15,000 meters) configurations. All of these had to be broken down into four or more loads for transport and took such a long time to reassemble and emplace that they were only suitable for siege or static defense roles.

The permanent force structure of the army as modified by the 1941 mobilization plan retained the three numbered heavy artillery regiments, the 1st, 2nd and 3rd, all armed with the 240mm Type 45 howitzer. These were supplemented by two special heavy artillery regiments raised by the Kwantung Army, the 2,191-man Mutanchiang and 1,548-man Tungning Heavy Artillery Regiments, also with 240mm howitzers.

Regiments in name, these units were actually battalions in firepower, having a total of eight 240mm weapons in two battalions. Each battery was provided with 17 heavy artillery prime movers and two 10-ton cranes to move and erect the guns. No ammunition was directly authorized for stockage at any level in the regiment, instead the unit relied on direct supply from the rear.

The 1st and a new 5th Regiments were mobilized to war strength in July 1940, followed by the 2nd, 3rd and the two named regiments a year later. Aside from a brief diversion of the 1st Heavy Artillery Regiment to the attack on Hong Kong in late 1941 and the siege of Bataan and Corregidor in early 1942, the five regular heavy artillery regiments spent the entire war in Manchuria.

Although the history of these regiments is simple, the issue is clouded by the raising of a small number of further heavy artillery regiments late in the war that were nothing of the sort. As fortresses were converted to field units their constituent "fortress heavy artillery regiments" were also renamed, simply adding a number and dropping the word "fortress" from the title.[5] In May 1944 the Amami Oshima Fortress HAR was redesignated the 6th HAR, the Nakagusku Wan Fortress HAR (on Okinawa) became the 7th HAR, and the Pescadores Fortress HAR became the 8th HAR. Substantial portions of the fortresses were used to form mixed brigades, and this was true of many of the fortress artillery elements as well. Thus, the 8th (former Pescadores) HAR regiment was reduced to a strength of 302 line personnel, 11 technical and 6 medical personnel. The 7th Regiment on Okinawa was configured as three batteries each with four 75mm Type 38 field guns, plus a further two 120mm guns attached by the Navy, for a total strength of 499 line, 18 technical and 6 medical personnel; hardly either "heavy" or a "regiment" of artillery.

In May 1944 the Chichi Jima Fortress HAR was split up to provide artillery components of the 109th Division plus forming the nucleus for the new 9th HAR. This was defined as consisting of a 70-man headquarters, a 140-man "A" battery and three 107-man "B" batteries. Each battery was provided with 30 rifles, while the battalion manned four 24cm Type 45 howitzers, two 15cm Type 45 guns, and four 10.5cm Type 38 guns, all emplaced on Chichi Jima in the coastal defense role.

The same process was used to form nominal heavy regiments elsewhere, with 12th and 13th Heavy Artillery Regiments being formed in January 1945 in Formosa and the 14th and 15th at the same time in the homeland, the 16th and 17th in February, and the 18th and 19th in mid-year, the last four all in the homeland. Of these only the 19th appears to have been a traditional heavy artillery regiment. In this process the Rashin Fortress HAR became the 14th HAR, the Nagasaki Fortress HAR the 17th HAR, and the Hoyo Fortress HAR the 18th HAR, for instance. The 19th HAR was anomalous, having been formed in Manchuria with 240mm howitzers, then shipped to the homeland in 1945.

Unit	Pers	Weapons
12th Hvy Art Regt	850	unk no of 75mm, 15cm & 24cm guns and 28cm how
13th Hvy Art Regt	850	unk no of 75mm, 15cm & 27cm guns and 28cm how
14th Hvy Art Regt	850	4x75mm G, 6x10cm G, 2x15cm G, 3x28cm H, 5x30cm H
15th Hvy Art Regt	1,221	4x75mm G, 2x10cm G, 2x12cm G, 7x15cm G, 8x24cm H, 6x28cm H

5 The fortress artillery units themselves are discussed under their parent fortress units in the infantry section (see pages 225–236).

Unit	Pers	Weapons
16th Hvy Art Regt	945	unk no of 75mm, 10cm & 15cm guns and 28cm how
17th Hvy Art Regt	387	6x75mm G, 4x10cm G, 4x28cm H
18th Hvy Art Regt	643	4x75mm G, 2x9cm G, 4x12cm G, 4x12cm H, 4x15G
19th Hvy Art Regt	1,208	n/a

As an example, the 18th Heavy Artillery Regiment consisted of an 83-man HQ, a 140-man heavy battery with 15cm guns, and four light/medium batteries each of 105 men, one with 12cm howitzers, one with 75mm field guns, one with 12cm AA guns, and one with 88mm AA guns.

A somewhat larger number of heavy artillery battalions were raised to utilize 150mm guns, 280mm howitzers and 305mm howitzers. The 1st through 9th Independent Heavy Artillery Battalions were part of the permanent force structure and the 1st–3rd were mobilized in mid-July 1940 and the remainder in July–August 1941 and dispatched to Manchuria with the exception of the 3rd Battalion, which appears to have been sent to China.

The only units to use the obsolete 280mm howitzers (called the "type C" organization by the 1941 mobilization plan) were the 1st and 2nd Heavy Artillery Battalions. The battery equipment for these units included four Type 94 special heavy artillery tractors and two 30-ton cranes to assemble and disassemble the howitzers. No basic load of ammunition was authorized for these weapons. Both apparently converted to 150mm gun units in late 1941 or early 1942.[6]

A more common weapon was the 305mm Type (Taisho) 7 howitzer. The short-barrel version (Type D1 organization) was used by the 4th Heavy Artillery Battalion, and the long-barrel version (Type D2 organization) by the 5th, 7th, 8th and 9th Heavy Artillery Battalions, although the 9th Battalion seems to have switched back and forth between 305mm howitzers and 150mm guns several times during the war. Most of these battalions spent the war in Manchuria, although the 4th Battalion was ordered to the Philippines in July 1944. The 9th Battalion participated in the invasion of the Philippines and was brought from Manchuria to Japan in March 1945.

Deploying these units was not something to be undertaken lightly, and like almost all other units deployed overseas the 4th Heavy Artillery Battalion was put on a diet before it moved out in 1944 to conserve shipping space. It landed in the Philippines with an authorized strength of 24 officers, 78 NCOs and 498 privates divided into a 176-man headquarters, two 173-man howitzer batteries and a 78-man service battery. Each battery was built around two 58-man platoons, each with 11 Type 95 low-speed 13-ton tractors with Type 94 tracked trailers to move a single Type (Taisho) 7 howitzer. The battalion headquarters included an 82-man signal/observation group and a 74-man baggage trains. The battalion service battery was provided with 10 tractors with trailers, 8 trucks and 2 repair trucks. In December 1944 the battalion reported it had a total of 44 Type 95 tractors, 23 Type 94 trailers, 41 trucks and three 4-legged 30-ton cranes. Of course, once the pieces were emplaced in their massive pits they tended to stay there, rendering the tractors superfluous most of the time. It did have three Mk 3C and four Mk 5 radios for communications with forward observers, but the unit does not appear to have accomplished anything of note

6 A single 280mm howitzer lobbed the occasional shell (most of them duds) into T'ung Kuan in Shansi in late 1944 and into 1945. It is not known what unit it belonged to.

Moving the 30cm Type 7 long howitzer was a massive undertaking. Here two of the trailers needed, on the top for the upper cradle and on the bottom for the baseplate.

against US forces.

Two battalions, the 3rd and 6th, were raised with 150mm guns (Type E organization). Because of their greater mobility than other heavy artillery the 1st, 2nd, and 9th Battalions converted to this weapon during the war as well. In addition, the three heavy artillery battalions raised in March 1942, 11th, 12th and 13th,[7] also used the 150mm gun, as did the 20th in the Philippines and 100th Heavy Artillery Battalions raised in September 1944, the latter for Okinawa.[8] The 3rd Battalion participated in the Hong Kong seizure and then moved to Rabaul, where it sat out the rest of the war. The 11th, 12th and 13th Battalions were all recalled to the homeland in March 1945.

Each Type E battalion had two gun batteries, each of which was provided with eight Type 92 eight-ton tractors to tow the four guns, along with 17 trucks for equipment. The service battery had ten more tractors for caissons, five trucks and two light repair vehicles. The service battery carried all the ammunition for the battalion, 900 rounds total, divided evenly among HE, HE long pointed and AP under normal circumstances.

During January–May 1945 a final batch of thirteen more heavy artillery battalions (numbered 33–45) were activated in the homeland to man whatever weapons were at hand. For the most part these were immobile units sited to fire out to sea or on likely landing beaches.

Unit	Bty	Str	Weapons
33rd Hvy Art Bn	4	602	2x12cm G, 2x15cm G, 8x28cm H
34th Hvy Art Bn	4	596	4x75mm T38 G, 3x10cm T38 G, 6x28cm H
35th Hvy Art Bn	5	705	4x75mm G, 3x10cm G, 2x12cm G, 2x15cm G, 4x28cm H
36th Hvy Art Bn	2	414	8x24cm T45 H
37th Hvy Art Bn	5	708	4x75mm T38 G, 8x12cm G, 2x15cm T45 G, 2x24cm T45 H
38th Hvy Art Bn	2	406	8x24cm H
39th Hvy Art Bn	n/a	384	6x28cm H
40th Hvy Art Bn	n/a	444	n/a
41st Hvy Art Bn	2	453	4x15cm T89 G
42nd Hvy Art Bn	2	453	4x24cm T45 H
43rd Hvy Art Bn	2	453	2x30cm T7 H
44th Hvy Art Bn	2	453	4x15cm T89 G

The majority of these weapons were antiques, as by mid-1945 the Japanese were bringing out of storage every weapon that could fire and for which ammunition was still available.

In addition, eight independent heavy artillery batteries were formed. The 1st (15cm gun) and 2nd (15cm gun and 24cm howitzer) were formed for the Kwantung Army in September 1941. The 4th (28cm howitzer) and 5th (12cm gun) were formed in January 1945 for the homeland, the 6th (28cm howitzer), 7th (10cm gun) and 8th (15cm gun) in July, while the 9th battery never completed its organization in Korea.

At least three separate heavy artillery batteries were formed in Manchuria in 1945, the

7 Raising of the 14th to 18th Heavy Artillery Battalions had been planned, but the process was suspended, apparently in 1942, and never resumed.

8 Probably typical, the 100th Battalion consisted of a 105-man HQ, two 158-man batteries, and a 78-man trains. Each battery manned four 15cm Type 89 guns pulled by 8-ton tractors. The 20th Battalion was unique in using captured 15cm guns (probably US 155mm).

1st with 150mm guns, the 2nd with 240mm howitzers, and the 6th with 305mm howitzers.

The fortress regiments had a varied organization.[9] Their strengths in mid-1945 were as follows:

Fort Art Regt	Str	Weapons
Tokyo Bay	1,825	8x10cm G, 12x15cm G, 4x20cm G*, 4x25cm G*, 4x30cm G*, 2x28cm H
Yura	1,184	4x9cm G, 6x12cm G, 6x15cm G, 10x24cm G, 11x27cm G, 8x28cm H
Maizuru	410	2x27cm G, 4x28cm H
Tsushima	1,766	2x75mm G, 22x15cm G, 4x30cm G*, 2x41cm G*
Iki	1,426	10x15cm G, 2x30cm G*, 2x41cm G*
Shimonoseki	1,025	2x75mm G, 5x12cm G, 10x15cm G
Fusan	999	19x75mm G, 4x9cm G, 7x15cm G, 2x41cm G*, 8x28cm H
Reisui	317	4x75mm G, 2x15cm G
Eiko Bay	n/a	12x75mm, 10x28cm H
Soya	603	4x10cm, 4x15cm, 6x12cm H
Tsugaru	1,350	2x10cm G, 12x15cm G, 2x30cm G*
**in naval turrets*		

As an example of internal organization late in the war the Tokyo Bay Fortress Artillery Regiment consisted of a headquarters and three battalions. These battalions were armed as follows:

1st Battalion
- 1st Battery: two naval turrets, each two 25cm guns
- 2nd Battery: four 15cm guns
- 3rd Battery: four 10cm guns
- 4th Battery: four 15cm guns
- "X" Battery: one naval turret, two 30cm guns (incomplete)

2nd Battalion
- 5th Battery: four 10cm guns
- 6th Battery: one naval turret, two 30cm guns
- 7th Battery: two naval turrets, each two 20cm guns

3rd Battalion
- 8th Battery: two 15cm guns
- 9th Battery: two 15cm guns
- 10th Battery: two 28cm howitzers

Separate from the fortress heavy artillery regiment were two additional units, the 1st Tokyo Bay Fortress Artillery Unit (603 men in 4 batteries with eight 28cm howitzers, two 15cm Type 89 guns and eight 75mm Type 38 field guns) and the 2nd Tokyo Bay Fortress Artillery Unit (483 men in 3 batteries with six 28cm howitzers and two 15cm Type 4 guns).

Similarly, the smaller Iki Fortress Artillery Regiment had the following organization at war's end:

9 A general discussion of the fortress artillery regiments can be found in the infantry section dealing with fortress units, pages 225–236.

1st Battalion
- Kurosaki Battery: one naval turret, two 41cm guns
- Wataraosima Battery: four 15cm guns
- Nagarasu Battery: four 15cm guns
- (mobile) battery: four 75mm guns
- (mobile) battery: five 75mm guns
- Irukabana Battery (incomplete): two 14cm guns
- Tansuura Battery (incomplete): two 14cm guns

2nd Battalion
- Oshima Battery: one naval turret, two 30cm guns
- Ikisuki Battery: two 15cm guns + (incomplete) two 14cm guns

In most cases fire control was primitive. The two exceptions were the naval turrets and the newer 15cm guns. The naval turret batteries were equipped with the Type 88 electrical fire control system, while the 15cm guns had the Type 98 electro-mechanical system. Both relied on depression-base rangefinders set into the roofs of concrete command posts as their key component, not surprising given Japan's rugged topography. The howitzers (usually actually mortars by Western definition) were usually only provided with a portable tripod-mounted depression rangefinder and plotting boards, while the other weapons generally fired off of on-carriage sights exclusively. No coast artillery fire control radar had been fielded by the war's end.

Artillery Mortars

A truly unique weapon was the Type 98 artillery mortar, which launched a 320mm spigot round from a short 250mm barrel. It was designed for one highly-specialized mission, the destruction or suppression of Soviet fortifications. The Soviets were assessed as having arranged their fortifications in three layers, the rearmost of which was out of range of Japanese fortress artillery. The Type 98s were "surprise weapons" that could be pack or hand carried forward at night, set up in 30 minutes or so, and then used to launch their massive 334kg shells the short remaining distance to the Soviets. They were organized into a new unit, the independent artillery mortar battalion/regiment (*Dokuritsu Kyūhō Daitai/Rentai*).

An artillery mortar battery was provided with four of these Type 98 mortars plus 20 rounds of HE ammunition, but no transport to carry either. The battery consisted of the command platoon, the firing battery and a large ammunition train. The firing battery had two platoons, each with two 10-man mortar sections. The ammunition trains was similar, providing an ammunition section to support each mortar section, but totaled 185 men to manhandle the massive ammunition into the firing positions. The battalion service battery carried a duplicate set of mortars and ammunition on a one-for-one basis, but had only four trucks for general transport. Such extra transport as was available was concentrated in the battalion HQ, which had six trucks. For communications each battery was provided with two Type 94 Mk 6 radios, as was the battalion HQ.

The 1st and 2nd Artillery Mortar Battalions were formed in 1938 and, in the absence of action in Manchuria, dispatched to China. Lacking the Type 98s, these first two units were actually equipped with 15cm mortars. The 1st Battalion served in China until moved to the 8th Area Army in the south Pacific in 1943, where it ended the war. The 2nd Battalion participated in the siege of Hong Kong and then that of Bataan and Corregidor before being dissolved. An initial three battalions (11th, 12th and 13th) of a second generation of artillery mortar battalions were formed in August of 1941 by the Kwantung Army, this time with Type 98s. Two more battalions, the 14th and 15th, were activated in September–November 1941 and used in the Philippines. The 15th was deactivated shortly thereafter, while the 14th was transferred to the Kwantung Army. All these units were apparently deactivated in late 1943.

April 1944 saw a major increase in artillery mortar strength. Four artillery mortar battalions (15th, 16th, 18th and 19th) were activated for duty with the 27th Army in northern Japan and remained there through to war's end, while three more (14th and 17th in Japan and 20th in Korea) were activated for duty with the 31st Army in the Central Pacific, each with a strength of 648 men. At this point their role seems to have shifted from offensive to defensive. The three battalions destined for the Pacific suffered heavy losses at sea and were finally landed as stragglers on Saipan, where they apparently fought as infantry.

These were followed in July by the 21st Independent Artillery Mortar Battalion for the 14th Area Army in the Philippines, and then by the 22nd Battalion for the homeland and the 23rd Battalion in October for the Kwantung Army. Finally, three more battalions (24th–26th) were activated in June 1945 for homeland defense and the 27th in July for the Kwantung Army. The wartime battalions had three slightly different strength authorizations, which each maintained to the end:

	Off	WO	MSG	SGT	PVT	Total
11th–20th & 27th Battalions	19	4	7	33	585	648
21st–23rd Battalions	19	4	7	33	594	657
24th–26th Battalions	20	3	8	50	569	650

All appear to have followed the standard pattern except the last batch (24th–26th) which abandoned the massive spigot mortar with its short range in favor of more conventional, if equally old, mortars. The 24th Battalion used the 150mm mortar, while the 25th and 26th Battalions used the 90mm mortar, with four mortars per company (twelve per battalion) in each case. These battalions do not appear to have had the service battery found in the earlier battalions. A 28th Battalion was formed in July 1944 and shipped to Manila, but left its heavy mortars on the ship and re-equipped with 120mm conventional weapons.

The Kwantung Army's original artillery mortar unit was one not envisioned under the IGHQ mobilization, the 1st Independent Artillery Mortar Regiment with Type 98s. It was raised on a peacetime reduced strength basis in August 1940 with an organization of a headquarters, six mortar batteries and a service battery and a total of about 1,500 men. In July 1942 it was mobilized and the 2nd, 3rd and 6th Batteries were detached from the regiment and formed into an improvised artillery mortar battalion and sent to Rabaul. The remainder of the regiment remained in Manchuria until August 1944, when it was transferred to the 32nd Army for duty on Okinawa.

The regiment, less its three batteries, was thus a close approximation of an artillery mortar battalion on mobilization. It left most of its motor transport (about 25 trucks), horses, and about 400 men in Manchuria, reducing regiment strength to about 400 men. The addition of local conscripts and Okinawan *Boei Tai* personnel served to bring regimental strength back to about 750, of which about 125 were Okinawans. Each battery had a HQ with 10–15 men, a command section of 25 divided into signal and observation sections, an ammunition trains of about 80 in two sections, two 320mm fire platoons of slightly over 40 each, and a 90mm section of about 20 formed from surplus personnel of the fire platoons. Each battery was provided with four 250mm mortars (plus four reserve weapons), each served by two guns teams of nine men. The two senescent 90mm Type 10 mortars were added locally and were fired against targets which the 250mm mortars were unable to reach. The 90mm weapons, however, were regarded as worthless and little use was made of them. Each battery was provided with two field telephones, a Mk 5 radio and a Mk 6 radio. Each battery was supplied with 150 rounds of ammunition. In addition the service battery held a further reserve of 70 rounds.

The 1st Artillery Mortar Regiment began firing against US troops on 8 April 1945, and by 24 April it had been destroyed as a unit, with only three weapons remaining. All the other 250mm mortars had been destroyed by US infantry mortar fire. Because of the short ranges involved none had apparently been hit by US artillery.

Rocket Artillery

Unlike the European powers, which used multiple rocket launchers for saturation fire, the Japanese used single-tubed launchers in much the way they used mortars. In fact, the rocket gun battalion was almost identical in organization to a mortar battalion and the latter could easily be re-equipped as the former.

Although several improvised rockets and launchers were used, the standard weapon was the 20cm Type 4 that used a mortar-type bipod and open-breech tube. The rocket had

A 20cm Army Type 4 rocket being fired.

a weight of 90kg, of which 43kg was the rocket motor and had a range of 2,500 meters.

The first unit formed was the 109th Division Rocket Company (later expanded to an extemporized battalion) on Iwo Jima. The large-scale raising came in July 1944 when the 1st–4th Rocket Artillery Battalions were ordered activated, the 1st for Korea, the 2nd for China (later moved to Kyushu), the 3rd for the Philippines, and the 4th for Kyushu. In addition, coastal divisions in the homeland in 1945 often included rocket artillery battalions.

A rocket artillery battalion was made up of a 65-man HQ, three 243-man batteries, and a 51-man ammunition trains. Each rocket battery consisted of a HQ (including a 5-man observation section and a 14-man signal section) and three 65-man platoons, each of which comprised four 16-man firing sections, each with a 20cm Type 4 rocket launcher. A company was provided with 12 of the Type 4 launchers, 84 rocket shells, and 60 Type 39 one-horse carts. The battalion ammunition trains was provided with ten trucks.

Self-Propelled Artillery

The IJA was slow off the mark in the development of self-propelled artillery. Although the Ho-Ni II 105mm SP howitzer and a conversion of the Type 97 to a 150mm SP howitzer were undertaken in 1944, few were built.

The initial provisional SP unit was the Independent SP Artillery Battalion, formed in November 1944 with four batteries and a total of 12 Ho-Ni I and Ho-Ni II guns. It was to be sent to the Philippines but was lost *en route*. The 1st Independent SP Artillery Battery (also known as the Sumi Unit) was formed in December 1944 with three 15cm SP howitzers and also sent to Luzon. One gun was lost *en route* and the others appear to have been captured by US forces in a vehicle park.

The decision to create a series of self-propelled artillery battalions was apparently taken in the spring of 1945 and the activation of such units began in mid-June. The plans called for the activation of ten such battalions, with the 5th Battalion activating in early June; the 1st in late June; the 2nd, 7th and 8th in July; the 3rd and 6th in August; and the 4th, 9th and 10th in September. It would appear that only the first five were actually formed.

The initial plans called for such a battalion to consist of a headquarters, three SP batteries and a towed battery. Surviving documents do not indicate the presence of a trains battery, but these may have been planned as well. The HQ battery included three armored OP vehicles and three tracked APCs. Each of the SP batteries had four SP guns, eight APCs (six for ammunition, one for fuel, one towing an AT gun), and two armored OP vehicles. The towed battery would have two armored OP vehicles and four tracked APCs towing 75mm Type 90 guns.

The presence of observation vehicles and light trucks to support the battery observation section forward clearly show that these units were meant to be primarily employed as indirect-fire units, rather than as anti-tank or assault gun elements.

By the time the Army began raising such units in earnest, however, the organization had been changed to three gun batteries and a service battery with a total strength of 27 officers and 451 enlisted. The battery was made up of a battery HQ (with a Type 100 observation vehicle and a field car), a forward observation group (with a Type 100 and an observation car), two platoons each of two squads, and an ammunition train (with two Type 1 full-track APCs for ammunition). Each firing section was made up of six men with a Ho-Ni SP piece and seven men with a Type 1 with trailer carrying 150 rounds of ammunition.

Artillery Intelligence

Also included in the artillery branch were a small number of artillery intelligence (target acquisition) regiments, primarily to support counter-battery efforts. Under the 1941 mobilization plan each of these was made up of a 258-man HQ, a 93-man survey company, a 225-man flash-ranging company and a 91-man sound-ranging company. The headquarters company consisted of balloon, photography and signal platoons. The survey company had three platoons, each with three survey sections and a computing section. The flash-ranging company provided three platoons each with three plotting stations, while the sound-ranging company had six listening posts. The regiment was motorized, provided with 38 trucks for the HQ, 10 for the survey company, 21 for the flash-ranging company and 9 for the sound-ranging company. In addition, other elements could be assigned as needed, including meteorological, aerial photo interpretation and chemical.

The initial such units, the 1st and 2nd Artillery Intelligence Regiments, were ordered formed on 9 February 1938 under the Kwantung Army and mobilized to a reduced war strength of 22 officers and 446 enlisted in July 1941. They remained there until March 1945 when they were returned to the homeland and enlarged to 631 men.

A more widely travelled unit was the 5th Artillery Intelligence Regiment, raised in July 1940 with no fewer than 27 officers and 738 enlisted, which participated in the siege of Hong Kong in December 1941 and was transferred to the Philippines in March 1942 to help soften up Bataan and Corregidor before being transferred to the Kwantung Army on the conclusion of that operation. The intelligence regiments also functioned as parent commands for detachments that could be sent out to active theaters. Such detachments operated in China, Burma and at Rabaul.

No further units were formed until February 1945 when the 3rd, 4th and 6th Artillery Intelligence Regiments were activated for defense of the homeland, each using the organization of the 1941 mobilization plan. A month later the 1st, 2nd and 5th Regiments were ordered transferred back to Japan so that all six regiments were in the homeland from March 1945.

Artillery Commands

To command all these disparate units the army formed a brigade-equivalent type of HQ, the artillery command (*Hōhei shireibu*)[10] to act as command and staff elements at army and area army levels. The 1st Artillery Command HQ was formed in 1940 and dispatched to China where it participated in several campaigns under the 23rd Army before being used in the assault on Hong Kong. In March 1942 it was transferred to the Philippines, along with a number of other artillery units, to coordinate the reduction of Bataan and Corregidor. In the meantime, three more artillery HQs were activated in July 1941 in order to provide one for each of the main subordinate armies of the Kwantung Army: the 5th Artillery HQ for the 4th Army, the 7th Artillery HQ for the 3rd Army, and the 8th Artillery HQ for the 5th Army. In May 1942 the 1st Artillery HQ was transferred to Manchuria to provide artillery coordination for the 4th Army.

10 Not to be confused with the artillery brigade HQ (*Hōheidan shirbeibu*) designation given to the artillery staffs of some of the infantry division HQs. The 1st–4th, 8th–12th, 14th, 19th, 24th, 25th, 28th and 51st Divisions all had nominal artillery brigade HQs as part of their make-up at some times. Presumably this facilitated control of attached artillery elements, although it otherwise essentially duplicated the function of the HQ of the divisional artillery regiment.

The number of subordinate elements varied among the units and over time as non-divisional artillery was shifted in and out of their superior armies. In Manchuria the 1st Artillery Command was usually the smallest, usually controlling the 1st Medium Artillery Regiment, the 1st Heavy Artillery Regiment, the 9th independent Heavy Artillery Battalion, and the 5th Artillery Intelligence Regiment. The 7th Command, on the other hand, usually shared the honor of the largest such unit with the 8th, with the following units often under command: the 4th Independent Mountain Artillery Regiment; the 9th, 17th and 22nd Medium Artillery Regiments, the 2nd Heavy Artillery Regiment; the 1st, 4th, 6th, 10th and 13th Independent Heavy Artillery Battalions, the 1st Independent Artillery Mortar Regiment, and the 2nd Artillery Intelligence Regiment.

Only two artillery commands were used outside of Manchuria before mid-1945. The 9th Artillery Command HQ was formed in July 1944 to control the 7th Medium Artillery Regiment, the 5th Independent Medium Artillery Battalion, the 3rd Independent Heavy Artillery Battalion, and the 1st Independent Artillery Mortar Battalion at Rabaul. In late 1944 the 5th Artillery Command HQ was moved to Okinawa to control the non-divisional artillery there (1st and 23rd Medium Artillery Regiments, 7th Heavy Artillery Regiment, 100th Independent Heavy Artillery Battalion, and 1st Independent Artillery Mortar Regiment).

As the threat to the homeland increased all the artillery command HQs in Manchuria (as well as most of their subordinate units) were recalled to Japan proper and on 1 June 1945 a new series of these HQs were formed, the 2nd, 6th, 10th, 11th and 12th, followed a month later by the 3rd Artillery Command HQ. Once again, the subordinate elements controlled by each HQ varied, but typical of the 1945 homeland Artillery Commands was the 4th:

4th Artillery Command HQ (119)
9th Independent Field Artillery Regiment (1,862)
20th Medium Artillery Regiment (700)
28th Medium Artillery Regiment (2,286)
44th Independent Heavy Artillery Battalion (453)
21st Trench Mortar Artillery Battalion (1,407)
24th Trench Mortar Artillery Battalion (1,407)
25th Trench Mortar Artillery Battalion (1,407)

6

Anti-Aircraft Artillery

Perhaps lulled into a false sense of security by its island homeland location, the IJA was somewhat slow off the mark in the development of air defense capabilities. It was a problem that was to plague them throughout the war.

Small numbers of extemporized AA guns were built during the 1920s, but it was not until 1928 that a standard AA weapon, the 75mm Type 88, was introduced. This weapon, of unexceptional design for its time, remained in service as the main Army air defense weapon to the end of the war, although with increasingly disappointing results as new, higher-performance aircraft were introduced. Exacerbating the problems presented by the mediocre performance of the Type 88 gun was the relatively poor performance of the standard Type 90 (1930) AA director. The introduction of the Type 2 Model 3 director in 1942/43 gave greater flexibility in firing under all conditions, but did little to improve accuracy.

In 1938 the IJA introduced its first autocannon, the 20mm Type 98. A lightweight, flexible weapon, powerful for its caliber, these guns used on-carriage iron sights exclusively. A centralized director to control all six guns of a battery was developed in 1942 and probably would have increased the Type 98's effectiveness had not production difficulties delayed its fielding.

Anti-Aircraft Units

For peacetime defense of the homeland the Army organized the 1st through 4th Anti-Aircraft Regiments in the homeland and the 5th and 6th in Korea, although it is not certain that these units were ever mobilized to full strength before their deactivation in 1940–42.

An anti-aircraft regiment of the 1930s consisted of a headquarters, a searchlight battery and two AA battalions. Each of the battalions was made up of two 4-gun batteries. A battery could be either "mobile" or "fixed" depending on the assignment of trucks, and consisted of a headquarters, an observation platoon, and a firing battery. The observation platoon held a range detail, a signal section and a machine gun section. The range detail manned a stereoscopic rangefinder, usually either a 2-meter Type 90 or a 3-meter Type 93, along with a speed and angle of path instrument. The machine gun section had two machine guns on AA mounts, either 6.5mm Type (Taisho) 11 or 13.2mm Hotchkiss, with the former more common. The firing battery consisted of a fire control section with a Type 90 fire control device with its 7-man crew, and two platoons, each of two gun squads with 75mm AA guns. Mobility, when authorized, was provided by 6x4 Sumida trucks. A searchlight battery had only two searchlight sections and a sound locator section.

The first air defense unit to serve outside the homeland was the 12th AA Battalion, raised under the Kwantung Army at the beginning of 1937 and redesignated a few months later as the 12th AA Regiment. This was followed in 1938 by the 9th, 10th, 11th, 13th and 17th AA Regiments, and the 9th–14th Field AA Units (extemporized batteries, later redesignated independent AA batteries) all also raised for duty in Manchuria. Each of these regiments consisted of a headquarters, two 75mm gun batteries, a 20mm machine cannon battery, a searchlight battery and a service battery.

Shortly thereafter, the Army began the process of providing air defense for the forces in China. The 1st through 8th Field AA Units were dispatched in early 1938, and by the end of the year no fewer than 43 field AA units had been sent, 30 of the Type A variety (four 75mm Type 88 guns) and 43 of the Type B (two 75mm Type 11 or Type 88 guns).

A Type A independent field AA unit had the following organization:

Battery HQ (2 officers + 3 NCOs + 4 privates) [9 pistols, 1 rifle] [1 car, 1 motorcycle]
  HQ Platoon [total platoon 8 pistols, 10 rifles, 2 machine guns]
  Platoon HQ (1 + 0 + 1) [1 motorcycle]
  Observation Section (0 + 2 + 17) [1 observation vehicle]
  Signal Section (0 + 1 + 14) [1 truck]
  AA Machine Gun Section (0 + 0 + 6)
Two Firing Platoons, each [total platoon 11 pistols, 8 rifles, 2 75mm AA guns]
  Platoon HQ (1 + 0 + 0)
  Two Squads, each (0 + 1 + 16) [1 tractor, 1 truck]
Ammunition Platoon [total platoon 22 pistols, 16 rifles]
  Platoon HQ (1 + 0 + 1) [1 motorcycle]
  Two Ammunition Squads, each (0 + 1 + 12) [3 trucks]
  Maintenance Squad (0 + 2 + 14) [2 trucks, 2 shop trucks]

The observation vehicle was built on a truck chassis and provided a Type 90 two-meter base optical height finder, two Type 90 target speed computers, and simple meteorological and survey instruments. The signal section carried ten field telephones and 40 small reels of field wire. The basic load of ammunition consisted of 32 rounds for each pistol, 30 rounds for each rifle, 4,000 rounds for each machine gun, and 200 rounds for each of the 75mm Type 88 guns.

The independent field AA units had relatively short lives, being quickly consolidated into AA regiments, the 15th and 16th (each 970 men) around Peking and Tientsin in July 1938, followed by the 20th and 21st for north China, the 22nd near Hankow and the 23rd in south China in November 1939 (each of these three 75mm batteries and a searchlight battery, totaling 570 men with twelve 75mm guns and 20 trucks). The 16th and 23rd were used in the invasion of Java, while the rest continued to serve in China through the war.

Prelude to War: the 1941 Mobilization Plan

The varying demands placed on these units, particularly with regard to mobility requirements, clearly called for a more flexible organizational framework. The result was a complete overhaul of air defense organization promulgated in the 1941 mobilization order. Henceforth, air defense units would be categorized into two groups: static units on the one hand, and mobile and semi-mobile units on the other, distinguished by the absence or presence, respectively, of the word "field" in their designations.

The mainstay of the static air defense was to be the AA Regiment, consisting of three binary battalions, two with 75mm guns and one with searchlights. The basic firing unit was the battery, which was made up of a HQ platoon, two firing platoons and a trains element. The HQ platoon included a signal section, an observer/FDC section with a Type

90 director and 3-meter optical rangefinder, and a 6-man machine gun squad with two 7.7mm Type 92 machine guns on high-angle mounts nominally for protection against strafing aircraft. The firing platoon was made up of two sections, each manning a Type 88 gun. The battery was provided with 200 rounds of Type 90 long pointed HE ammunition for each of its 75mm guns and 2,160 rounds for each machine gun. A searchlight battery consisted of two platoons, each with three 150cm searchlights each with an associated sound detector/locator. The AA regiment was completely immobile, having not a single truck in the entire unit. It had no basic load of ammunition other than that on hand in the batteries, and relied on transport of its supported unit to keep it resupplied.

With the approach of war the air defenses of the homeland were mobilized, including the 1st to 6th Air Defense Regiments and 1st and 2nd Independent Air Defense Battalions for the Tokyo area, the 11th to 14th Regiments and 11th and 12th Battalions for the Osaka area, the 41st and 42nd Regiments for Korea and the 51st and 52nd Regiments for Formosa, all in November 1941. Notably, the regiments were almost all significantly larger than envisioned by the 1941 mobilization plan.

Table 6.1: Air Defense Regiments Mobilized November 1941

Regt	Off	Enl	Total
1	59	1,557	1,636
2	48	1,149	1,197
3	53	1,316	1,369
4	47	1,324	1,371
5	46	1,270	1,316
6	4	129	133
11	?	?	?
12	69	1,860	1,929
14	41	1,054	1,095
21	67	1,897	1,964
22	59	1,658	1,717
23	51	1,320	1,371
41	87	2,490	2,577
42	59	1,673	1,732
51	34	875	909
52	42	1,103	1,145

Fixed AA defenses were obviously inappropriate in the wide-open spaces of Manchuria and in July 1941 the AA regiments of the Kwantung Army were disbanded and, with one exception, used to form the more mobile field AA battalions and field searchlight battalions.

Back in the homeland, for cases where a full regiment was not required or where greater flexibility in assigned assets was needed, the mobilization plan provided for independent AA battalions and batteries and independent searchlight battalions. These used the same building-block components as the AA regiment with only minor changes. These various units were subordinated as needed to AA defense unit HQs and strategic location AA defense unit HQs, which performed limited liaison functions.

Only one independent (non-field) AA battalion was sent to combat in the Pacific, the 50th Battalion, formed from the 10th AA Regiment in Manchuria in 1941. It consisted

Table 6.2: Anti-Aircraft Units, from October 1940

	Personnel	Pistols	Carbines	Rifles	AA Machine Guns	20mm Auto Cannon	75mm Type 88 AA Guns	150cm Searchlights	Riding Horses	Pack Horses	Cars	Trucks	Tractors	Repair Vehicles
Mobile & Semi-Mobile Units														
Field Anti-Aircraft Defense Unit HQ	75	0	11	49	0	0	0	0	0	0	2	4	0	0
Field Anti-Aircraft Battalion (Type A)														
Battalion HQ	134	11	45	100	0	0	0	0	0	0	2	20	0	0
Three AA Batteries, each	149	14	31	0	2	0	4	0	0	0	2	9	6	0
Service Battery	86	6	46	0	0	0	0	0	0	0	2	17	0	2
Field Anti-Aircraft Battalion (Type B)														
Battalion Headquarters	102	11	32	150	0	0	0	0	0	0	2	6	0	0
Three AA Batteries, each	121	14	4	0	2	0	4	0	0	0	2	0	0	0
Service Battery	57	6	29	0	0	0	0	0	0	0	1	8	6	0
Ind Field Anti-Aircraft Battery	179	14	48	40	2	0	4	0	0	0	2	6	6	1
Field Searchlight Battalion														
Battalion HQ	89	8	14	100	0	0	0	0	0	0	1	6	0	0
Two S/L Batteries, each	182	14	12	0	0	0	0	6	0	0	0	0	0	0
Field Machine Cannon Battery (Pack)	245	14	2	60	0	6	0	0	21	136	0	0	0	

Table 6.2: Anti-Aircraft Units, from October 1940

	Personnel	Pistols	Carbines	Rifles	AA Machine Guns	20mm Auto Cannon	75mm Type 88 AA Guns	150cm Searchlights	Riding Horses	Pack Horses	Cars	Trucks	Tractors	Repair Vehicles
Field Machine Cannon Battery (mot)	121	13	34	60	0	6	0	0	0	0	2	15	0	0
Static Anti-Aircraft Units														
Anti-Aircraft Defense Unit Headquarters	55	0	3	0	0	0	0	0	0	0	2	0	0	0
Strategic Location AA Defense Unit HQ	50	0	3	0	0	0	0	0	0	0	2	0	0	0
Anti-Aircraft Regiment														
Regiment HQ	54	0	5	0	0	0	0	0	0	0	3	0	0	0
Two Anti-Aircraft Battalions, each														
Battalion HQ	81	0	2	0	0	0	0	0	0	0	1	0	0	0
Two AA Batteries, each	123	0	2	0	2	0	4	0	0	0	1	0	0	0
Searchlight Battalion														
Battalion HQ	64	0	6	0	0	0	0	0	0	0	3	0	0	0
Two S/L Batteries, each	161	0	2	0	0	0	0	6	0	0	1	0	0	0
Independent Anti-Aircraft Battalion														
Battalion HQ	101	0	2	20	0	0	0	0	0	0	1	0	0	0
Two AA Batteries, each	123	0	2	0	2	0	4	0	0	0	1	0	0	0
Independent Anti-Aircraft Battery	117	0	2	0	2	0	4	0	0	0	1	0	0	0

Table 6.2: Anti-Aircraft Units, from October 1940

	Personnel	Pistols	Carbines	Rifles	AA Machine Guns	20mm Auto Cannon	75mm Type 88 AA Guns	150cm Searchlights	Riding Horses	Pack Horses	Cars	Trucks	Tractors	Repair Vehicles
Independent Searchlight Battalion														
Battalion HQ	67	0	2	20	0	0	0	0	0	0	1	0	0	0
Two S/L Batteries, each	158	0	0	0	0	0	0	6	0	0	0	0	0	0
Sound Locator Battery	210	0	2	0	0	0	0	0	0	0	1	0	0	0
Anti-Aircraft Observation Unit	170	0	0	0	0	0	0	0	0	0	0	0	0	0
Signal Unit (Type A)	200	0	0	0	0	0	0	0	0	0	0	0	0	0
Signal Unit (Type B)	105	0	0	0	0	0	0	0	0	0	0	0	0	0

Note: rifles in headquarters are undistributed weapons for unit self-defense.

of three firing batteries that prior to deployment were enlarged to six guns each, and it left Manchuria in December 1942 for Rabaul and then the Solomons.

Due to their cumbersome organization and lack of mobility the AA regiments were not widely used during the war outside of the homeland. Two regiments were transferred south, the 16th AA Regiment winding up in Java and the 23rd to Indochina and then Java. Three more regiments were raised in mid-1942: the 24th in April for homeland duties (where it remained), and the 25th and 26th in June by the Kwantung Army.

Two regiments later raised for the Kwantung Army diverged from the standard set by the 1941 mobilization plan by having three batteries in each AA battalion instead of the normal two. Of these, the 25th AA Regiment was destined to become the only such unit to see heavy combat. In early 1944 this unit was transferred to the command of the 31st Army in the Central Pacific. It was split up on arrival, with the bulk of the regiment assigned to Saipan, while the 2nd Battalion HQ and 5th (gun) and 9th (searchlight) Batteries went to Truk, the 3rd (gun) Battery to Nomoi (later moved to Truk), and the 4th (gun) Battery to Puluwat. In partial compensation the 44th Field Machine Cannon Company was added to the regiment's strength on Saipan on 28 March 1944, followed by the 43rd Independent Field AA Battery on 23 April.

For the move to this combat theater the 25th AA Regiment was reorganized. The gun batteries were expanded to eight officers and 164 enlisted, divided into three gun platoons (each two 75mm Type 88 guns), an 18-man ammunition platoon, and a 61-man HQ platoon. In this new organization the HQ platoon was comprised primarily of a 13-man spotting section, an 18-man observation/FDC section, a 10-man signal section with telephones, and a 9-man MG section. The new type battery thus had two Type 92 MGs and six Type 88 AA guns. Each gun battalion consisted of three such batteries and a 43-man HQ. The searchlight battalion was identical to a gun battalion but equipped their line squads with searchlights instead of 75mm guns. Unusually, the searchlight batteries included machine gun sections like the gun batteries. For mobility each AA gun battery had one tractor, so that it could change positions, but not quickly. The regiment was allotted 13 trucks to support repositioning and normal supply.

The Army calculated at that point that one day's ration of fire for the Type 88 gun was 800 rounds and that three such units of fire should be stockpiled on the islands with the guns. In fact, only 955 per gun actually arrived with the regiment. It is not known whether the deficiencies were corrected.

The authorized and actual equipment of the 25th, the only AA regiment to deploy to the Central Pacific, on its arrival was as shown in Table 6.3.

Shortly after its arrival in the Pacific the 25th AA Regiment was caught in the 31st Army reorganization of May 1944. The 2nd Battalion HQ and 3rd, 5th and 9th Batteries were redesignated the AA Unit of the 51st IMB on Truk, while the 4th Battery became the AA Unit of the 11th IMR on Puluwat. The remainder of the Regiment, reinforced by the 43rd Independent AA Battery, was reorganized into a reformed 25th AA Regiment, consisting of a regiment HQ, 1st through 4th AA batteries, and 5th and 6th searchlight batteries, with no battalion HQs. The new regiment was destroyed in the US invasion of Saipan in June 1944. Its revised organization on Saipan was:

Table 6.3: 25th AA Regiment Equipment Status early 1944

		Type 14 Pistols		Type 99 Short Rifles		Type 92 Machine Guns		Type 10 Grenade Dischargers		75mm Type 88 Guns			Type 96 Searchlight		Type 90 small Sound Locators		Cars		Trucks		Type 96 Tractor Trucks		Repair Vehicles	
		TO&E	on hand	TO&E	on hand	TO&E	on hand	TO&E	on hand	TO&E	on hand		TO&E	on hand	TO&E	on hand	TO&E	on hand	TO&E	on hand	TO&E	on hand	TO&E	on hand
Saipan	Regiment HQ	0	31	0	32	0	2	0	0	0	0		18	18	18	18	3	0	4	2	0	0	0	0
	1st Bn HQ	0	7	0	10	0	0	1	1	0	0		0	0	0	0	2	0	2	1	0	0	0	0
	1st Battery	0	16	0	55	2	2	0	0	6	6		0	0	0	0	1	0	1	1	6	1	0	0
	2nd Battery	0	16	0	65	2	2	0	0	6	5		0	0	0	0	1	0	1	1	6	1	0	0
	6th Battery	0	17	0	30	2	2	0	0	6	5		0	0	0	0	1	0	1	0	6	1	0	0
	3rd Bn HQ	0	9	0	10	0	2	1	1	0	0		0	0	0	0	2	0	2	0	6	1	0	0
	7th Battery	0	26	0	35	2	2	0	0	0	0		0	0	0	0	1	0	1	0	0	0	1	1
	8th Battery	0	0	0	0	0	0	0	0	0	0		0	0	0	0	0	0	0	0	0	0	0	0
Truk	2nd Bn HQ	0	0	0	0	0	0	1	1	0	0		0	0	0	0	2	0	2	1	0	0	0	0
	5th Battery	0	0	0	0	2	2	0	0	6	5		0	0	0	0	1	0	1	2	6	2	0	0
	9th Battery	0	0	0	0	0	0	0	0	0	0		1	1	0	0	1	0	1	1	0	0	0	0
Nomoi	3rd Battery	0	0	0	0	2	2	0	0	6	6		0	0	0	0	1	0	1	1	6	1	0	0
Puluwat	4th Battery	0	0	0	0	2	2	0	0	6	6		0	0	0	0	1	0	1	1	6	1	0	0
Total Regiment		169	122	360	237	12	18	3	3	36	33		18	18	18	18	20	0	66	11	42	8	1	1

Note: pistols and rifles were undistributed within the regiment by TO&E.

	Off	WO	NCO	Other	Total
Regiment HQ	6	0	9	104	119
4 Gun Batteries, each	4	1	13	126	144
2 Searchlight Batteries, each	4	1	13	148	166
Attached medical	2	0	5	16	23
Attached intendance & technical	3	0	8	0	11

It was equipped with 14 Type 92 machine guns and 19 Type 88 guns for armament, 10 cargo trucks and 12 searchlight trucks for mobility, and two Type 94 MK 3C radios and 53 telephones for communications.

The three AA regiments remaining in China by the end of the war had been formed in 1939 as small units with 570 men each, but reorganization in early 1945 yielded divergent organizations. The 15th AA Regiment was the largest, consisting of 1,613 men in five gun batteries (each four 75mm), two machine cannon batteries (each six 20mm), and two searchlight batteries. The 21st AA Regiment was slightly smaller with 1,456 men in six gun batteries, one machine cannon battery, one searchlight battery, and a service battery. The 22nd AA Regiment, in contrast, retained its original organization with 571 men in three gun batteries, a searchlight battery and a service battery.

One anti-aircraft regiment completely out of any numerical sequence, the 171st, was raised in Manchuria in September 1942. This unit was anomalous as well in its organization, which approximated the pre-war pattern rather than the 1941 mobilization plan. It had a TO&E strength of 1,515, but its authorized strength was only 887. The 171st Regiment consisted of a headquarters, four gun batteries, a machine cannon battery and a searchlight battery.

This regiment was probably raised to ameliorate the loss of GHQ air defense assets felt that year, since the remnants 10th AA Regiment was reorganized into the 1st Armored Division AA Unit and that of the 11th AA Regiment into the 2nd Armored Division AA Unit that same month.

While all this was going on a different series of units was being raised for homeland defense, spurred by the Doolittle raid of mid-1942. In December 1941 the homeland had been protected by only 300 AA guns, mostly 75mm. These were for the most part concentrated in the pre-war AA regiments, with about half the weapons grouped into two AA brigades (Eastern and Central) in early 1942. The level of training was low, equipment was old, and coordination poor. Shortly after the Doolittle raid the air defense forces in the homeland were completely reorganized. In contrast to the anti-aircraft regiments (*Koshabo Rentai*) deployed to China and Manchuria, the homeland forces were grouped into air defense regiments (*Boku Rentai*). These were similar to the anti-aircraft regiments but were completely static units lacking all but the most basic of administrative facilities.

On 19 November 1942 the new homeland air defense organization was implemented, which provided three regional air defense commands and several smaller independent units as follows:

Eastern Air Defense Headquarters
- 1st–7th Air Defense Regiments
- 7th Anti-Aircraft Regiment
- 1st and 2nd Independent Air Defense Battalions

A 75mm Type 88 battery training during the war. Note the empty cartridge cases on the ground and the new round being brought out on the right.

1st Machine Cannon Battalion
Central Air Defense Headquarters
11th–13th and 15th Air Defense Regiments
11th and 12th Independent Air Defense Battalions
12th Independent Anti-Aircraft Battalion
2nd Machine Cannon Battalion
Western Air Defense Headquarters
21st–23rd and 25th Air Defense Regiments
Independent Units
41st, 42nd, 51st and 52nd Air Defense Regiments

The air defense regiments were deployed outside Japan in only one instance, to protect the oilfields at Palembang in Sumatra, perhaps the primary objective in Japan's decision to go to war in the first place. In March 1943 the Palembang Air Defense Headquarters was formed, consisting of the 101st, 102nd and 103rd Air Defense Regiments, as well as the 101st Machine Cannon Battalion. Each of these regiments consisted of 1,368 men with five batteries (20 guns) of 75mm AA guns, probably complemented by a machine cannon battery and a searchlight battery. In February 1944 the three 100-series air defense regiments were redesignated anti-aircraft regiments, the last AA regiments to be formed outside Japan for the rest of the war.

By the end of the war the Palembang units had been reconfigured significantly. By that time the 101st Regiment had been merged with the machine cannon battalion to form, in essence, a light AA regiment, the only such unit in the IJA, this probably spurred by the carrier raids on the refineries in February 1945. The 101st Regiment was configured as a headquarters with seven officers and 46 enlisted, and six batteries each with four officers and 126 enlisted. The heavier regiments each had an authorized strength of 1,321, divided into a headquarters (11 officers plus 116 enlisted), six gun batteries (each four plus 126) and three searchlight batteries (each four plus 134).

Table 6.4: AA Regiments at Palembang, August 1945

	101st Regiment	102nd Regiment	103rd Regiment
120mm naval	0	6	6
76mm naval	0	12	18
75mm Type 88	0	24	27
40mm Vickers	0	0	4
40mm Bofors	12	0	0
20mm Type 98	32	0	0
20mm Type 2	6	0	0
81mm AA mortars	100	0	0
70mm AA mortars	30	0	0
fire control radars	0	5	8
Japanese searchlights	0	11	13
captured searchlights	0	16	16
Japanese sound locators	0	19	21
Dutch sound locators	0	2	3

Within the homeland, however, the US bomber offensive began to make itself felt. Following the Doolittle raid the development of 88mm and 120mm guns accelerated. This did result in additional AA guns being supplied for defense, although it was diluted by the continually escalating requirements for such weapons from the combat theaters. By 1 June 1944 the total number of AA guns in the homeland had risen to 550, comprising 300 guns in the Eastern District, and 125 guns each in the Central and Western Districts.

As the inventory of guns increased the AA regiments in the homeland were enlarged, initially, from September 1942, as "type B" units with two gun battalions (each of three 6-gun batteries) and one searchlight battalion (with three batteries each with three lights). As equipment became available they were upgraded to "type A" regiments that doubled the number of batteries in each gun battalion and doubled the size of the searchlight batteries.

The Type A regiment was adopted in May 1943. This 3,142-man unit operated seventy-two 75mm Type 88 AA guns and thirty-six 150cm Type 93 searchlights, each of the latter paired with a Type 95 sound locater, along with 36 Type 92 machine guns on high-angle mounts. The regiments were immobile, with only enough transport for routine resupply, and relied entirely on telephones for communications.

Table 6.5: Type A Anti-Aircraft Regiment, from May 1943

	Officers	Warrant Off	NCOs	Privates	Carbines	Medium MGs	75mm AA Guns	Searchlights	AA Directors	Cars	Trucks
Regiment Headquarters	15	0	17	52	6	0	0	0	0	2	9
Two AA Battalions, each											
Battalion HQ	4	0	8	68	6	0	0	0	0	1	0
Six Batteries, each	4	1	13	128	6	2	6	0	1	0	0
Searchlight Battalion											
Battalion HQ	3	0	6	60	6	0	0	0	0	1	0
Three Batteries, each	4	1	13	150	12	2	0	6	0	0	0

In April 1944 a massive reorganization was ordered that redesignated the earlier air defense regiments and added to them with a new series of anti-aircraft regiments to yield a total of 17 regiments, numbered in blocks. Anti-Aircraft Regiments 111 through 118 were raised for and by the Eastern District, Regiments 121 through 125 by and for the Central District, and Regiments 131 through 134 by and for the Western District. At the same time the 3rd, 13th and 23rd Independent AA Battalions were formed in April 1944, followed by the 4th, 21st, 22nd and 24th in October, with the first group conforming to the 1941 mobilization plan, but the latter group being larger with four batteries per battalion.

Nevertheless, production of new weapons was agonizingly slow, the penalty for having ignored the development and industrialization of modern AA guns during the 1930s.

In mid-1944 IGHQ decided to place all AA artillery in Japan under the command of the 6th Air Army, which also controlled the fighters. Further study of this plan indicated, however, that since the air intelligence, signals and searchlight assets were under local control that coordination would prove impossible and this plan was dropped.

In October 1944 US bombers began their offensive against Japan from bases in the Marianas. By this time AA artillery strength had reached about 1,200 in the homeland.

Continuing shortages of equipment meant that some regiments were still organized as Type B units at the end of the war, and others were somewhere in between. These shortages plagued the air defenses to the end of the war:

> Newly organized anti-aircraft units frequently waited 2–4 months before receiving their equipment. No anti-aircraft directors were available to the anti-aircraft defenses of the industrial city of Nagoya, with its important manufacturing plants, including the Mitsubishi airplane works, until March, 1945; cables for these directors were not furnished until May, 1945. In Kyushu, where our invasion was anticipated, several anti-aircraft gun defenses could not fire at night, because they were without radars or searchlights, nor fire by day, under "unseen" conditions, because of the lack of radars.[1]

1 GHQ USAFPAC, AAA Research Board, *Survey of Japanese Anti-aircraft Artillery*, 1 February 1946, p. 68.

Although the regiments had a fixed organization, they rarely followed it, being organized to man whatever weapons were on hand in a given area. As a result their composition varied widely, as can be seen from an examination of the strengths of the 122nd and 123rd AA Regiments, both Central District units, at the time of the surrender:

	122nd AA Regt	123rd AA Regt
Personnel	4,948	1,900
75mm Type 88 guns	40	14
88mm Type 99 guns	36	18
120mm Type 3 guns	0	6
150cm searchlights	32	14
Sound locators	55	n/a
Type 1 radars	2	1
Type 3 radars	1	1
Height-finders (optical)	28	13
Air speed calculators	35	26
Predictors/directors	26	11
Radios	4	5
Telephones	431	140
Trucks	5	3

Most striking of all, of course, is the complete absence of mobility in both regiments, lacking even the transport to resupply themselves *in situ*, and relying entirely on telephones for the dissemination of warning and liaison information.

It had been planned early on to provide each gun battery with a fire control radar. The IJA planned to make the German Wurzburg their standard fire control radar, but by the end of the war only a single example had been placed in service. As an interim measure some elements of Japanese radars and the Wurzburg were incorporated into a new model, the Tachi 31, and 85 of these were built between October 1944 and the end of the war.

Due to a very poor distribution network it seems unlikely that more than about a hundred of the various radars were actually placed in service. This was clearly insufficient for 292 batteries in the AA divisions and as a result the fire control radars were actually usually assigned to the battalion HQs. Priority went to the 1st AA Division, which had 75 radars in its possession at the end of the war for 23 gun battalions (127 batteries), or three per battalion. The 2nd AA Division, on the other hand, had only six radars for its eight battalions (44 batteries). In any event, the performance of the radars was so poor that they were very rarely used for fire control, but instead served to locate targets for hand-off to the optical fire control equipment. To minimize maintenance burdens they were usually turned on only after receiving an alert from the early-warning network.

Commanding the Homeland AA Units

A variety of HQs were formed to control the burgeoning number of AA units in the homeland. In December 1944 the 1st Anti-Aircraft Division was formed from the units in the Eastern District, specifically the 111th through 118th AA Regiments, 1st through 4th Independent AA Battalions, 95th and 96th Field AA Battalions, 1st and 4th Machine Cannon Battalions, and the 1st Searchlight Regiment.

In April 1945 the Central District's air defense assets were split into two commands: the Nagoya AA Group and the Central AA Group. The former consisted of the 124th and 125th AA Regiments, 5th and 12th Independent AA Battalions, 97th Field AA Battalion, 12th and 106th Machine Cannon Battalions, and the 11th Independent Searchlight Battalion. The Central AA Group was made up of the 121st, 122nd and 123rd AA Regiments; 11th, 12th, and 22nd Independent AA Battalions; and the 11th Machine Cannon Battalion.

At the same time the Western AA Group was formed, combining the 131st through 134th AA Regiments; 21st, 23rd and 24th Independent AA Battalions; 21st Machine Cannon Battalion; 21st Independent Searchlight Battalion; and 55th Special Machine Cannon Unit.

On 6 May 1945 the three AA groups were reinforced and redesignated AA Divisions. The 2nd AA Division was the Nagoya AA group reinforced by the 47th Independent AA Battalion, three independent AA batteries, and six independent machine cannon batteries. The 3rd AA Division was the Central AA Group with the 45th Independent AA Battalion, two independent AA batteries, and five independent machine cannon batteries added. Similarly, the Western AA Group plus the 43rd Independent AA Battalion, two independent AA batteries, and seven independent machine cannon batteries formed the 4th AA Division.

The only part of the homeland left out of this plan was the northern island of Hokkaido, and in July 1945 the 5th AA Unit HQ was activated, along with the 67th and 68th Independent AA Battalions, two independent field AA batteries and a searchlight battery to provide air defense coverage.

Smaller Static AA Units

For operations where a full anti-aircraft regiment was not required, independent AA battalions were authorized. Since the AA regiments themselves were not very large and in any case the "non-field" AA units lacked the mobility needed to be useful in most theaters, relatively few of these units were actually raised and they served almost exclusively in the homeland.

The 1st, 2nd, 11th, 12th and 31st Independent AA Battalions were activated in November 1941. The 41st Independent AA Battalion was activated in April 1942 to provide air defense for Najin Fortress, but no other battalions appear to have been raised until 1944.

At that time the 3rd, 13th and 23rd Independent AA Battalions were formed in April 1944, followed by the 4th, 21st, 22nd and 24th in October, with the first group conforming to the 1941 mobilization plan, but the latter group being larger with four batteries per battalion. March 1945 saw the mobilization of the 5th Independent AA Battalion, in April the 47th, in June the 43rd and 47th, and July the 67th and 68th, this pair being for the 5th Area Army in the north.

One battalion served overseas, the 27th Independent Anti-Aircraft Battalion under the 32nd Army on Okinawa.

The original battalions were divided into a headquarters and two batteries. Each battery was allotted two Type 92 machine guns on high-angle mounts, four 75mm AA guns, a passenger vehicle and a light observation vehicle. The battalion HQ had one more passenger vehicle and one more light observation vehicle.

The strengthening of the AA defenses in the homeland, however, brought significant changes to this structure. Only the 13th and 22nd Battalions appear to have retained their two-battery organization, and they were each reinforced by a searchlight battery. By the end of the war most of the battalions had three or four batteries, each of four or six guns, with common configurations being four 6-gun batteries (total 753 men) and three 6-gun batteries (528 men).

Transport for such a battalion was limited to a single car and one to three trucks, all held in the battalion HQ. Deliveries of ammunition and most routine deliveries of food came directly to the batteries, bypassing battalion HQ, so little transport was needed. Normal communication was by phone. The battalion HQ had a radio transceiver and each battery a receiver, but these were used only when the phone lines were down.

The 67th and 68th Battalions each had a strength of 670 men, while the 27th Battalion had 505 in three batteries of six 75mm Type 88 guns each.

A number of independent AA batteries were also raised during the war. The majority of them were activated in early 1942: the 24th to 27th Independent AA Batteries in the Western District, the 31st to 35th Batteries in the Northern District, and the 40th to 45th Batteries by the Korea Army. The bulk of those raised by the Korea Army were eventually shipped overseas, the 42nd (Pagan), 43rd (Saipan) and 45th (Guam) Batteries being transferred to the Central Pacific in May 1944. Shortly after arrival they were absorbed by other units as part of the 31st Army reorganization, with the 42nd Battery becoming the AA unit of the 9th IMR, the 43rd Battery becoming the 3rd Battery of the reorganized 25th AA Regiment, and the 45th Battery the 1st Company of the reorganized 52nd Field AA Battalion.

Each of these batteries had a strength of 105 built around two 38-man gun platoons, each with two gun squads with 75mm guns, plus a command platoon with a 15-man fire control section, a 13-man signal section and an eight-man lookout/observer section. It had been planned to convert these Pacific-bound batteries to six-gun configuration, but it would appear that only the 42nd Battery was actually so reinforced.

A second wave of AA batteries was activated in early 1945, with the Eastern District mobilizing the 48th through 51st Independent AA Batteries for the 1st AA Division, the Tokai Army the 52nd through 54th Batteries for the 2nd AA Division, the Western District the 55th and 56th Batteries for the 4th AA Division, and the Central District the 57th and 58th Batteries for the 3rd AA Division. Each of these had a nominal strength of 4 officers and 140 enlisted and was armed with six guns. The 59th–61st Batteries were formed in Burma, each to man four captured 3.7" British guns.

Static anti-aircraft defense also included light AA weapons, although the Army was a bit slow off the mark in the organization of such units. The first of these units, the 1st, 2nd and 3rd Machine Cannon Battalions, were ordered activated on 14 May 1943 for assignment to the Eastern, Central and Western AA Brigades respectively, with the 4th Machine Cannon Battalion following shortly in the Eastern Brigade. Each of these had 535 men in three batteries. These were followed later that year by the 101st and 102nd Machine Cannon Battalions raised in Sumatra. Each of these East Indies-based battalions had a TO&E strength of 861 men.

In July 1944 the 103rd, 104th and 105th Machine Cannon Battalions were activated in Japan for the 32nd Army. Each of these had a 40-man HQ and three 100-man machine cannon companies, each with six 20mm autocannon, giving the battalion a total strength of 13 officers, 3 warrant officers, 42 NCOs and 266 privates of the line branch, plus 6 technical and intendance personnel, and 10 medical personnel. All three of these battalions were stationed on Okinawa and were destroyed there during the US invasion.

At about the same time the 12th Machine Cannon Battalion was raised for Homeland defense, along with the 106th Battalion which was, presumably, activated for service overseas but which stayed in Japan.[2] These were followed by the 4th Battalion in October 1944.

The TO&E for these machine cannon battalions called for a 27-man headquarters and four 108-man batteries. The battery was to man six twin-mount 20mm Type 4 guns in three platoons, although in fact the single-mount Type 98 appears to have been common. The battery was also provided with a single Type 94 one-meter base rangefinder and 7,200 rounds of 20mm tracer ammunition. The guns used on-carriage sighting equipment exclusively. A battery director capable of controlling and firing all six guns remotely had been developed and appeared effective, but production difficulties prevented significant fielding of the system before the war ended.

In practice the homeland machine cannon battalions varied widely in strength, from 304 men in three batteries in the case of the 23rd and 106th Battalions to 864 men in eight batteries in the 1st Battalion.

To complement these units sixty independent machine cannon batteries were ordered activated in April 1945, each of 138–144 men divided into three platoons, each with two Type 4 pedestal twin-mount 20mm AA guns.[3] A first group of twelve were activated, with the Eastern District providing the 1st through 6th, the Central District the 7th through 10th, and the Western District the 11th and 12th Independent Machine Cannon Batteries. This was followed by a second batch comprising the 13th through 19th (Eastern District), 20th and 21st (Korea Army), 22nd through 25th (Tokai Army), and 26th through 33rd (Western District). The third and final group was made up of the 35th through 41st (Eastern District), 42nd through 46th (Tokai Army), and 47th through 60th (Central District).

Six of these batteries (1st, 2nd and 13th–16th) were assigned to the 1st AA Division, six (23rd–25th and 42nd–44th) to the 2nd AA Division, five (47th–51st) to the 3rd AA Division, and eight (11th, 12th, 26th–28th, and 31st–33rd) to the 4th AA Division. Seven of the batteries (18th–21st and 35th–37th) were assigned to field armies for protection of tactical targets, while the rest were assigned to the 1st and 6th Air Armies for the defense of air bases.

Two anomalous machine cannon batteries, the 78th and 79th, were formed in Singapore to man captured 40mm Bofors guns. Each of these batteries was built around two platoons, each of which commanded three 10-man gun sections. In addition, the battery had a 21-man aerial lookout/warning section and a 9-man observer section for a total strength of 4 officers and 98 enlisted. The batteries included no central fire control equipment, as that had been destroyed before capture.

Field Anti-Aircraft Units

While these AA and machine cannon units were adequate, if only barely, to defend static targets their complete lack of mobility rendered them useless for the defense of tactical targets. For this role a separate series of units, field AA and machine cannon units, were raised.

2 At that time the 2nd Battalion was redesignated the 11th, and the 3rd Battalion the 21st.

3 A few batteries were issued Navy twin 25mm guns in lieu of the Type 4s.

For defense of maneuver units the mobilization plan called for the creation of "field" AA battalions and batteries. The field AA battalions came in two varieties: mobile (Type A) and semi-mobile (Type B). A type A field AA battalion consisted of a HQ, three batteries and a service battery. The battery of such a battalion consisted of a 42-man HQ platoon, two 31-man gun platoons, and a 43-man trains element. The HQ platoon included a 6-man AAMG section, a 13-man signal section and a 17-man observer/FDC section, the latter with a specially equipped truck known as a Type 94 Mk 1 observation vehicle. A gun platoon consisted simply of the platoon leader, two 13-man gun sections and four drivers. It was provided with two 75mm Type 88 guns, two gun tractors and two ammunition trucks. The trains element was provided with two more ammunition trucks, two baggage trucks, and two reserve tractors. The battery carried a total of 200 rounds for each 75mm gun and 2,160 rounds for each MG.

The field AA battalion (Type A) also included a service battery with 17 trucks carrying baggage, rations and 800 rounds of 75mm ammunition, along with a maintenance section with two repair trucks. The battalion HQ operated its own FDC section with a Type 94 Mk 1 observation vehicle.

The type B field AA battalion was similar to the type A battalion, but with a much reduced scale of transport. A gun platoon was identical to that of the "A" battalion, but without the drivers and vehicles. The battery HQ platoon included a 14-man observation section (manning the height-finder), an 8-man director section, a 13-man signal section, a 7-man AAMG section, a 6-man lookout section and a single driver with a car. Gun section size was reduced to twelve men and the platoon's drivers eliminated. The battery HQ itself had the battery commander, two runners, two buglers, two medics and a second driver with a car. The battery ammunition trains had only a warrant officer, an ammunition specialist and two medics. In this case, the motor vehicles were concentrated in the service battery, which could move only one battery at a time. Ammunition allocation remained the same as in the Type A battalion, but this (along with baggage and rations) almost certainly could not be moved by the battalion in a single lift.

Experience worldwide was to soon show that single batteries of heavy AA guns had little value, except for morale. Nevertheless, the Army formed such batteries throughout most of the war. The mobilization plan specified an independent battery as similar to the battery in the Type A battalion, but reinforced with its own maintenance section.

The first mobilization of field AA battalions did not occur until July of 1941. At that time the 32nd through 36th, 40th, 44th and 45th Field AA Battalions, all Type B, were activated and assigned to the Kwantung Army. A month later a second, more mobile, wave was ordered activated for the Kwantung Army as well, this consisting of the short-lived 46th and the 49th through 56th Field AA Battalions, all Type A; and the 48th Type B Field AA Battalion. One further field AA battalion, the 47th (Type B) was also raised at the same time but kept in GHQ reserve in the homeland. As war approached two further battalions, the 57th and 58th, were activated in Japan. These two battalions were the last Type A field AA battalions to be mobilized. The army, short of motor vehicles for every application, simply could not afford to equip further units of this type. Thus, at the outbreak of the Pacific War there were two distinct groups of Field AA Battalions, those in the 32 to 48 number block as semi-mobile "B" units, and numbers 49–57 as mobile "A" units. Three more "B" battalions, 38th, 39th and 41st, were mobilized in January 1942.

The thrust to the south included the reassignment of the 30th, 31st and 45th Field AA

Battalions to the 14th Army (Philippines); the 33rd, 34th, 35th and 51st Battalions to the 25th Army (Malaya); and the 44th Battalion to the 16th Army (NEI) in November 1941. The 16th Army was further reinforced by the 45th and 48th Battalions in January 1942.

When units took up position in the operational southern regions a few changes to organization became necessary. The battery of a "B" battalion retained the same strength, at 121 men, and its gun platoons remained unchanged, but about nine men were customarily moved from the HQ platoon to the ammunition trains to deal with the combat demands for ammunition supply.

As the air portion of the Allied offensives in the south Pacific picked up in intensity all the remaining "A" and "B" type field AA battalions were pulled out of Manchuria and shipped out. Two remained on the mainland, the 46th (B) in Korea and the 55th (A) in China. The others, however, were sent to New Guinea and the Solomons. In some cases the battalions of this second wave to the Pacific were reorganized to increase firepower and reduce shipping. The most noticeable change was the addition of a third gun to each firing platoon of the batteries.

The type B (semi-mobile) units sent overseas found themselves in static duties and some were reconfigured locally with six Type 88 guns per battery while retaining the original manpower authorizations. This was accomplished by reducing the number of motor vehicles authorized for the battalion from 25 to 12 and otherwise reducing overhead and using the manpower so saved as gun crews. As thus reorganized the AA battery was reconfigured into two platoons each of three squads.

Similarly, many of the batteries of the Type A (mobile) battalions sent to the south Pacific were now built around two platoons, each consisting of a platoon leader and three 15-man gun sections, with each section provided with five rifles and carbines, a Type 88 gun and a tractor. The battery ammunition section consisted of seven men with a single truck. The battery HQ had a 7-man observation section, a 7-man signal section, a 6-man air warning section (two of them also doing duty as a tractor crew), a 6-man AA MG section with a truck, and a 9-man command section. Gone were almost all of the trucks, so that while the battery still had a limited ability to shift firing positions, it could not undertake tactical or operational moves without the attachment of motor transport assets from outside.

The one anomalous unit in the mix was the only field AA battalion sent to the Central Pacific. This unit, the 52nd (A) Battalion, was dispersed with its HQ and one battery on Guam and one battery each on Yap and Woleai. Although the unit does not appear to have taken its motor vehicles with it to the Pacific, the batteries remained armed with two Type 92 MGs and four Type 88 guns apiece.

With Japanese air supremacy largely unchallenged for the first half-year of the Pacific War the original group of "A" and "B" battalions proved sufficient. Thus, it was not until February 1943 that a new group of field AA battalions were activated: the 58th–60th for the 8th Area Army in New Guinea and the Solomons. These were followed in April by the 61st to 63rd, all destined for the East Indies and in late June by the 64th–66th Battalions for the 18th Army in New Guinea. In October the 67th–73rd Battalions were activated, to be scattered about the southern regions from New Guinea to Burma.

These wartime field AA battalions (numbered 58 and higher) followed a new special TO&E that increased firepower without increasing manpower. A third gun was added to each firing platoon to bring the battery up to six 75mm guns and the battalion to eighteen.

A platoon thus consisted of three 14-man squads plus two men in the HQ. The battery HQ comprised an 8-man command group, an 8-man lookout (early warning) squad, a 10-man observer squad, a 7-man machine gun squad and a 7-man signal squad. The battery also included a 10-man ammunition platoon. The amount of transport authorized for the battalion is not clear, although it seems likely that it was supposed to be somewhere between the "A" and "B" models, seemingly closer to the former. At full strength these battalions were apparently to have six tractors and eight trucks in each battery, plus 17 more trucks in the battalion ammunition trains and a further 19 in the battalion HQ. In practice, of course, these figures were probably rarely reached.[4] These elements combined to give the new field AA battalions a total strength that varied slightly from 545 (in the earlier 1943 battalions) to 528 (in the later 1943 battalions).

January 1944 saw the activation of the 74th and 75th Field AA Battalions in Japan for duty in China, followed by the 76th to 78th Battalions in May for the Philippines. In July the 79th to 81st Battalions were activated for the 32nd Army and the 82nd and 83rd Battalions for the Formosa Army. These were followed in August by the 84th to 88th Battalions, the 89th Battalion in September, the 90th to 91st Battalions in October and the 95th to 98th Battalions in December, with most of these units destined for the homeland and the Kwantung Army. As the ring constricted around Japan, however, the need for mobile AA unit diminished and efforts were concentrated instead on static units. As a result only two field AA battalions were raised in 1945, the 94th (for Singapore) and the 99th (for China). These 1944/45 battalions generally adhered to the 1943 organization with eighteen 75mm guns, six machine guns, and 84 rifles each. Once again, authorized strength varied slightly, the normal baseline strength being 521 (17 officers, 3 WOs, 45 NCOs and 440 privates of the line branch and 2 officers, 7 NCOs and 7 privates of the medical branch) for the units assigned to the Pacific, while those assigned to the Kwantung Army and China had an additional six sergeants of the line branch, bringing strength to 527 men.

At the same time the field AA battalions were activated, a number of independent field AA batteries were also being mobilized. Such a unit was made up of a 70-man HQ, two 35-man gun platoons and a 35-man trains element. Each gun platoon consisted of a HQ and two gun sections, each manning a 75mm Type 88 gun.

As activated in Manchuria, a gun section was made up of a section leader NCO, twelve gun crew and four drivers and assistant drivers. The section had one 75mm gun, a tractor and a truck. The platoon HQ was made up solely of the lieutenant platoon leader. The trains element consisted of a warrant officer platoon leader, two 13-man ammunition squads (each with three trucks), and a 16-man maintenance squad (with two cargo and two shop trucks). The HQ platoon was made up of a 19-man observation section, a 15-man signal section (with 10 field phones and no radios), and a six-man MG squad. The battery HQ itself had a total of nine men. The HQ and HQ platoon were provided with three field cars, a truck, a tractor and an observation vehicle for transport. As thus configured, the independent field AA battery of the Kwantung Army was somewhat more mobile than the standard envisioned by the 1941 mobilization plan, but stayed within the overall guidelines for manpower, weapons and equipment.

The first eleven units of this type (21st to 31st Independent Field AA Batteries) were

4 In fact in the 68th independent Field AA Battalion on New Guinea five of the six guns in each battery were static pedestal mounts that had to be set onto concrete bases.

activated in July 1941, with six going to the Kwantung Army. Three more (32nd to 34th Batteries) were activated in January 1942. The 35th through 41st Batteries were activated in late December 1942, followed by the 42nd in June 1943 and the 43rd–54th in November, mostly for the southern regions. This group found itself almost without exception in the East Indies, New Britain, and New Guinea. In July 1944 two more batteries (53rd and 54th) were activated for duty in China, and three batteries (55th to 57th) were mobilized in Manchuria by the Kwantung Army. Finally, December 1944 saw the creation of three batteries (61st to 63rd) by the Southern Army, followed by the 65th Battery in Manchuria and the 66th and 67th Batteries in Hokkaido in 1945.

The independent field AA batteries that served in Manchuria and the central Pacific retained their four-gun organization. Units sent to the south, however, were often reconfigured. An example was the 31st Independent Field AA Battery, activated in July 1941 in Japan and sent to Manchuria in August. In December it joined the Philippine invasion forces and in February 1942 participated in the invasion of Java before returning to the homeland in May. On 20 November 1942 it was reorganized to consist of 162 men, 6 Type 88 AA guns (in two 3-gun platoons) and 5 motor vehicles and within a month had been reassigned to the Southern Army and sent to Rabaul. It January 1943 it was moved again, to Kolombangara. Such a reorganization, of course, reduced the battery from a fully-mobile to a semi-mobile unit. On the jungle islands of the south Pacific, however, the distinction was often academic.

Field Machine Cannon Units

The elimination of autocannons from the organization of the anti-aircraft regiment in October 1940 freed a number of these weapons for use in independent batteries. The 1941 mobilization plan provided for thirteen of these batteries, numbered 15 to 27. Six of these batteries (16th, 21st, and 25th–27th Field Machine Cannon Batteries) were motorized units, while the remainder used pack transport.

Both motorized and pack-type field machine cannon batteries were equipped with six 20mm Type 98 autocannon. In the case of the motorized battery, transport was provided by 12 cross-country and three road-bound trucks, while in the pack-type battery similar duties were performed by 36 pack horses to carry the guns, 42 more for ammunition, 29 for rations and the balance for general supplies and equipment. A motorized battery had a basic load of 3,660 rounds of HE and 4,340 rounds of AP ammunition, while the pack-type battery was provided with only 2,400 rounds of HE and 1,200 rounds of AP.

The first four batteries (15th to 18th) were activated in July 1941, followed by the remainder of this group plus two more motorized units, the 28th and 29th, the following month.

The batteries relied on telephones both for contact with outside units and between battery HQ and platoon HQs for warning, there were no radios.

The field machine cannon batteries did not participate in the initial expansion to the south. In late 1942 and early 1943, however, the bulk of them were dispatched to New Guinea and New Britain. For this change of theater the machine cannon batteries were reorganized.

The new organization mandated for the pack-type units for the southern areas by November 1942 Army Order A-99 reconfigured the batteries as semi-motorized formations. Such a unit had a strength of four officers, 16 NCOs and 137 privates divided into a 13-man

HQ (including a 3-man sanitation squad), a 24-man command platoon (comprised of a 12-man lookout squad, a 7-man observation squad, and a 5-man gas squad), three gun platoons, and 21-man trains with a single truck. Each platoon was made up of a platoon leader, two 9-man gun squads (each with a 20mm autocannon and a truck) and two 7-man machine gun squads, each with a 13.2mm heavy machine gun. With seven trucks the battery was mobile to the extent that they could carry most of their heavy gear, but the troops obviously would have had to walk.

Units deployed to Rabaul apparently kept their trucks as long as they were there, but those deploying to less-developed areas quickly had theirs taken away as useless. At the same time, strengths were reduced and, in some cases, the battery commanders chose to reorganize. An example was provided by the 15th Field Machine Cannon Battery, which converted to the semi-motorized configuration in Japan in November 1942, soon moved to Rabaul and then, in May 1943, moved to Finschafen in New Guinea, without its trucks and at a reduced strength.

The new organization of this battery had 111 men with the battery commander, no HQ unit, and two "units," one with 20mm Type 98 autocannon and the other with 13.2mm MGs. The 20mm group consisted of a 20-man HQ and two gun platoons, each with a commander and three 7-man squads. The machine gun group was made up of a 7-man HQ and six squads (no platoon HQs) each of five men. In addition, the battery included a 10-man ammunition trains (7 men for 20mm and 3 for 13mm). With the battery HQ personnel, such as medical, gas, observers and spotters, decentralized to the new "unit" headquarters, this was clearly optimized for separate operation in two locations. Organizational formats used by other units included three autocannon platoons (each two 20mm) and one machine gun platoon (six 13mm MGs), and the original mixed organization of three platoons each with two 20mm and two 13mm weapons.

The motorized-type field machine cannon batteries lost most of their motor vehicles when they deployed to the south Pacific but increased their original manpower allocation to accommodate extra weapons. An example of such a unit was the 27th Battery in the Solomons, which consisted of three platoons, each with a platoon leader and four 9-man squads, two equipped with a 20mm gun and a truck and the others with a 13.2mm MG. The battery also included a 19-man trains and a 33-man HQ, the latter including a 7-man spotting detail and a 12-man lookout section.

In August 1943 a new group of field machine cannon batteries was activated, the 30th to 39th, all for the 18th Army. Being destined for New Guinea and environs to start with they adopted a new organization of 105 men with no transport. As originally raised, each of these batteries was organized into a 34-man HQ (including an 18-man observation group), two 25-man platoons, and a 21-man ammunition trains. Each of these platoons was made up of the platoon leader and three 8-man squads each manning a 20mm gun. Another block of batteries, the 40th to 47th, was activated in October and November 1943, all but one for the south Pacific.

The one battery dispatched to the central Pacific, the 44th, seems to have retained its original organization. Other units deployed to the south Pacific, however, were invariably given six 13.2mm MGs to augment their firepower, although with no additional manpower. This 105-man organization immediately became the standard for all subsequent field machine cannon batteries raised, with machine guns added only for those destined for the south Pacific.

In February 1944 two batteries (48th and 49th) were raised for China, followed by three more (50th to 52nd) in May 1944 for the south Pacific. This was followed by a large group of eighteen batteries (53rd to 70th) raised in July 1944, this being subdivided in blocks with the first three going to the Philippines, the next six to the Formosa Army, the last three to the Kwantung Army, and the remainder to Japan and China. A further seven batteries (71st to 77th) were activated in October 1944 for the Kwantung Army. Following this the 78th and 79th Batteries were raised in Singapore and the 80th in Burma. The final mobilization of field machine cannon batteries came in February and March of 1945, when the China theater raised five batteries (81st to 85th) and the 10th Area Army (formerly the Formosa Army) raised eight more (86th to 93rd).

Special Machine Cannon Units

The one remaining type of air defense unit resulted from an inter-service marriage of convenience. Delays in shipbuilding and an increasing shortage of fuel for the remaining ships left the IJN with a considerable surplus of their 25mm light AA guns in both single and twin mounts. The Navy proposed fitting these weapons to extemporized land mountings and turning them over to the Army, with the understanding that a substantial portion of them would be used to defend naval bases and airfields. The Army agreed and on 22 August 1944 officially activated the 1st through 55th Special Machine Cannon Units, followed a month later by the 56th to 62nd Units, and then the 63rd to 72nd in February 1945.

Each of these battery-size units had a TO&E strength of 85 men divided into a 19-man HQ and six 11-man sections. Each section was provided with either a twin-mount 25mm or two single-mount 25mm. There were three standard configurations for the unit: one with six twin mounts, a second with six single mounts and three twin mounts, and a third with twelve single-mounts. A battery was authorized 12,000 rounds of ammunition as organic holdings. Nominally, these units were motorized formations with each authorized about 20 trucks. It is not known, however, how many actually received their motor vehicles. Certainly, many were completely immobile once deployed.

The 1st through 6th Special Machine Cannon Units were raised in the East Indies, the 7th through 19th in the Philippines, and the 20th and 21st for Iwo Jima. The 22nd through 30th Units were raised on Mindanao for the defense of Navy airfields. The 31st to 39th were activated on the Palau islands. The Navy claimed the 41st, 42nd, 45th–50th, and 55th Units for the defense of their airfields in the Ryukyus. The 45th and 46th Units were assigned to the Bonins, the 51st–54th to Luzon, the 58th–62nd to Palembang, and 63rd–72nd to the Philippines.

An example would be the 21st Special Machine Cannon Unit, activated in May 1944 by the Western District Army in the homeland, based on a cadre of one officer and 37 enlisted from the 3rd Independent Machine Cannon Battalion, followed by instruction in the 25mm gun by Navy officers and NCOs. In June they received their three twin-mount Model 1 and six single-mount Model 3 weapons and immediately embarked for Iwo Jima, where the unit was later destroyed in combat.

Field Searchlight Units

Only a small number of field searchlight battalions were activated during the war. Such a unit was divided into an 80-man headquarters and two 185-man searchlight batteries. Each searchlight battery was in turn divided into two platoons, each of which had three sections. A section was provided with a 150cm searchlight and a Type 90 sound locator. When units converted to the Type 1 searchlight the separate generator trailer would have been added to each section. The unit was semi-mobile, with the battalion HQ holding all of the motor vehicles, consisting of a passenger car and six trucks. The unit had no radios, further reducing its capacity for mobile operations, relying instead on 30 telephones. For local defense the battalion was provided with 98 carbines and 100 rifles.

The 1st Field Searchlight Battalion was included in the Anshan Defense Unit of the Kwantung Army pre-war, while the 2nd, 3rd and 4th Battalions were activated in August 1941 for the 3rd, 4th and 5th Armies, respectively, with the 3rd moving to the south Pacific in 1942. In May of 1944 the 2nd Battalion was reassigned to the Southern Army and moved to the Philippines, followed by the 4th Battalion, and in October the 5th and 6th Battalions were activated, followed shortly by the 7th Battalion, all for service with the 3rd Army in Manchuria.

In addition, the 1st through 4th Independent Field Searchlight Batteries, each of 185 men, were activated in December 1942, the 7th and 8th for New Guinea in April 1943, the 9th through 12th for the Indies in November 1943, and the 19th in July 1945.

Higher Commands

Varying levels of control were exercised over both static and field anti-aircraft units by assorted command echelons. The Kwantung Army formed the Kwantung Field AA HQ in 1939 to control its air defense assets and in December 1940 this was redesignated the 3rd Field AA HQ, controlling the 26th AA Regiment and the 1st Manchukuoan AA Regiment. In July 1941 the 15th and 16th Field AA HQs were formed in Manchuria to control a variety of units of battalion size and lower, and in May 1942 the 14th Field AA HQ was activated by the Kwantung Army to control the 10th and 25th AA Regiments. In addition, the 11th, 12th and 13th Field AA HQs were sent to Manchuria in August 1941, but do not appear to have been used.

While these controlled the field AA defenses for the Kwantung Army the defense of strategic targets was entrusted to three Kwantung Army Air Defense Commands (KAADC) raised in December 1940, with the 1st KAADC controlling the 12th and 19th AA Regiments, the 2nd KAADC the 9th and 18th Regiments, and the 3rd KAADC the 10th, 13th and 17th Regiments.

As the war expanded abroad, further field AA HQs were raised for overseas duty: the 17th HQ for the 25th Army's invasion of Malaya and the 18th HQ for the 16th Army's invasion of the East Indies, both in late 1941. The 15th Field AA HQ moved from Manchuria to New Georgia in December 1942 to command the 41st and 58th Field AA Battalions, the 31st Independent Field AA Battery, the 22nd, 23rd and 27th Field Machine Cannon Batteries, and the 3rd Field Searchlight Battalion, with the 12th Field AA HQ following for the 18th Army, the 19th at Rabaul, and the 20th for the 2nd Army, although the 15th and 19th were abolished in June 1944. The 21st HQ was formed in August 1944 to control the AA defenses of Okinawa.

For defense of the vital Sumatra oil fields the Army formed mixed air defense units under the command of the 9th Air Division. In addition to the 8th Air Brigade, this division also contained the Palembang Defense Unit and the Pangkalanbrandan Defense Unit. The Palembang unit was made up of the 21st and 22nd Fighter Regiments; the 101st, 102nd, and 103rd AA Regiments; the 101st AA Balloon Regiment, and the 101st Machine

Cannon Battalion. The Pangkalanbrandan Unit was comprised of the 71st Fighter Squadron, the 104th AA Regiment, the 102nd Machine Cannon Battalion, and the 14th Air Intelligence Regiment.

In total, few such higher HQs, however, appear to have been needed, perhaps due to the dispersed nature of fighting in the Pacific and limited need for air defenses in China.

Shipping Artillery

Although not directly relevant to the subject some mention should be made of the Shipping Artillery. These provided the guns and crews to protect Army-controlled transports and cargo vessels, including the large number of civilian charters, mainly against air attack. At the start of the Pacific War the Shipping Artillery Regiment consisted of two AA battalions (each of three companies), a machine cannon battalion (three companies) and a depot. By early 1944 it had split into two regiments: the 1st, based in Japan, and the 2nd, based in Singapore until July 1944, then Manila. Each consisted of fifteen AA batteries, three light AA batteries, three surface gun batteries, a mortar company, a machine gun company, a depth charge company, a hydrophone company, an air watch company and two sea watch companies. Each of the AA batteries manned six 75mm AA guns, each light battery sixteen 20mm AA guns, and each surface battery six 75mm Type 38 field guns. The regiments and batteries were primarily administrative HQs, as the gun crews were formed into detachments for each ship as needed. As an example, the 9,800-GRT Ayatosan Maru in 1942 was provided with a detachment consisting of a 4-man spotting squad, a 2-man signal team, and three 20mm platoons each of two 6- or 7-man gun crews.

For the defense of smaller vessels the Army raised the 1st and 2nd Shipping Machine Cannon Regiments, each of two light AA battalions and a machine gun company, each of three companies, for a total of 4,920 men.

On a related, but smaller-scale, it is worth noting that the Navy's Central Pacific Area Fleet felt obliged in April 1944 to follow suit and order each of its base commands to activate five shipboard air defense squads, three Type A (each 14 men with two 13mm MGs), a Type B (15 men with two 25mm single-mount guns) and a Type C (ten men with two 7.7mm MGs). These would be assigned to shipboard defense as needed by the base force commander.

Air Intelligence

Providing early warning of the approach of enemy aircraft was the responsibility of the air intelligence formations. Part of the Army Air Force, these units primarily served to alert fighters, but were also sometimes tied into the AA artillery net as well. The first-generation, known as air intelligence units, were battalion-size and consisted of a headquarters and two companies, each company having two platoons. A platoon would have four sections, each of an NCO and seven men each with a Mk 3 radio or a telephone, to act as spotters. The 1st and 2nd Units were formed in September 1938, and the 3rd in March 1940, all under the 2nd Air Army in Manchuria. The 15th and 16th Units were formed in 1938 for China.

The 16th Unit participated in the invasion of Malaya in December 1941 then moved to Burma in March 1942. The dispatch of this unit was made possible by the activation in China of a new type of formation, the air intelligence regiment. This was similar to the air intelligence unit, but could have three or four subordinate companies. The first of these, the 6th Air Intelligence Regiment, was formed in June 1942. This was followed in June 1942 by the 1st Regiment, a unique training unit in the homeland.

The 20th Unit was formed in August 1942 for Sakhalin, then the 14th Unit in October for the southern areas. The most active of these formations, the 4th Air Intel Unit, was formed in November 1942 and shipped to Rabaul the following month. At the same time the 7th Unit was formed for the southern regions and the 8th to support the 7th Air Division on New Guinea.

Air intelligence elements continued to expand, retaining the "unit" designations for the homeland, Korea and Formosa to the end of the war, while switching to the "regiment" name overseas. There were two drivers for the adoption of the larger units. The first was simply the geographical dispersion of the air war which was such that, for instance, the 4th Air Intelligence Unit had to provide a large number of observation posts around the Bismarks and New Guinea, necessitating the addition of a third company in mid-1943.

The second was the introduction of the new Type B air warning radars, initially the Tachi-6, then the Tachi-7, which prompted the incorporation of new radar companies into the organizations. A radar company was usually composed of four 50-man platoons, each of which was responsible for a single radar set. One or two such companies were added to most of the air intelligence units in mid-1943.[5] The 4th Unit, for example, received a radar company in mid-1943 and a second in early 1944, necessitating its redesignation as the 4th Air Intel Regiment. Similarly, the 2nd Air Intel Regiment was formed for Burma in August 1943, the 15th Unit in China was expanded and redesignated the 6th Regiment in November 1943, the 5th Air Intel Regiment formed in China and the 7th for Singapore and the 14th Unit in Sumatra redesignated a regiment in April 1944,[6] the 1st and 2nd Units in Manchuria were disbanded to form the 11th Regiment in October 1944.

The 6th Air Intelligence Regiment comprised a radar company, a signal and signals intercept company and a lookout company, with the signals elements talking directly to Tokyo and Manchuria to provide early warning.

Where the chain of command tended to be more compact, as in the homeland, the early warning units retained the "unit" designation regardless of size. The largest of these were those formed for the home islands, the 32nd, 35th, 36th and 37th Units in April 1945, followed by the 31st and 33rd Units in June.[7] Strengths varied from 2,584 to 6,414, with each including a small HQ, a signal unit, and a warning broadcast unit. Each also included one or two visual observation groups (each 450–550 men) and three to five radar companies (each about 350 men).

Also formed were the Fifth Area Army Air Intel Unit in April 1944, the Formosa Army Air Intel Unit (later the 10th Area Army Air Intel Unit) and the Korea Army Air Intel Unit (later the 37th Unit) in July 1944, and the 32nd Army Air Intel Unit on Okinawa in April 1945. The Korea unit had 3,369 men in eight visual platoons, seven small radar companies, and the usual base units, and was probably representative of the others.

---

5 Such was the importance of radar that the 9th Unit was actually formed in March 1944 for the Philippines with just a 27-man HQ and two 260-man radar companies.

6 Interestingly, the 14th Regiment in Sumatra continued to use its bistatic radars as well as the newer Type B radars to achieve very good coverage of the approaches to Palembang.

7 The 35th Unit was a redesignation of the Chubu Army Air Intel Unit, formed in June 1944.

7

Other Army Units

Airborne

There must have been some thought of using airborne forces in the late 1930s, for development of the take-down version of the Type 99 light machine gun is said to have begun in 1938. In any event, the training of Army paratroopers began in December 1940. On 15 November 1941 the Army was able to the form the 1st Raiding Regiment, followed by the 2nd Regiment on 1 December.[1]

On 1 December 1941 the 1st Airborne Brigade (*Teishindan*) was activated at Nittabaru Airfield on Kyushu. This was made up of a HQ, a signal unit and the two parachute regiments, with the following strength:

	Off	WO	MSG	SGT	PVT	Total
Brigade Headquarters	9	0	12	0	32	53
Signal Unit	n/a	n/a	n/a	n/a	n/a	n/a
Two Parachute Regiments, each	33	24	7	140	556	760

Each of the parachute regiments, actually battalions, was divided into three 135-man rifle companies and an engineer company. A rifle company consisted of three rifle platoons (three rifle sections each),[2] a machine gun platoon (two squads) and an anti-tank section with a single gun (37mm AT, 20mm ATR or 37mm infantry gun). The engineer company had three platoons equipped with flamethrowers and demolition charges.

On 8 December the 1st Airborne Brigade was reassigned from the direct control of the Inspectorate-General of Air to the Southern Army. The brigade, less the 1st Parachute Regiment, was assigned the mission of capturing intact the oil refineries and airfield at Palembang. The 2nd Regiment was dropped in two waves on 14 February 1942. The airfield was taken after a two-day battle, but the attempt to take the refineries in a *coup de main* was only partially successful, due in part to the fact that all heavy equipment, dropped separately, got lost. No further major airborne operations were carried out until December 1944, and in May 1942 the 1st Airborne Brigade reverted to the command of the 51st Air Training Division in the homeland.

On 31 July 1942 the 1,058-man 1st Parachute Training Regiment was reconfigured as an operational unit and redesignated the 3rd Parachute Regiment to expand the 1st Airborne Brigade. At the same time a new 4th Parachute Regiment was activated. The force was again reorganized on 10 August 1943 when the 3rd and 4th Parachute Regiments were

1 Designated *Teishin* (raiding or commando) units but called airborne or parachute here to avoid confusion with later, non-parachute, raiding/commando units.

2 Initially paratroopers were armed only with pistols and grenades for the drop, with the Type 99 short rifles and heavy weapons being dropped separately in containers. This left the troops very vulnerable to immediate counterattack, and starting in 1943 paratroopers were equipped with Type 2 take-down rifles that could be strapped to the individual.

placed on slightly larger TO&Es and the 1st Airborne Tank Unit (*Teishin Senshatai*) and the 1st Airborne Engineer Unit (*Teishin Kokeitai*) added. The Airborne Brigade then had the following strength:

	Off	WO	MSG	SGT	PVT	Total
Brigade Headquarters	9	0	12	0	32	53
1st & 2nd Parachute Regiments, each	33	24	7	140	566	760
3rd & 4th Parachute Regiments, each	38	31	18	116	615	818
Airborne Tank Unit	20	10	8	43	220	301
Airborne Engineer Unit	17	6	6	36	192	257

Added at the same time was the 1st Airborne Flying Brigade, which consisted of the 1st and 2nd Airborne Flying Regiments (each about 40 Type 100 transports)

The larger parachute regiments had three rifle companies, a heavy weapons company and an engineer company. Such a regiment was armed with 769 pistols, 455 rifles, 27 light machine guns, six medium machine guns, four 70mm infantry guns and four 37mm AT guns or 81mm mortars.

The airborne engineer unit was a glider-borne outfit with two engineer companies and an equipment platoon.

The airborne tank unit was formed in anticipation of the entry into service of the large Ku-7 glider. It was organized into a headquarters, a light tank company and a motorized company and was originally equipped with about 24 Type 95 and Type 98 light tanks, some of these later being replaced by Type 2 light tanks. In the event, the Ku7 glider was never placed into production and the Airborne Tank Unit never achieved actual airborne status.

In late 1943 the 1st Airborne Brigade was assigned to the 3rd Air Army in the southern area, but on 20 June 1944 the brigade (less 3rd and 4th Parachute Regiments) was reassigned back to the 1st Air Army in the homeland. The two wayward parachute regiments, instead, were sent to the Philippines.

With the new parachute weapons, particularly the take-down versions of the rifle and light machine gun, it proved possible to drop troops and their weapons together. The November 1943 load guidance provided that each paratrooper would carry two days of rations plus, in the case of a rifleman, his rifle, 120 rounds of ammunition and three hand grenades. A light machine gunner would drop with his machine gun, two 30-round magazines and two hand grenades, while his assistant would jump with a pistol, 180 rounds of MG ammunition, gun accessories and three grenades. A grenadier would drop with a pistol, his Type 89, two hand grenades and six discharger grenades. Other heavy weapons operators would carry a pistol and three hand grenades.

A major reorganization of the airborne forces came on 21 November 1944. On this date the 1st Parachute Group (*Teishin Shudan*) was formed, along with a second brigade HQ, two glider infantry regiments (*Kakku Hohei Rentai*) (actually battalions) and a machine cannon unit (*Teishin Kikanhotai*). With new TO&Es in force for all units, the Parachute Group had the following strength:

	Off	WO	MSG	SGT	PVT	Total
Group Headquarters	33	0	40	0	86	159
Two Brigade Headquarters, each	15	0	44	0	81	140

	Off	WO	MSG	SGT	PVT	Total
Four Parachute Regiments, each	49	33	30	121	661	894
Two Glider Infantry Regiments, each	49	28	30	109	710	926
Airborne Tank Unit	25	16	8	69	324	465
Airborne Machine Cannon Unit	17	8	5	34	244	308
Airborne Engineer Unit	19	9	56	0	341	425
Airborne Maintenance Unit	9	1	28	2	300	340

Each of the glider infantry regiments, like the parachute regiments, were actually battalions in strength, consisting of a 38-man HQ, three line companies, an anti-tank company, an infantry gun company, and a 30-man signal unit. The three line companies were identically organized, each consisting of three rifle platoons (each 3 light MG and 3 grenade dischargers) and a machine gun platoon (4 medium MGs). Because each platoon had not only a lieutenant platoon leader, but also a warrant officer deputy, the company had a strength of 7 officers, 6 warrant officers, 23 NCOs and 127 privates (including attached). Two of these companies were denominated rifle companies while the third received special training and was called an engineer demolition company. The anti-tank company had 156 men in a HQ, two gun platoons (each two 47mm) and an ammunition platoon. The infantry gun company had 180 men in a HQ, two gun platoons (each two 75mm regimental guns) and a supply platoon. The regiment was to be carried in Type 4 (Ku8) gliders, each of which could carry 20 men or a mountain gun with crew.

The tank unit was now a battalion consisting of a headquarters with two light tanks, a tank company with twelve more, an infantry company, an AT company (Type 94 tankettes towing four 47mm guns), a motor transport company with light trucks and a maintenance unit. The light tanks appear to have been Type 2 models.[3] The engineer unit retained its two-company organization.

In late November 1944 the Airborne Group HQ and the 2nd Airborne Brigade HQ flew south to take control of the airborne forces in the Philippines. For the most part the airborne forces fought as ground infantry. The nominal order of battle for the airborne forces at the beginning of 1945 was thus:

- 1st Airborne Group (Philippines)
  - 2nd Parachute Brigade
    - 3rd Parachute Regiment
    - 4th Parachute Regiment
  - 2nd Glider Infantry Regiment
  - Airborne Engineer Unit
- 1st Parachute Brigade (Japan)
  - 1st Parachute Regiment
  - 2nd Parachute Regiment
  - Airborne Tank Unit
  - Airborne Maintenance Unit

3 In addition, an extemporized 5th Airborne ("raiding") regiment was formed on Luzon in late October 1944, consisting of one infantry battalion (with three line companies, a heavy weapons company and a labor company) and a heavy weapons battalion (with mountain artillery, anti-tank and machine cannon companies).

1st Glider Infantry Regiment (Japan)
Airborne Machine Cannon Unit (Japan)

The 2nd Parachute Brigade undertook small-scale parachute and air landing raids, of platoon or company size, on Leyte and Ormoc in late 1944 and early 1945 without much effect.

On 16 February 1945 the 1st Parachute Brigade (plus 1st Glider Infantry Regiment) was officially detached from the 1st Airborne Group and assigned to the new 6th Air Army in the homeland. Meanwhile, the Airborne Group HQ was also functioning as the Kembu Group HQ and thus quickly lost its airborne characteristics, and the troops were integrated into the the Kembu Group itself. The Airborne Engineer Unit was disbanded in mid-1945. The 1st Parachute Brigade remained in Japan to the end of the war and took no further part in combat.

The IJA parachute arm had been initially used in the classic airborne assault role against Palembang on Sumatra in early 1942. During the next few years two glider infantry battalions were added to provide real line strength in airborne operations with their heavier weapons, but no attempt was made to develop artillery or heavy mortar support units for the airborne forces. By the end of 1943 the IJA's airborne arm was suitable only for commando and raiding operations. In the face of Allied air supremacy delivery of large airborne forces became problematical, and vastly improved Allied command and control and mobility of ground forces rendered the life expectancy of those units that did land rather short.

Commando Units

As the IJA was pushed back onto the defensive a number of HQs formed extemporized commando or raiding units from organic assets, both as an expression of the traditional Japanese martial spirit and in an attempt to devise a tactic not vulnerable to the crushing American superiority in firepower. Eventually, some of the commando units achieved a semi-permanent status with fixed TO&Es. A commando manual published in 1944 by "Eastern 33" force that standardized their organization as 191-man commando companies provided an example of such an organization. Such a company consisted of a 43-man HQ squad, a 25-man signal squad (with two Type 94 Mk 3A radios), and three 41-man commando platoons. A platoon consisted of the platoon leader, platoon sergeant, a three-man medic team, a three-man engineer team, three runners and three nine-man squads. Individual armament for the company totaled 39 Type 14 pistols, 72 Type 38 carbines, 10 Type 100 submachine guns, and 334 Type 97 hand grenades. The company HQ was also assigned four portable inflatable boats, four Type 89 heavy grenade dischargers and four Type 96 light MGs, along with demolition equipment, for distribution as needed. It was noted that these units could be landed by parachute, glider, submarine, small boats or infiltrated by land.

The theoretical advantages of these types of units attracted the attention of IGHQ and in December 1943 organization of the 1st–10th Commando (*Yugeki*) Companies and the 1st and 2nd Commando Group HQs began, followed by official IGHQ sanction in the IJA's order of battle in January 1944. All these units were destined for the 2nd Area Army in the south Pacific. The 3rd–6th Commando Companies were raised locally by the 2nd Area Army, the 1st and 2nd Commando Companies were raised by the Formosa Army using Japanese officers and NCOs and Formosan aborigine troops, and the other units in

Japan proper. These companies had TO&E strengths of 196 men and were organized into two platoons each and relied on rifles and demolition charges for their missions. The 2nd Commando Group, commanding the 2nd, 3rd and 4th Commando Companies, was sent to Morotai where it was destroyed later in 1944. The 1st Commando Group, controlling the rest of the commando companies, was split up over a wide area, with the 1st Commando Company being diverted to Luzon, the 5th, 7th, 9th and 10th Commando Companies assigned to northwest New Guinea, and the 6th and 8th to the lesser Sundas. For the most part the commando companies operated independently.

The exception was the 1st Commando Company, which was joined by two locally-formed raiding companies in the extemporized "Gigo Force." The two local Gigo Force companies had been given five months training, and were thus not ad hoc units in the usual sense of the phrase. Each consisted of five platoons of two 12-man sections, and each section was equipped with six rifles, three submachine guns, a Type 96 light MG and a grenade discharger with eight rounds. Total company strength was two officers, 25 NCOs and 110 privates.

Three further commando companies (15th–17th) were encountered in Burma in late 1944, but it is not known if they were part of a second series, or simply locally-raised units.

The Kwantung Army formed its own elite formations. In July 1940 a nominal Mobile (*Kido*) Regiment had been formed, composed of two field-grade officers, nine company-grade officers, 26 master sergeants and 244 privates of the line branches, along with 17 medical and technical personnel, with the mission of interdicting the Trans-Siberian Railway and otherwise disrupting the massing of Soviet Forces. The "mobile" in its title presumably referred to its tactics rather than apparent capabilities, for it was provided with only 15 horses and no motor vehicles.

In early 1944 IGHQ gave permission for the expansion of the Mobile Regiment into a three-regiment brigade. On 10 May 1944 the unit was expanded and retitled the 2nd Mobile Regiment as part of the 1st Mobile Brigade formed the same day. The Mobile Brigade was officially added to the Kwantung Army's order of battle in June 1944 and the organization was completed in August. They remained in Manchuria to the end, but there is no evidence they had any impact on Soviet operations.

The brigade had the following composition:

	Off	WO	MSG	SGT	PVT	Total	Horses
Brigade Headquarters	20	0	29	0	40	89	19
Three Mobile Regiments, each	75	0	650	0	1,531	2,256	136

Little is known of the internal organization of the 1st Mobile Brigade, but the totals given by IGHQ records are striking on several counts. The first is the use of master sergeants as first-line leaders. Normally, master sergeants were technicians attached for medical, pay or maintenance duties. In the mobile regiments there were no fewer than 627 master sergeants of the line branch, to supervise 1,492 privates of the line branch. A second anomalous feature was the presence, according to the IGHQ organizational files, of horses in the brigade, but no vehicles. Since the Trans-Siberian Railway ran close to the border in several places, raiding vehicles may have been considered unnecessary.

While the mobile brigade was designed for theater-level penetration missions, a number of battalion-sized raiding units were organized by the Kwantung Army for tactical commando operations. About 12 such units were activated in mid-1944 for assignment to divisions and armies in Manchuria. Two of these, the 3rd and 4th Raiding Units, were transferred to the 32nd Army in the Ryukyus in September, but the remainder served in Manchuria to the end of the war. These raiding units apparently had varying authorized strengths, for the 3rd Raiding Unit had a TO&E strength of 5 officers, 22 NCOs and 373 privates in 1945, while the 4th Raiding Unit's authorized strength was 5 officers, 22 NCOs and 284 privates. Both units were at 98 percent strength in December 1944 and the 3rd Raiding Unit, at least, was well above strength by the time the two were destroyed on Okinawa in 1945.

Airfield Units

Close to the other end of the spectrum from the elite commando units were the rather sedentary airfield battalions and companies of the Army Air Force. These units provided maintenance, administrative and security support for the IJA's flying units. The IJA Air Force raised about 250 airfield battalions numbered 1 through 258. They were activated in nonsequential blocks, the largest of which were the 100th–107th in July 1943, the 131st–138th in May 1944, the 147th–157th in June 1944, the 139th–146th and 158th186th in July 1944, the 187th–200th in October 1944, the 208th–226th and 232nd–239th in February 1945, and the 249th–258th in June 1945.

Strength and organization of an airfield battalion varied somewhat according to the types of aircraft it was planned to support, with single-engine fighters requiring less support than multi-engine aircraft.

The early war airfield battalions had as their major components two maintenance companies and a guard company. A guard company had a strength of about 230 men and consisted of a HQ, a chemical section, three ground defense rifle platoons and an anti-aircraft platoon. The six-man chemical section was responsible not only for chemical defense, but also for the operation of smoke generators used to screen the airfield from enemy attack. The 50-man ground defense platoons were similar to the regular army's rifle platoons. The 70-man anti-aircraft platoon had nine squads, each equipped with a 20mm autocannon or a machine gun.

In December 1943 a new TO&E was developed for airfield battalions that called for a total strength of 418 men. This was not only applied to all airfield battalions raised after that date, but the existing battalions were reorganized onto this standard during December 1943 to February 1944. The revised organization split off the maintenance companies into separate units to increase flexibility and replaced it with a supply company. The guard company was reduced to 183 men, keeping the AA platoon but probably losing one of the ground defense platoons.

Prewar doctrine called for the operation of air regiments from single bases, but in practice many regiments were split up among several airfields, sometimes quite a distance apart. In some cases these units were supported by similarly parceling out the supporting airfield battalions, but this was not always practical and in 1940 the Army Air Force began raising independent airfield companies in substantial numbers. Airfield Companies numbered 1 through 88 were raised during the war. By and large, they were raised as needed and not in large blocks, the only major activations being the 41st–44th in July 1943, the 45th–48th in February 1944, the 61st–66th in May 1944, and the 76th–81st in April 1945.

The original organization for a typical airfield company called for a 50-man HQ, a 50-

man maintenance platoon, an 80-man supply/signal platoon, a 63-man guard platoon, and a 30-man AA platoon. The guard platoon consisted of a platoon leader, a liaison sergeant, a runner, and four 15-man rifle sections each with a Type 11 light MG. The AA platoon was usually made up of two sections, each with two 20mm AA guns. A larger formation was the 32nd Airfield Company in the Philippines in September 1944, consisting of a 60-man HQ (including 7 medics), a 40-man signal platoon that also included a 5-man meteorological section and an 8-man photo section, a 68-man maintenance platoon, an 88-man guard platoon, a 38-man materiel platoon, and a 65-man transport platoon, for a total of 360 men. The guard platoon was made up of three rifle squads, a light MG squad, and two 13-man AA machine cannon sections. The reduction in size that was imposed on the airfield battalions was similarly applied to the airfield companies, and a TO&E strength of 226 men (probably representing a reduction in maintenance) was imposed in February–March 1944, although apparently not always to existing units.

8

Naval Ground Forces

The Imperial Japanese Navy was divided for administrative purposes into four homeland naval districts: Yokosuka, Sasebo, Kure and Maizuru, each headquartered from the port of the same name. The naval districts were omnipresent in naval organization, being responsible for recruiting and conscription within their areas, the training of seamen and specialists, pay and all manner of other personnel functions. Each was also responsible for ship's complements and naval units ashore as laid out in the Navy Administrative Orders Manual. Thus, the Kure Naval District was responsible, for instance, for forming and maintaining the personnel strength of the submarine I-174, the impressed merchantman Nachi Maru, the 938th Naval Air Group, and the shore-based 121st Construction Battalion. Each also, of course, operated substantial fleet support activities within their geographical bounds, such as shipyards, fuelling depots, ammunition points, etc. The naval districts also commanded a variety of tugs, patrol boats, harbor service craft, etc.

The actual combat operations of the navy on the high seas were directed through a different chain of command. Numbered fleets were regionally defined, with, for instance, the 4th Fleet responsible for the mandated islands of the Pacific, while the bulk of the carriers and battleships were concentrated into a central striking force, the Combined Fleet. Each fleet had its own support facilities, such as repair shops, etc. at its headquarters base, but in other locations relied on "base forces" subordinate to that fleet to supply the support they needed.

Naval Landing Forces

Japanese sailors landed by small boats from their warships had participated in ground combat in both the Russo-Japanese War and the seizure of German possessions in WWI. These had been short-term expedient measures and they did not prove a useful precedent for operations in China. Protection of Japanese treaty rights and the concomitant necessary intimidation of the local Chinese placed a heavy strain on the ships' crews, who were being called on to perform double duties. Increasing Japanese belligerency exacerbated this problem and on 1 October 1932 a Navy administrative order was issued authorizing the formation of semi-permanent landing parties to be known as Special Naval Landing Forces. The order provided for two TO&Es, a larger one of 1,979 men and a smaller one of 1,434. At least one such was used in the fighting in Shanghai in 1932, but was deactivated shortly thereafter.

Although a number of SNLFs were apparently envisioned only one, the Shanghai SNLF, was formed. This unit used the larger TO&E and was built around three rifle companies and a rifle/heavy weapons company. Each of the rifle companies consisted of six rifle platoons and a machine gun platoon. The fourth company had three rifle platoons and a heavy weapons platoon. This latter had four weapons, either 3-inch naval guns or a mixture of 75mm regimental guns and 70mm battalion howitzers. The unit also included, at one time or another, a tank company and other supporting units. By the time of the battle of Shanghai in 1937 it had expanded to six rifle companies and a large defense battalion, the latter equipped with four each of 15cm and 12cm howitzers, 12 mountain guns, 4 AA guns and eight 15cm mortars. The force also had four tanks and some armored cars. The Shanghai SNLF served in China throughout the war, engaging in coastal and riverine operations. By 1 April 1945 it had a strength of 53 officers, 117 warrant officers, 648 petty officers and 2,827 sailors. By that point it included a large defensive component with no fewer than twenty 12cm AA guns, eight 8cm AA guns and 120 AA machine cannon.

Japanese doctrine at the time called for naval landing parties to seize landing beaches in advance of army units and with the increasing likelihood of combat the Navy began planning for SNLFs that would be raised by the naval districts, without having to call on ships complements. These units were to be formed at the four major naval bases (Kure, Maizuru, Sasebo and Yokosuka) and were given the rather convoluted generic designation of "specially established naval district special naval landing forces." Individually, such a unit would carry the name of the district where it was raised, i.e., the Sasebo Naval District 2nd Special Naval Landing Force, although for convenience they will be here referred to by the more common Western form: i.e., Sasebo 2nd SNLF.

The order authorizing the naval district SNLFs was issued on 15 October 1936 and it provided for a new TO&E. The tables called for a total of 539 men organized into a battalion headquarters, two rifle companies, and signal, engineer, ammunition, medical and supply units. The battalion HQ consisted simply of the battalion CO, his adjutant, a color-bearer, and an HQ platoon with a nine-man runner squad and a nine-man scout squad.

A rifle company consisted of the company CO, a HQ (identical to the battalion HQ platoon), four rifle platoons, and a machine gun platoon. Each rifle platoon was made up of a platoon leader, two runners to carry his commands, and four squads each of a petty officer and eight sailors. The machine gun platoon consisted of the platoon leader, three orderlies, three MG squads (each of a petty officer and seven sailors) and an ammunition squad (of a PO and five sailors).

The battalion signal platoon consisted of a warrant officer and eleven enlisted, including six telephonists and four radio operators. The medical unit was made up of the medical officer, a warrant officer assistant, a 14-man stretcher-bearer squad and a nine-man medic squad. The battalion was completed by a 20-man pioneer unit, a 17-man ammunition unit and a 22-man supply unit.

The 1936 order represented an improvement to the earlier improvised formations but reflected a doctrine that stressed landing away from enemy opposition followed by overland marches by follow-on army units to the objective. The primary means of internal communications within the battalion was the runner, there were no heavy weapons other than the three medium MGs per company, and transport for supplies was almost nonexistent.

The decision to seize Shanghai at the outset of the war with China spurred the creation of seven SNLFs (Sasebo 1st–4th, Kure 1st and 2nd, and Yokosuka 1st) in July–August 1937. These were followed by two more (Sasebo 5th and 6th) for Tsingtao in December, then by six (Sasebo 7th, Kure 3rd–5th, and Yokosuka 2nd and 3rd) in 1938, four (Sasebo 8th and 9th, Kure 6th, and Yokosuka 4th) in 1939, and finally a single contribution from Maizuru, the 1st, in 1940.

These "first-generation" SNLFs had short lives, being deactivated once the specific operation they were earmarked for had completed, usually 2–6 months. The organization of the early ones varied, but the Kure 1st may have been typical, with 700 men in a HQ, two rifle companies (each four rifle platoons and an MG platoon) and an infantry gun company. The exceptions were the three that were sent to the island of Hainan, the Sasebo 8th, Yokosuka 4th and Maizuru 1st, which remained on garrison and pacification duties there until the Japanese surrender in 1945. These were large, but relatively light, formations; with the Yokosuka 4th consisting of five rifle companies (each six platoons), the Maizuru 1st of six rifle companies (each five platoons), and the Sasebo 8th of five rifle companies and a 4-platoon "gun company." This gave the Maizuru unit an authorized strength 21 officers, 23 warrant officers, 170 petty officers and 1,102 other ranks on activation in June 1940.

Anticipating that several SNLFs might have to act in concert, a December 1937 administrative order authorized a TO&E for a combined SNLF headquarters that could be set up to command two or more naval district SNLFs. This small formation totalled only 8 officers and 30 other ranks and contained no signal or service support elements whatsoever. Two such HQs were formed for operations in China, the 1st (December 1937 to November 1939) and the 2nd (April to November 1938).

Operations highlighted a number of problems, and on 12 July 1939 a set of Naval Landing Force Regulations was issued that incorporated changes to the SNLF organization. The regulations turned the SNLF into a combined arms team and authorized a larger combined SNLF HQ that was at least marginally capable of controlling and supporting them.

The new SNLF had a basic organization consisting of a headquarters, three rifle companies, an artillery unit (company), and supporting units. The basic unit, the rifle platoon, was analogous to the Army's rifle platoon, consisting of three rifle squads and a grenadier squad, although slightly smaller in personnel. As with the Army, doctrine called for the grenadier squad to be employed as a unit under the command of the platoon leader, rather than parceled out as individual weapons.

The company was made up of three rifle platoons, a machine gun platoon, and a headquarters platoon. The company HQ platoon consisted of three squads each of a sergeant or corporal and ten privates. One of the squads performed scout duties and was equipped with a light MG, one provided runners and general dutymen, and the third provided signallers to operate a portable radio, six field telephones, a signal lamp, and six bugles.

Fire support was provided by the artillery unit, consisting of two platoons with 70mm Type 92 infantry howitzers and two with 75mm Type 41 regimental mountain guns. Each of the platoons included an ammunition element to carry 500 rounds per gun. The artillery unit HQ platoon included two HQ squads (each a petty officer and eight men with a truck) for general duties, and two FDC squads (with the same strength). One of the FDC squads could plot the fire for the mountain guns while the other did the same for the infantry howitzers or, if the infantry howitzers used direct or semi-indirect fire, one FDC squad could support each mountain gun platoon. To facilitate communications the unit HQ platoon was provided with three portable radios and six field telephones.

The SNLF signal unit consisted of two petty officers assigned to code duties and four radio teams each of a petty officer and three privates. Each of these teams manned a large radio and one had the additional duty of working an airground liaison radio. The unit had no organic vehicles and relied on trucks held elsewhere in the force to tow their radio trailers.

The engineer unit was divided into three squads, each of a petty officer and ten enlisted. One squad specialized in laying and removing mines, one was made up of carpenters and

Table 8.1: Special Naval Landing Force, from July 1939

	Officers	Warrant Officers	Petty Officers	Other Ranks	Pistols	Rifles	Light MGs	Medium MGs	Grenade Launchers	70mm Inf Howitzers	75mm Mountain Guns	Flamethrowers	Cars	Trucks	Carts
Force Headquarters															
Command Group	2	0	1	0	3	0	0	0	0	0	0	0	0	0	0
Attached Personnel	4	0	2	5	1	7	0	0	0	0	0	0	0	0	0
Headquarters Platoon	1	0	3	30	2	30	1	0	1	0	0	0	0	0	0
Signal Unit	1	0	6	12	19	0	0	0	0	0	0	0	0	0	0
Three Rifle Companies, each															
Company Commander	1	0	0	0	1	0	0	0	0	0	0	0	0	0	0
Attached Personnel	1	0	2	0	1	2	0	0	0	0	0	0	0	0	0
Headquarters Platoon	0	1	3	30	2	30	1	0	0	0	0	0	0	0	0
Three Rifle Platoons, each															
Platoon Headquarters	1	0	0	2	3	0	0	0	0	0	0	0	0	0	0
Three Rifle Squads, each	0	0	1	10	1	9	1	0	0	0	0	0	0	0	0
Grenadier Squad	0	0	1	10	0	11	0	0	4	0	0	0	0	0	0
Machine Gun Platoon															
Platoon Headquarters	1	0	0	2	3	0	0	0	0	0	0	0	0	0	0
Four MG Squads, each	0	0	1	7	0	7	0	1	0	0	0	0	0	0	0
Two Ammunition Squads, each	0	0	1	10	1	1	0	0	0	0	0	0	0	0	0
Artillery Unit															
Unit Commander	1	0	0	0	1	0	0	0	0	0	0	0	0	0	0
Attached Personnel	1	0	2	4	1	6	0	0	0	0	0	0	0	2	0

Table 8.1: Special Naval Landing Force, from July 1939

	Officers	Warrant Officers	Petty Officers	Other Ranks	Pistols	Rifles	Light MGs	Medium MGs	Grenade Launchers	70mm Inf Howitzers	75mm Mountain Guns	Flamethrowers	Cars	Trucks	Carts
Headquarters Platoon	1	1	4	32	6	32	0	0	0	0	0	0	0	2	5
Two Mountain Gun Platoons, each															
Platoon HQ	1	0	0	2	3	0	0	0	0	0	0	0	0	0	0
Two Gun Squads, each	0	0	1	12	13	0	0	0	0	0	1	0	0	1	0
Two Ammunition Squads, each	0	0	1	10	2	9	0	0	0	0	0	0	0	1	1
Two Infantry Gun Platoons, each															
Platoon HQ	1	0	0	2	3	0	0	0	0	0	0	0	0	0	0
Two Gun Squads, each	0	0	1	10	11	0	0	0	0	1	0	0	0	1	0
Ammunition Squad	0	0	1	10	4	7	0	0	0	0	0	0	0	1	1
Engineer Unit	1	0	3	30	34	0	0	0	0	0	0	0	0	0	0
Transportation Unit	1	0	4	40	45	0	0	0	0	0	0	0	5	13	6
Medical Unit	4	0	5	32	41	0	0	0	0	0	0	0	0	0	0
Supply Unit	3	0	10	44	57	0	0	0	0	0	0	0	0	0	0
Undistributed Equipment	-	-	-	-	20	250	0	0	0	0	0	2	0	14	0

metalsmiths, and the third consisted of electricians and mechanics. The transport unit was made up of four squads (each a petty officer and ten enlisted) and operated a total of five passenger cars, ten cargo trucks, two water sprinkler trucks, an ambulance, six hand carts and four portable generators.

The medical unit was built around two stretcher-bearer squads (each a petty officer and twelve enlisted) and a medic squad (a petty officer and eight enlisted). The unit also included three officers and two POs from the medical corps and one officer to supervise the medics and handle administrative matters. Finally, the supply unit consisted of a headquarters with three officers and six POs and four supply squads each with a PO and eleven enlisted. The supply unit had no transport assets.

Lacking the regimental and divisional trains elements of the army infantry the SNLF tended to carry slightly more ammunition at this battalion level than their army counterparts. Thus, each line soldier carried 180 rounds for his rifle and each rifle squad 1,440 rounds for its light MG. The grenadier squads carried 26 rounds (all HE) for each of their Type 89 weapons, while the MG platoons carried 4,200 rounds for each of their Type 92 MGs. As mentioned, the artillery unit carried a total of 2,000 rounds of 70mm and 2,000 rounds of 75mm ammunition. In addition, the SNLF had an authorized ammunition reserve that provided a further 180 rounds per rifle, 4,000 rounds per LMG, 8,000 rounds per medium MG, and 50 rounds per grenade discharger. As compared to the army's infantry this allocation was about 50 percent higher in medium MG, 100 percent higher in grenade discharger, and 200 percent higher in mountain gun ammunition, even including regimental reserves in the army. The SNLF's defined reserve holdings also included 250 reserve rifles, 50 rifle silencers, six 30cm searchlights, two flamethrowers, and carrier pigeons as required.

Simultaneously, a new TO&E was published for the headquarters of a combined SNLF. The HQ itself consisted of a core of eight officers to be supplemented by specialists as required. Also included in the overall headquarters designation were a headquarters company, a signal platoon, and engineer, transport, medical and supply sections. The HQ company consisted simply of a 12-man HQ and two rifle platoons. These men could be used as scouts, HQ security guards, runners, etc., as needed and provided a valuable pool of reserve manpower. The signal platoon was similar to that of the SNLF signal unit but had an additional two POs for code duties. The engineer section (8 men), transport section (18 men), medical section (13 men) and supply section (10 men) supported only the CSNLF headquarters and did not represent supporting echelons above the subordinate SNLFs.

Also published at the same time were TO&Es for three "special" units that, although not ordinarily part of an SNLF, could be raised for attachment as necessary: the mountain gun company, the tank company, and the chemical warfare company.

The 163-man mountain artillery company could be used either to replace the artillery unit in an SNLF or as a centralized fire support unit for a CSNLF. The larger size of the gun squad meant that the mountain gun company could be equipped with the more powerful 75mm Type 94 mountain gun, as opposed to the 75mm Type 41 regimental gun in the SNLF artillery unit. As was the case with the artillery unit 500 rounds per gun were carried and the unit belied its "mountain" designation by using trucks for transport. Communications were slightly weaker than in the SNLF artillery unit, the signal and FDC squads having only two radios and six telephones between them.

The tank company was built around a light tank platoon (nominally the HQ platoon)

Table 8.2: Mountain Gun, Tank & Chemical Companies, SNLF, from July 1939

	Officers	Warrant Officers	Petty Officers	Other Ranks	Pistols	Rifles	Light MGs	75mm Mountain Guns	Armored Cars	Light Tanks	Medium Tanks	Cars	Trucks	Carts	Bicycles
Mountain Artillery Company															
Company Headquarters	2	0	2	2	4	2	0	0	0	0	0	1	1	0	0
Headquarters Platoon															
Platoon Headquarters	1	0	0	0	1	0	0	0	0	0	0	0	0	0	0
Signal Section	0	0	2	17	19	0	0	0	0	0	0	0	0	0	0
FDC Section	0	0	2	18	20	0	0	0	0	0	0	0	1	0	0
Two Gun Platoons, each															
Platoon Headquarters	1	0	0	2	3	0	0	0	0	0	0	0	0	0	1
Two Gun Sections, each	0	0	1	14	15	0	0	1	0	0	0	0	2	0	0
Ammunition Platoon															
Platoon Headquarters	0	1	0	2	3	0	0	0	0	0	0	0	0	0	0
Four Ammunition Squads, each	0	0	1	11	0	12	0	0	0	0	0	0	1	1	0
Tank Company															
Company Headquarters	2	0	3	9	14	0	0	0	0	0	1	2	0	0	0
Headquarters Platoon	1	0	4	12	17	0	0	0	0	4	0	0	0	0	0
Three Tank Platoons, each	1	0	3	9	13	0	0	0	0	0	3	0	0	0	0
Trains Platoon	0	1	5	50	23	33	2	0	0	0	0	0	12	1	0

Table 8.2: Mountain Gun, Tank & Chemical Companies, SNLF, from July 1939

	Officers	Warrant Officers	Petty Officers	Other Ranks	Pistols	Rifles	Light MGs	75mm Mountain Guns	Armored Cars	Light Tanks	Medium Tanks	Cars	Trucks	Carts	Bicycles
Chemical Company															
Company Headquarters	2	0	1	2	2	4	0	0	0	0	0	1	0	0	0
Headquarters Platoon	0	1	2	20	1	22	0	0	0	0	0	0	0	0	2
Two Chemical Platoons, each	1	0	3	35	5	33	0	0	0	0	0	0	2	0	1
Mobile Decontamination Platoon	1	0	3	42	44	0	0	0	3	0	0	0	4	0	0

for scouting and liaison, and three medium tank platoons. The trains platoon included a maintenance detachment of two POs and twenty other ranks with a parts truck and a supply truck, and three resupply squads each with a PO and ten other ranks with a total of ten cargo trucks carrying 3,200 AP and 1,800 HE rounds of 57mm ammunition (for the medium tanks); 2,500 rounds of 37mm AP ammunition (for the light tanks); and 200,000 rounds of MG ammunition. The company HQ included a field car, a radio car, and a medium tank. There were five radios in the company: one in the radio car, one in the company commander's medium tank, and one in each of the medium tank platoon leaders' tanks.

The chemical company was a defensive unit that included a headquarters platoon (with a Type 1 weather observation instrument), a mobile decontamination platoon (with three tankettes fitted with sprinkler systems for front line decontamination, and four trucks loaded with bleach), and two chemical platoons each with six gas detector kits and 50 small decontamination sets.

Shortly thereafter, a second set of tables of organization for "special" units was issued, considerably less detailed than the first. They provided for a howitzer company (four 15cm field howitzers), a heavy mortar company (four 15cm mortars), a light mortar company (twelve 81mm mortars), an infantry gun company (four 70mm infantry howitzers), an anti-tank company (six 37mm AT guns), an autocannon company (six 25mm AA guns), a machine gun company (twelve medium MGs), a dual-purpose gun company (two searchlights, six 13mm AA machine guns, four 76mm DP guns), and a field anti-aircraft company (two searchlights, two 13mm AA machine guns, four 75mm AA guns). The dual-purpose gun company manned static pedestal-mounted weapons, but the rest of the units were mobile formations with motor vehicles to transport the heavy equipment while the troops marched.

The addition of these "special" units to the potential order of battle of the naval ground forces gave the force a great deal of organizational flexibility. Nevertheless, when the naval districts actually began mobilizing SNLFs they tended to use the TO&Es only as a general guide.

As war with the United States grew closer the Navy began raising the landing forces needed for a sweep through the Pacific. They already had the Shanghai SNLF performing littoral duties in China, and three more from the first-generation of SNLF (Maizuru 1st, Yokosuka 4th, and Sasebo 8th) that would spend the entire war engaged in brutal pacification campaigns on Hainan Island, that being a Navy responsibility.[1] The Navy had begun experimenting with parachute units in June 1941, and in September the Navy Ministry ordered the raising of two parachute-trained SNLFs. The strength authorizations for 844 men each were issued on 15 October and the Yokosuka 1st and 3rd SNLFs actually formed a month later. A strength table for 927 men in the non-airborne Yokosuka 2nd SNLF was also issued on 15 October.

On 5 November strength tables were issued for the Sasebo Combined SNLF HQ (109 men), the Sasebo 1st (1612) and 2nd (1441), Kure 1st (1404) and 2nd (1394), and Maizuru

1 The 742-man Yokosuka 4th SNLF, along with the Kure 6th and the 746-man Sasebo 8th SNLFs had participated in the capture of Hainan Island in February 1939. The Kure unit was deactivated and replaced by the Maizuru 1st. One company of the Yokosuka 4th was detached for duty on Kwajalein Island in October 1942, leaving three rifle companies plus signal, transport, engineer and "police" units, for counter-guerilla operations on Hainan.

2nd (1071) SNLFs, all conventional ground units.

As mentioned two SNLFs (Yokosuka 1st and 3rd) were raised and trained as parachute units. They thus had a unique TO&E authorized on 15 October 1941 that called for 844 men in a HQ company, three rifle companies and an anti-tank company. The rifle platoon of the parachute SNLF was similar to that set forth in the 1939 TO&E, but added a second light MG to each rifle squad and relieved the four grenade discharger men of their cumbersome rifles. At the company level the machine gun platoon lost two of its MG squads and one ammunition squad, while the company HQ lost the general-duty squad. These changes reduced the rifle company strength to 188 men.

The anti-tank company was a very austere version of the 1939 "special" unit, consisting of two 2-gun platoons and having a strength of about 60 men. The guns were carried by flying boat to be landed near the drop zone.

The headquarters company was comprised of a 22-man headquarters, a 43-man intendence unit, a 52-man medical unit, a 31-man transport unit, a 31-man engineer/maintenance unit, a 47-man signal unit, and a 12-man demolitions unit. Heavy weapons within the HQ company consisted of two light MGs in the demolitions unit and two more light MGs and a grenade discharger in the engineer/maintenance unit.

Both parachute units were used in the invasion of the Dutch East Indies, the Yokosuka 1st SNLF in the capture of the Celebes port of Menado, and the Yokosuka 3rd SNLF in the conquest of Timor. These were the only two airborne assaults by SNLF personnel. In the Menado operation the HQ, two rifle companies were dropped by parachute the first day from 28 "Tina" transports, while ten men and an AT gun were carried by seaplane and landed on a nearby lake. The third rifle company was dropped on the second day. At Koepang, Timor, two rifle companies were dropped the first day from 28 "Tina" and "Thalia" transports, and the third on the second day.

In April 1943 the Yokosuka 1st SNLF absorbed the Yokosuka 3rd and its authorized strength was increased to 1,326. The new force, consolidating all the Navy's nominal parachute units, was now built around five 189-man rifle companies, each of three rifle platoons and a two-squad MG platoon. The rifle platoon had a four-man HQ (with a Type 2 AT grenade launcher) and four 9-man squads, three of which each had two Type 99 light MGs and a Type 100 rifle grenade launcher and a grenadier squad with four Type 89 grenade dischargers. The new organization also introduced the Type 100 submachine gun to the SNLF, although on a small scale, two per company and rifle platoon HQ. The AT platoon was expanded slightly to five 13-man squads, each with a Type 1 AT gun. In September 1943 it was sent from the Homeland to Saipan,[2] but four months later two companies were detached to Rabaul and the unit returned to an approximation of its original configuration, but with two 70mm and two regimental guns replacing the AT guns.

The Yokosuka 1st and 3rd SNLFs were unique among the prewar formations not only in being parachute units, but also for the fact that they were the only two to share a common TO&E. For the rest, authorized strengths varied from 927 to 1,612.

Representing the low end of the spectrum was the Yokosuka 2nd SNLF, which was organized using the July 1939 TO&E but with an anti-aircraft machine gun platoon replacing the artillery unit. Representing something of the middle ground was the Maizuru 2nd SNLF, with a reduced artillery unit.

2 About half the unit's parachutes were taken to Saipan, but since no parachute jumps had been made since the invasion of the Dutch East Indies, this probably represented optimism rather than a planned mission.

These conventional SNLFs were fully utilized in the advances of late 1941 and early 1942 through the Pacific. Only one unit was used in the drive east, the 3rd Company of the Maizuru 2nd which captured Wake Island. All the others joined the parachute SNLFs in southern operations; the Yokosuka 2nd capturing Miri and Kuching, the Maizuru 2nd Rabaul, the Kure 1st Ambon, the Kure 2nd Tarakan, and the Sasebo 1st and 2nd (as the Sasebo Combined SNLF) Menado, Kendari and Makassar.

It should be noted here that these SNLFs were not considered elite formations by the Japanese, even within the Navy. Rather, they were generally regarded simply as sailors doing shore duty. Reflecting this, US intelligence noted that the early SNLFs when facing determined resistance "exhibited a surprising lack of ability in infantry combat." In fact, once an SNLF had captured an objective it usually stayed there as a garrison and defense unit rather than being reassigned for further offensive actions.

In February 1942 the Maizuru 2nd SNLF was converted to the 8th Special Base Force to guard Rabaul, and on 10 March 1942 the Kure 1st, Kure 2nd, Sasebo 1st, and Sasebo 2nd were all converted to special base forces and guard units in the Dutch East Indies.

Even as the existing units were being converted to their new, more sedentary duties, a second wave of SNLFs was being raised for operations around New Guinea and the Solomons. These comprised the Kure 3rd and Sasebo 3rd in February 1942; and the Maizuru 3rd, Kure 5th, Sasebo 5th, and Yokosuka 5th in May (Kure and Sasebo naval districts did not raise SNLFs designated 4th).

The February 1942 SNLFs initially adhered to the July 1939 TO&E, the only two SNLFs to completely do so. In light of increases in Allied aircraft and armor strength, however, the Kure 3rd, at least, was quickly reorganized by early June into two rifle companies (each 5 platoons), an artillery company and a field AA company plus the standard support elements (signal, engineer, transport and supply platoons). All these used the standard TO&Es except the artillery company, which consisted of a 32-man HQ, a 55-man ammunition platoon and two 41-man anti-tank platoons, each with a light MG and two 37mm Type 94 guns.

Two of the May ("5th") SNLFs were configured this way to start with. The Kure and Yokosuka units were combined with the 647-man 2nd Combined SNLF HQ for the invasion of Midway, but with the failure of this operation the HQ was abolished the next month and the two SNLFs were placed directly under the Combined Fleet. The Sasebo 5th SNLF was similar, but included a lookout company that provided observation posts to detect enemy air and sea movements and 48 enlisted radar specialists.[3] The Kure 3rd and Sasebo 3rd SNLFs were earmarked for the invasion of Port Morseby, but were turned back at the battle of Coral Sea. Shortly thereafter the three May 1942 SNLFs arrived in the area and HQs and rifle companies of these three plus the Kure 3rd SNLFs were sent to assault Milne Bay, where they took heavy losses before being evacuated in September.

On 15 August 1942 two more SNLFs were officially raised: the Yokosuka 6th and Sasebo 6th. These were organized as in the July 1939 TO&E, but with a field AA company added and a small tank section (two Type 95 light tanks) formed from existing personnel. In addition, the 70mm infantry howitzers in one infantry gun platoon of each SNLF were replaced by 37mm AT guns. The Sasebo 6th SNLF was sent to Bougainville and Buin, where the surviving remnants finished the war.

3 The Kure 5th SNLF was deactivated in October 1942, the Yokosuka 5th in February 1943, and the Sasebo 5th in February 1944.

The Yokosuka 6th SNLF, on the other hand, was immediately dispatched to the Gilberts in reaction to the US raid on Makin in August. One rifle company (plus the AT platoon and the tank section) was sent to Makin, and the rest of the force to Tarawa. In December two of the Makin company's platoons were returned to Tarawa. In the meantime, the main body of the SNLF had been considerably reinforced in light of its defensive role. These additions consisted of a coast artillery battalion, an anti-aircraft battalion, and a seagoing unit. The coast artillery battalion consisted of an 8" gun battery (two guns), two 14cm gun batteries (each two guns), and a 76mm gun battery (three guns). The AA battalion consisted of the previously formed field AA company, two heavy batteries (each two 127mm twin mounts), a 13.2mm AA machine gun battery (five twin and three single mounts), and a searchlight battery. The seagoing unit consisted of three torpedo boats and a minesweeping unit. Other additions included a second tank section, an infantry gun platoon and an AT platoon. On 15 February 1943 the Yokosuka 6th SNLF was redesignated the 3rd Special Base Force.

These units were followed in September 1942 by the 963-man Maizuru 4th SNLF, probably combining two rifle companies with a coast artillery unit. This, in turn, was followed by the 8th Combined SNLF and its two constituent units, the Kure 6th SNLF and the Yokosuka 7th SNLF in November. The 8th CSNLF completed the transformation of the SNLFs from offensive assault formations to defensive units specializing in coastal and anti-aircraft defense.

As originally organized the Yokosuka 7th SNLF consisted of a HQ, a rifle company, two heavy coast defense artillery companies, a light CD artillery company, a field AA company, an AA machine gun company, and the usual supporting services.

The rifle company differed from the normal organization in that the grenade discharger squads were deleted from the rifle platoons and one of the 11-man squads was deleted from the HQ platoon. The field AA company used the standard TO&E, but the other units of the SNLF used unique organizations. A heavy coast defense artillery company had 156 men and was divided into a HQ platoon (including a 15-man searchlight squad and a rangefinder squad), two 24-man gun platoons (each manning two 12cm naval guns), and a 48-man ammunition platoon. The light coast defense gun company was by far the largest component of the force, totaling 278 men. It was built around four gun platoons, each with four 5-man squads manning 76mm L/40 low-angle guns. The company also included an ammunition platoon with four 17-man sections to support the platoons, and a headquarters platoon with two 7-man fire control squads, four 6-man searchlight squads, four 5-man signal squads, and four 5-man general-duty squads. The AA machine gun company had a strength of 165 men and was divided into a machine cannon platoon (two 40mm twin-mount guns) and two MG platoons (each five twin 13.2mm MGs).

The Kure 6th SNLF was similar in overall structure, but substituted a dual-purpose gun company for the light CD artillery company, replaced the 12cm coastal guns with 14cm models, and added a lookout company. The DP gun company was equipped with four 12cm guns and used the standard TO&E. The rifle company followed the standard TO&E except that the machine gun platoon was replaced by an infantry gun platoon and a mountain gun platoon. The machine cannon platoon of the AAMG company had three twin and two single-mount 40mm. This SNLF also included a tank section with two Type 95 light tanks in the HQ. The lookout company of about 100 men manned optical observation stations and two radar stations.

Within a very short period of time it was realized that the 8th CSNLF had been shortchanged with regard to self defense, so on 31 December 1942 a rifle company and infantry support units were added to both constituent SNLFs. The new rifle companies were organized almost identically to the standard naval rifle company but were much more heavily armed, albeit less mobile. Each eleven-man rifle squad now had two Type 99 light MGs, to give the company a total of eighteen. The grenade discharger squads were much heavier units, each of the eleven men being equipped with his own Type 10 light grenade discharger. These weapons, which were ordinarily used at company and battalion HQ for signalling purposes, had a considerably shorter range than the Type 89 grenade dischargers usually employed, but were also quite a bit lighter and handier. In the machine gun platoon each of the eight-man squads was now responsible for two Type 92 medium MGs instead of one. The only change in manpower to the company was the addition of a fourth general-duty squad to the company HQ platoon. These changes gave the rifle company a total strength of 246 men, with 18 light MGs, eight medium MGs and 33 light grenade dischargers. One other innovation was the provision of a Type 100 submachine gun for the two runners in each platoon HQ.

For fire support of the infantry each of the two SNLFs received an anti-tank platoon and a mortar platoon. The 48-man AT platoon consisted of two firing sections (each 13 men with a 37mm gun and a truck), two ammunition squads (each six men with a truck), and a ten man HQ. The mortar platoon had 76 men divided into four 8-man gun squads (each with an 81mm Type 94 mortar and a truck), two 12-man ammunition squads (each a truck), an 8-man FDC and a 4-man HQ. In the case of the Yokosuka 7th SNLF these platoons were simply attached directly to the force, but in the Kure 6th a gun company was formed that consolidated these two platoons plus the infantry gun and mountain gun platoons from the (1st) rifle company.

The 8th Combined SNLF HQ itself was a large formation with an authorized strength of 1,465. It included 112 men to man 16 special cargo boats, a 162-man lookout company (including two radar sets), and a large ordnance detachment to maintain the guns of the subordinate SNLFs. The 8th CSNLF was sent to the New Georgia area where it was gradually reduced in strength through attrition until it was redesignated the 14th Base Force (CSNLF HQ), 88th Guard Unit (Kure 6th SNLF) and 89th Guard Unit (Yokosuka 7th SNLF) in December 1943.

In February 1943 the Sasebo 7th and Kure 7th SNLFs were formed. These were identically organized with 1,660 men each, using slightly modified versions of the basic 1939 TO&E. The HQ, three rifle companies and artillery unit all followed the standard organization except that the 70mm howitzers in one infantry gun platoon per SNLF were replaced by 37mm AT guns. These SNLFs also each included a field AA company, which differed from the standard TO&E in that it included two AA machine gun platoons, each of 37 men with two gun sections (each 12 men with two 13.2mm MGs) and an 8-man ammunition squad. The transport unit was also different in that it comprised a 51-man boat platoon with 8 boats, a 46-man wagon platoon, a 50-man patrol platoon (four 12-man squads), and a 15-man maintenance platoon. Each of these SNLFs also included a 40-man "special duty unit" that manned four light tanks and a handful of flamethrowers. The Sasebo 7th SNLF arrived on Tarawa in March 1943 and it was destroyed there about a year later. The Kure 7th SNLF was moved to Buin, and there served out the war.

These were followed by the Kure 8th in April 1943. This had an authorized strength of 668 men and was stationed in the Nicobar Islands until September 1943, when it was converted to the 14th Guard Unit. In January 1944 the Sasebo 101st SNLF was created with a strength of 218 men, a submarine-borne commando unit taken largely from a company of the Yokosuka 1st SNLF, followed shortly by the similar 102nd. In early 1944 authorization was granted for the organization of a few semi-provisional SNLFs in the Pacific from the stranded crewmen of sunken vessels.

As the Allies drove towards the Japanese homeland frantic efforts were made to mobilize as many troops as possible. At the same time, the Navy found itself with a large number of sailors made surplus by the sinking of a good portion of their fleet and the immobilization of most of the remainder for lack of fuel. As a result, the Maizuru naval district was able to raise a 5th SNLF (3,131 men) and a 6th SNLF (3,778 men) in July 1945. Similarly, the Sasebo ND was able to raise the Sasebo Combined SNLF HQ and its 11th–14th SNLFs. The organization of these regimental-sized formations is not known, but some indication of their strength can be gauged from their armament.

The Sasebo 12th SNLF was detached to Ainoura area and had a total strength of 2,735 men with 27 rocket launchers (with 140 rockets), 20 mortars, 8 machine guns and 940 rifles. The Sasebo 13th SNLF was detached to the Hario area with 2,466 men with 26 rocket launchers (with 92 rockets), 20 mortars, 55 machine guns, two howitzers and 489 rifles. The Sasebo CSNLF HQ (499 men) remained at the base, commanding the Sasebo 11th (2,635 men) and 14th (2,528 men) SNLFs, with a total armament of 43 machine guns, 2 heavy grenade dischargers, 49 rocket launchers, one 37mm infantry gun, one 70mm infantry howitzer, twenty 81mm mortars, two 150mm mortars, nine 47mm AT guns, 4 tanks, two 76mm naval guns, 3 short 12cm naval guns and 3 long 12cm naval guns. Clearly, these extemporized SNLFs were organized with whatever weapons could be found. In addition, the Sasebo ND organized one small SNLF, the Sasebo 105th, with 600 men.

The Yokosuka Naval District had formed a Combined SNLF in mid-1943, apparently as a training and holding unit. It consisted of a 32-man HQ, five battalions, a 13-man tank unit with three light tanks, an 11-man anti-tank gun with one AT gun, a 21-man signal unit with telephones, a 13-man pioneer unit, a 52-man medical unit and a 35-man supply unit. Each of the battalions was made up of a 45-man headquarters and two companies. A company had a 38-man HQ, three 48-man rifle platoons (each with three light machine gun and one grenadier squads) and a 56-man machine gun platoon with four 8-man machine gun squads and two 11-man ammunition squads.

In July 1945 the Yokosuka Combined SNLF was expanded and strengthened into a combat force, largely using personnel from the Tateyama Gunnery School's landing school as cadre, to defend the Miura Peninsula. It consisted of a headquarters, a 200-man rifle company, a heavy gun unit, a chemical unit, and the 11th through 16th Yokosuka SNLFs. The heavy gun unit had two batteries, one with eighteen 12cm guns and the other with eighteen 12cm rocket guns. The 11th–15th SNLFs were similar, each of about 100 officers and 3,000 enlisted divided into a HQ and three battalions. Each battalion was made up of three infantry (rifle and MG) companies, a special attack company, a mortar company with eighteen 81mm mortars, a rocket gun company with eighteen 12cm and 8cm rocket launchers, and a gun battery with six 12cm guns. The special attack company was a 180-man suicide unit in which the men were provided with 30kg bombs they were to carry down the beach at landing craft, 60kg shaped charges strapped to the body with the carrier to throw himself at a tank, and hand grenades for night attacks against gun positions and command posts.

The Yokosuka 16th SNLF was a tank unit, made up of a headquarters with four light tanks, a tank battalion with 40 medium tanks, and an amphibious battalion with 20 Type 2 and 20 Type 3 amphibious tanks.

The primary concern of the Yokosuka CSNLF appears to have been American tanks. These were to be the primary targets of the special attack companies, and in addition shaped charge grenades and pole-mounted charges were distributed generously through the infantry units, while tanks were also the primary target of the 12cm guns.

Naval Ground Units in Japan

Japan proper was divided into four naval districts (Sasebo, Maizuru, Kure and Yokosuka) for administrative purposes, but into six naval sectors for defense. Two of the sectors, Maizuru and Sasebo, were geographically identical with the districts of the same name. The other two districts, however, each comprised two sectors: Yokosuka and Ominato sectors in the Yokosuka district, and Kure and Osaka sectors in the Kure District.

Most of the districts/sectors, plus the Chinkai District in Korea, included a defense unit (*bōbitai*) with coastal batteries and underwater mines. With war approaching mobilization tables were issued on 5 November 1941 for the following strengths:

Defense Unit	Off	WO	PO	Other	Total
Yokosuka	22	8	123	287	440
Saeki (Kure)	22	9	126	322	479
Sasebo	22	9	133	342	506
Maizuru	16	8	98	236	358
Ominato	25	9	108	254	396
Chinkai	21	9	111	239	380

As combat approached the homeland, of course, the defense units were expanded. An example was the Saeki Defense Unit. Its combat elements were found in the mine branch, the gunnery branch and the defense brigade. By the end of the war the mine branch operated three minelayers, four tug boats, three special patrol boats, three special sub chasers, two torpedo boats and 14 one-man submarines, along with ten shore-based stations with sonars and, in a few cases, magnetic detectors. The gunnery branch manned a highly diverse collection of weapons in eight batteries:

Hebino Saki 1: 1 off + 35 EM with 2 12cm L/40 DP guns & 1 2-meter rangefinder
Hebino Saki 2: 2 + 35 with 3 75mm Type 88 AA guns
Onyushima: 2 + 35 with 3 75mm Type 88 AA guns
Oshima: 2 + 95 with 5 15cm L/40 Type 41 guns & 1 2-meter rangefinder
Senzaki: 2 + 46 with 3 15cm L/40 Armstrong guns, 1 4.5-meter rangefinder, 1 S/L
Yura: 2 + 77 with 4 14cm L/50 Type 3 guns & 1 2-meter rangefinder
Ukuru: 2 + 43 with 3 15cm L/40 Armstrong guns & 1 4.5-meter rangefinder
Nookayama: 1 + 14 with 5 single 25mm AA & 1 quad 13mm AA

The defense brigade was hastily organized to utilize the local supply of sailors for ground combat. It consisted of three 1,300-man battalions, each barely armed with 24 pistols, 295 rifles, 10 light machine guns, 6 medium machine guns, 12 mortars, and 12 rocket launchers. The Defense Unit did not include the AA command in the Saeki area, a separate unit.

In addition, each naval sector (except Osaka) formed guard units (*Keibitai*) with cadre-strength coast and anti-aircraft artillery units for specific areas of coastline within the sectors that could be filled in and expanded as needed for the defense of the district. Being cadre units they tended to have small personnel strengths for most of the war, usually 700–800 men. These units, with their initial TO&E strengths in parentheses were Ise (guarding Nagoya)(163), Maizuru (358), Onagawa (123), Oshima (202), Saeki (479), Sasebo (506), Shimonoseki (249) and Yokosuka (440). A defense unit was usually organized into a mine department, a gunnery department (with subordinate batteries), and administrative and support departments.

By the end of the war the Sasebo Guard Unit had increased to no fewer than 10,490 men and the Sasebo Defense Unit to 4,447; and the Maizuru Guard Unit had risen to 4,293. Additional units were also formed in 1945, with those in the Sasebo Naval District comprising the Hakata Bay Guard Unit (359 men), Nagasaki Bay Guard Unit (289), and the Tsushima Guard Force (1,020).

The guard and defense unit HQs in the homeland, however, were largely administrative units. Each of the six naval defense sectors of the homeland was divided for command purposes into a mobile defense force, a water defense force and a land defense force. The water defense force provided patrol boats, minesweepers and a minelaying command. The land defense force consisted of a variable number of anti-aircraft commands, seacoast artillery commands, local security guard detachments and lookout posts. Anti-aircraft and coast defense batteries were assigned directly to these commands, with no intervening battalion or group HQs.

The water defense command of each sector was divided into water defense groups, with the Yokosuka Water Defense Command, for instance, controlling the Ise, Yokosuka, Onajawa and the rather distant Chichi Jima Water Defense Groups. A water defense group, in turn, consisted of an anti-submarine flotilla, a casemate unit, an observation detachment, and a maintenance battalion. The anti-submarine flotilla operated the command's boats, mainly patrol boats, submarine chasers, minelayers and minesweepers. The casemate unit operated the controlled minefields and detectors from shore-based installations. The observation detachment manned lookout posts to warn of approaching ships and aircraft. The maintenance battalion had five specialized 200-man companies to perform repairs and upkeep on the group's equipment.

As an example of a casemate unit, that in the Yokosuka Water Defense Group controlled one casemate Type A (mines, acoustic detectors and magnetic detectors), three casemates Type B (acoustic detectors only) and one casemate Type C (magnetic detector only). The Type A casemate consisted of a 3-man generator crew, a 3-man telephone crew, a 3-man communications maintenance crew, six observers and visual signallers, a 3-man searchlight crew, a cook, a medic, and a 26-man detection/mine section with three plotters, six magnetic detector operators, seven acoustic detector operators, and ten controlled minefield operators.

A sector's land defense command was responsible for anti-aircraft and coastal defense and local security guard detachments. This was exercised through its anti-aircraft and seacoast artillery commands. There were a total of 43 anti-aircraft commands at the end of the war with the following weapons:

Sector	Command	127mm	120mm	100mm	80mm	76mm	75mm	40mm	25mm	13mm
Sasebo	Karatsu	10	20	0	0	4	0	4	74	86
	Sasebo	32	16	8	0	8	0	8	73	61
	Kumamoto	0	4	0	0	0	0	0	20	28
	Kagoshima	30	26	8	0	20	0	8	363	140
	Saeki	0	13	0	0	0	12	0	122	12
Kure	Senzaki	0	0	0	0	0	2	0	19	0
	Bofu	24	8	4	0	4	0	7	63	19
	Iwakuni	0	8	0	0	0	6	0	60	4
	Matsuyama	0	8	0	0	0	0	0	9	0
	Oimizuri	0	0	0	0	0	0	0	6	0
	Kochi	0	4	0	0	0	3	0	28	2
	Kure	54	30	8	0	7	6	6	74	29
	Kurashiki	0	4	4	0	0	0	0	24	0
	Kannonzi	0	4	0	0	0	0	0	12	0
	Idamada	0	0	0	0	0	0	0	4	0
	Niho	0	0	0	0	7	0	0	28	20
	Tokushima	0	8	0	0	0	0	0	36	16
	Muroto	0	0	0	0	0	0	0	3	2
	Usa	0	8	0	0	0	12	0	48	0
Maizuru	Maizuru	16	12	0	4	21	0	4	103	47
	Fukachiyama	6	0	0	0	0	0	0	12	6
	Komatsu	0	0	0	0	2	0	0	18	6
	Otsu	0	0	0	0	0	0	0	4	1
Osaka	Osaka	0	6	0	0	0	6	0	136	38
	Suido	0	0	0	0	0	4	0	38	4
	Nara	0	8	0	0	0	4	0	131	36
Yokosuka	Ise Bay	12	18	4	0	0	12	4	329	74
	Enshu	0	0	4	0	0	22	8	70	16
	Izu	4	0	0	0	0	0	0	76	45
	Boso	8	12	0	0	0	0	0	111	91
	Idirizuka	22	8	4	0	0	0	6	112	96
	Yokosuka	96	76	20	0	15	6	12	561	419
	Keinin	0	8	0	0	0	0	0	80	51
	Kasumigaura	12	46	0	0	2	0	2	170	39
	Kanto	0	0	0	0	0	0	0	24	0
	Kooriyama	0	0	0	0	0	0	0	24	7
	Sendai	12	4	0	0	0	0	4	96	45
	Idimezi	0	0	0	0	0	0	0	30	2
	Ku	0	4	0	0	0	0	0	19	7
Ominato	Ominato	16	22	0	0	8	0	2	143	22
	Chitose	8	4	4	0	0	0	0	48	10
	Hokkaido East	0	16	0	0	7	0	0	64	38
	Hokkaido North	0	0	0	0	0	0	0	22	2

Note: figures represent gun barrels.

An AA command headquarters was relatively small, usually consisting of 3 officers, a warrant officer, 10 petty officers and 50 sailors, although there was considerable variation in size to handle widely varying numbers of batteries.

There were four types of batteries: gun, auto weapons, composite and searchlight. There were no fixed TO&Es for the batteries, but the norm was for each gun battery to have four or six guns with one officer and 75 enlisted; an auto weapons battery to have twelve 25mm guns with one officer and 80 enlisted; a composite battery to have a gun platoon (four guns, usually 127mm), an auto weapons platoon (four guns) and a searchlight squad; and a searchlight battery to have one to three lights and a sound locator with one officer and 60 enlisted.

An example of an AA battery would be the 5th AA Battery of the Sasebo Defense Unit in 1943, which manned six 12cm Type (Taisho) 10 dual-purpose guns each with a 7-man crew, a 150cm Type 93 searchlight with 5-man crew, a sound locator, a comparator and a rangefinder.

Coast artillery was primarily an army responsibility and the navy manned only small numbers of guns of relatively small caliber. Although the administrative structure had been in place, very little appears to have been done before the panic of 1945. By the end of the war such weapons as were available were concentrated in eight areas with the numbers of batteries and guns as shown:

	Bties	**20cm**	**15.5cm**	**15cm**	**14cm**	**12.7cm**	**12cm**	**8cm**
Nagoya	3	0	0	0	6	0	0	0
Mutsu Bay	n/a	10	0	0	0	4	25	7
Tokyo-Sagami Bay	15	1	2	11	4	3	13	11
Yura	4	0	0	8	1	0	0	0
Hoyo (Bungo) Strait	7	0	0	12	4	0	8	0
Kagoshima	6	0	0	4	4	0	4	0
Tachibana Bay	12	0	0	4	0	0	14	0
Sasebo	5	0	0	2	1	2	0	0

The battery varied widely in size, as can be gauged from the figures above. It was generally organized into four sections: a command section, a range section, a gun section and a searchlight section. The gun section manned from one to four guns, while the searchlight section manned two lights. The personnel strength was not fixed and generally represented the minimum needed to operate the guns, with the use of skeleton crews common right to the end of the war.

The guns and searchlights were navy models that had been removed from, or intended for, non-operational vessels. It had been planned to install shipboard-type electrical fire control equipment at the batteries, but this had not been accomplished by the end of the war. Generally the only fire control equipment available to the batteries were on-carriage sights, 1- or 2-meter base stereoscopic range finders and plotting boards with speed and course angle calculators.

Local security was provided by the sector's guard unit until 1945, when extemporized elements were added. An example of this can be found in the Yokosuka District's security forces (*keibi butai*) in the spring of 1945, where the regular guard unit's three rifle companies (each of three platoons) was to be supplemented by no fewer than 15 numbered security companies. These companies, which were to be drawn from the various elements available, including the gunnery school, the machinists school, air group personnel, etc., would be constituted only in emergency and each was to consist of a rifle company (four platoons each of four squads) reinforced by a machine gun platoon of four squads. A further 22 "security units" of unspecified size (presumably platoon to company equivalents) were to be formed by the naval air groups in the area.

An exceptional deployment was of Naval forces to the Ryukyus chain, considered part of Japan proper, but distant from it. Under the October 1944 Army/Navy agreement the IJN's artillery contribution consisted of one AA defense unit at Yontan Airfield with thirty-six 25mm machine cannon, plus various coastal batteries. Those coastal forces for Okinawa were to consist of one battery with four 20cm short guns, two batteries with a total of five 15.5cm guns, two batteries each with two 15cm guns, one battery with four 15cm howitzers, one battery with two 14cm guns, five batteries each with two 12cm guns, and three batteries each with two 12cm short guns. For the other, smaller, islands of the chain the Navy contributed one battery of two 20cm short guns, six 2-gun and one 3-gun batteries with of 15cm guns, two 2-gun batteries of 14cm guns, one battery of two 12cm guns, five 2-gun batteries of 12cm short guns, and four 2-gun batteries of 8cm guns.

By the time of the 1945 invasion the Okinawa Base Force consisted of:

Area A: 3,000 men, 80 7.7mm MGs, 25 13mm MGs, 30 25mm AA, 6 12cm guns, 3 15.5cm guns
Area B: 800 men, 8 7.7mm MG, 25 13mm MG, 3 25mm AA
Area C: 3,000 men, 8 7.7mm MG, 25 13mm MG, 3 25mm AA
Area D: 1,500 men, 20 7.7mm MG, 15 13mm MG, 10 25mm AA, 5 12cm guns, 6 15cm guns, 5 20cm guns
Kunigami District: 600 men, 8 7.7mm MG, 10 13mm MGs, 18 25mm AA

Another distant portion of Japan was the Ogasawaras (Bonins) so their naval forces were also named rather than numbered. The detachment of the Chichi Jima Guard Unit deployed to Iwo Jima was strengthened and redesignated as the Iwo Jima Guard Unit in March 1944. The initial force of about 500 men was provided with four 12cm AA guns, four 8cm AA guns, and 26 twin-mount 25mm AA guns. The reinforcing increment that arrived from the Yokosuka was the same size but brought with them four 14cm CD guns, six 12cm CD guns, four 12cm AA guns, and four twin-mount 25mm.

Two naval sectors (also referred to as guard districts) were also established outside Japan proper, in Japanese territories: Chinkai (now Chinhae, near Pusan) and Takao (on Formosa). The Takao sector included a guard unit of the same name, while the Chinkai sector included both a guard unit and a defense unit of that name, as well as the Rashin (now Najin, in northeastern Korea) Defense Unit. As with the similar units in the homeland, these were rather small formations, with the unit strengths being about 300 men in each case.

Naval Ground Units Outside Japan

The blockade of the long China coast created the need for local base facilities, and this was met by the formation of Amoy, Hankow, Shanghai and Tsingtao area special bases forces. These survived to the end of the war providing service support to the China Fleet.

The vast Japanese Mandated Islands, sprawling across the central Pacific, created a different set of problems. Japan was forbidden to fortify these islands by the terms of the League of Nations mandate under which she administered them. Even after Japan pulled out of the League and withdrew from the various naval treaties in 1937 she vowed not to construct military installations there.

On the other hand the Japanese administrators, the Nan'yo-Cho, engaged in aggressive commercial development, including the construction of dual-use facilities, such as airfields, coaling facilities and radio stations. These civil-funded efforts were relatively noncontroversial within Japan.

The strategic use of these islands stirred considerable debate within the Navy, however, for there were elements who opposed spending any of the tight naval budget on anything but warships. Further, there was disagreement even among those who favored fortifications as to which islands to fortify. One element wanted to use the Marianas for an imperial shield, another favored a more forward defense using the Marshalls.

The 1928 South Seas Defense Plan was something of a compromise, calling for the creation of both infrastructure and skeletonized forces that, while adhering to the naval convention ban on fortifications, would permit the militarization of selected islands within 60 days of the decision to proceed.

Under this plan the navy authorized the formation in September 1935 of the 1st through 6th "special defense units" (*Tokubetsu Bobitai*). Since fortification of the mandated islands was prohibited by treaty, and in any event no decision could be reached on which islands to fortify, they did not include any gun batteries or underwater detection sections. Instead, they provided guards, administrative overhead, signal units etc. in preparation for the time when Tokyo could finally decide whether to fortify the islands and, if so, which islands to invest in. The strengths of the defense units as authorized in the 1935 order were:

	Off	WO	PO	Sailors
1st Defense Unit	13	4	51	144
2nd Defense Unit	13	4	49	136
3rd Defense Unit	12	4	40	116
4th Defense Unit	16	8	74	218
5th Defense Unit	12	4	40	114
6th Defense Unit	14	10	36	144

Of these, 3rd Defense Unit was activated at Palau to garrison those islands, the 4th at Truk (with a detachment at Ponape) and the 5th at Saipan. In accordance with this plan a rather modest program was carried out in 1938 that saw the transfer to the inner islands of thirty-two 15cm L/40 coastal guns (eight each to Saipan and Palau, twelve to Truk and four to Ponape), six 12cm L/40 coastal guns (four to Saipan and two to Palau), and fourteen 8cm AA guns (ten to Truk and four to Ponape). The guns were delivered, but not mounted.

On 15 May 1939 the 4th Fleet was activated and made responsible for the defense of the mandated islands. Not a seagoing command, the fleet supervised all of the local defense units, air, land and sea, assigned to the mandates.

Part of any forward defense strategy was the creation of forward bases for the fleet. No thought was apparently given to establishing large, full-service bases, along the lines of Pearl Harbor and Cavite for the USN, but there would clearly be a need for fueling, provisioning and minor repairs without returning to the homeland. This was answered by the creation of a numbered series of units called "base forces" (*Konkyochitai*). These were to be assigned to the numbered fleets to operate the smaller forward bases needed by the navy. These base forces had service responsibilities (limited engineering and maintenance for ships, supply, and construction), administrative responsibilities for local ground and surface units, and tactical responsibilities to defend the base itself.

Where the administrative and service requirements were limited, as where a base force was co-located with a numbered fleet HQ with its own service elements or where harbor and port facilities were minor, the occupying unit was designated a "special base force" (*Tokubetsu Konkyochitai*) and took on more of the characteristics of a purely defensive unit.

The 1st and 2nd Base Forces were formed for use on the China coast. In October–November 1940, three were formed under the 4th Fleet for the mandates, the 3rd at Palau, 4th at Truk and the 5th on Saipan with small detachments through the Carolines. In January 1941 the 6th Base Force was activated on Kwajalein, responsible for operations in the Marshalls. Each of these assumed command of the co-located existing defense unit with the same numerical.

The base forces themselves tended to be rather small units, with strengths between about 200 and 1,500 men. With the decision to add weapons to the pre-war base forces additional personnel had to be added when the time came to mount the weapons. The batteries had a variable organization depending on the number and type of weapons manned. In addition to the gun crews (size unspecified in the organization table), each battery was authorized six men per sound locator (for AA batteries), 3 men for each 90cm searchlight, 4 for each 110cm searchlight and 5 for each 150cm searchlight, along with one man per telephone, four per radio, four per generator, three per command post, etc. Defense stations were used in conjunction with minefields and came in two varieties: a Type A with 38 men with three mine control sets and three hydrophones; and Type B with 27 men and two hydrophones.

Fortification of the mandated islands apparently began seriously in mid-1941. Little seems to have been done at Palau, except to mount the existing stored weapons. Elsewhere, however, a substantial armaments program was launched. The first to be armed were Truk (and its administrative dependency, Ponape) and Saipan, which received the following armament in August 1941:

	Truk	**Ponape**	**Saipan**
15cm coastal batteries (guns)	4 (10)	2 (4)	0
14cm coastal batteries (guns)	0	0	1 (4)
12cm coastal batteries (guns)	0	0	3 (10)
12.7cm dual-purpose twin-mount batteries (mounts)	3 (12)	1 (4)	1 (2)
8cm AA batteries (guns)	7 (20)	4 (8)	3 (8)

The decision to arm the more forward Marshall Islands, which would presumably be the first line of defense against the US Navy, had not yet been taken. Then, in October 1941 Kwajalein was given four batteries of 12.7cm dual-mount guns, totalling twelve of these twin systems. In early November the process of arming the Marshalls was truly begun with the dispatch of seven batteries, each with three twin-mount 12.7cm dual-purpose guns, three batteries for Jaluit, and two each for Maloelap and Wotje. The personnel to man these weapons were drawn from the 6th Defense Unit, which was now divided into four gun units, the 1st at Jaluit, the 2nd at Maloelap, the 3rd at Wotje and the 4th at Kwajalein.

Distant administration and control of these units proved difficult, however, and they were quickly redesignated as examples of a new kind of unit, the "guard unit" (*keibitai*), roughly analogous to the US Marine defense battalions. The forces on the outlying islands were given guard unit designations in the 50-series. The defense forces of Kwajalein itself remained denominated the 6th Defense Unit.

In early December two more guard units were formed to defend the expected early captures in the central Pacific, the 54th Guard Unit for Guam and the 65th Guard Unit for Wake.

These actions completed the preparations for war in the mandated islands. At the time of the Pearl Harbor attack the naval ground forces in the Pacific had the following strengths:

	Off	WO	PO	Sailors	Total
3rd Base Force (Palau)					
Headquarters	12	3	15	19	49
3rd Defense Unit	12	4	40	116	172
4th Base Force (Truk & Ponape)					
Headquarters	12	3	15	19	49
4th Defense Unit	16	8	74	218	316
5th Base Force (Saipan)					
Headquarters	12	3	15	19	49
5th Defense Unit	12	4	40	114	170
6th Base Force (Marshalls)					
Headquarters (Kwajalein)	12	3	15	19	49
6th Defense Unit (Kwajalein)	14	11	36	144	204
51st Guard Unit (Jaluit)	13	8	45	332	398
52nd Guard Unit (Maloelap)	13	8	45	332	398
53rd Guard Unit (Wotje)	13	8	45	332	398
54th Guard Unit (for Guam)	9	5	34	267	315
65th Guard Unit (for Wake)	24	14	87	841	966

The 8th Special Base Force was formed for Rabaul on 1 February 1942 to support the 8th Fleet in that area, and it was followed quickly by the 9th and 10th for the Malay Peninsula, 11th for Saigon, 12th for the Andamans, and the 21st to 25th Special Base Forces for the Dutch East Indies. Others followed as needed for the various ebbs and flows of the war in the Pacific.

The expansion into the mandated islands created a template for the garrisoning of far-flung locales once the war began. On 10 April 1942 the Navy forces in the Pacific were completely reorganized. Base forces would henceforth be responsible not only for the operation and defense of their own local facilities, but would also exercise command over guard units set up in their regions. The guard units in the Marshalls were redesignated in the 60-series (presumably to reflect their subordination to the 6th Base Force on Kwajalein), while two new guard units were formed under the 4th Base Force. The new organizational framework for the 4th Fleet was:

3rd Spec Base Force	Palau
4th Base Force	Truk
41st Guard Unit	Truk
42nd Guard Unit	Ponape
5th Spec Base Force	Saipan
54th Guard Unit	Guam
6th Base Force	Kwajalein
61st Guard Unit	Kwajalein
62nd Guard Unit	Jaluit
63rd Guard Unit	Maloelap
64th Guard Unit	Wotje
65th Guard Unit	Wake
8th Base Force	Rabaul
81st Guard Unit	Rabaul
82nd Guard Unit	Lae

Further changes to the organization, of course, were made on a continuing basis in response to changing circumstances. With the fall of the Philippines Palau had become a naval backwater and on 15 June 1942 the 3rd Special Base Force was abolished, while on 15 September the 51st Base Force was formed for the Kuriles. Six months later a new 1st Base Force was formed at Buin under the 8th Fleet to control the new 87th Guard Unit and the Sasebo 6th SNLF.

Strengths of the wartime special base forces varied from 520 to almost 1,500 men depending on the tactical requirements, but they were consistently organized into a fleet support department, a sea defense department, a land defense department, and harbor, signal, medical and supply departments. The further forward a special base force was deployed, the greater its tactical responsibilities, the extreme being represented by the 3rd Special Base Force on Tarawa.

In the central Pacific the US raid on the Japanese air base at Makin in August 1942 had immediate effects. The new Yokosuka 6th SNLF was dispatched to Jaluit in September and its first company (rifle, reinforced) with a strength of 365 men was immediately dispatched to Makin to provide security. The rest of the SNLF was sent to Tarawa, where they were joined in December by the 111th (Navy) Pioneers. On 15 February 1943 the Yokosuka 6th SNLF was redesignated the 3rd Special Base Force (the same designation earlier used at Palau). Being the most forward of the base forces, and given its derivation, it is not surprising that it was organized almost solely for defense. At the time of the redesignation it consisted of the following elements:

Base Force HQ
Land Defense Group
Three Rifle Companies (one on Makin), Gun Company
Coast Artillery Group
One 8" Battery (2 guns), two 140mm Batteries (each 2 guns), one 8cm Battery (3 guns)
Anti-Aircraft Group
Two 12.7mm Batteries (each 2 twin mounts), one 75mmBattery (4 guns), one MG Battery (5
Water Defense Group

Table 8.3: Numbered Base Forces of the Pacific War

	Activated	Deactivated	Location	Guard Units
1st	Pre-war	Mar 42	Formosa	
1st	Nov 42	Sep 45	Buin	84th (Nov42–Apr43), 87th (Sep43–Sep45)
2nd	Pre-war	Mar 42	Formosa	
2nd	Nov 42	Mar 44	Wewak	90th (Feb44–Mar44)
3rd	Pre-war	Jun 42	Palau	
3rd Sp	Feb 43	Nov 43	Tarawa	
4th	Pre-war	Sep 45	Truk	41st & 42nd (Apr42–Sep45), 43rd (Jun42–Jan44), 44th (Mar44–May44),47th (Mar–Sep45), 48th (Apr–Sep45), 49th (May–Sep45), 67th (Jan44–Sep45), 69th (Feb44–Mar44)
5th	Pre-war	Aug44	Saipan	44th (May–Jul44), 54th (Dec41–Oct44), 55th (Mar–Dec44), 56th (Mar–Dec44)
6th	Pre-war	Sep 45	Kwajalein	51st, 52nd & 53rd (Dec41–Apr42), 61st (Apr42–Jul44), 62nd, 63rd & 64th (Apr42–Sep45), 65th (Dec41–Sep45), 66th (Jun43–Sep45), 68th (Dec43–Mar44)
7th	Pre-war	Jun 42	Chichi Jima	
7th	Jul 42	Mar 44	Lae	82nd (Dec42–Mar44), 85th (Aug43–Mar44)
8th	Feb 42	Dec 44	Rabaul	81st (Apr42–Dec44), 82nd (Apr42–Dec42), 83rd (Mar–Dec43), 84th (Jul–Oct42), 86th (Aug43–Dec44), 89th (Dec43–Dec44)
9th	Pre-war	Sep 45	Cam Rahn	91st (Dec41–Feb42)
10th Sp	Feb 42	Sep 45	Singapore	1st (Mar–Aug42), 7th (Jul–Sep45), 9th (Aug43–Jun44), 101st (Aug–Sep45)
11th Sp	Pre-war	Sep 45	Saigon	10th (May–Sep45), 11th (Jan–Sep45), 110th (Aug–Sep45)
12th Sp	Jan 42	Sep 45	Port Blair	12th (Jun–Oct43), 13th (Sep–Oct43), 14th (Oct43–Sep45, 25th (Dec43–Sep45)
13th	Oct 43	Sep 45	Rangoon	12th (Oct43–Aug45), 13th (Oct43–Sep45), 17th (Oct43–Sep45)
14th	Dec43	Sep 45	Kavieng	83rd (Dec43–Sep45), 88th (Dec43–Dec44), 89th (Dec44–Sep45)
15th	Jun 44	Sep 45	Penang	9th (Jun44–Sep45)
21st Sp	Mar 42	Sep 45	Surabaya	3rd (Jan44–Sep45)

Table 8.3: Numbered Base Forces of the Pacific War

	Activated	Deactivated	Location	Guard Units
22nd Sp	Mar 42	Sep 45	Balikpapan	2nd (Mar42–Sep45), 4th (Sep–Dec44), 6th (Sep–Dec44)
23rd Sp	Mar 42	Sep 45	Makassar	3rd (Mar42–Jan44), 4th (Dec44–May45), 6th (Oct42–Nov43 & Dec44–May45), 8th (Dec44–Sep45)
24th Sp	Jan 42	Sep 44	Ambon	4th (Mar42–Nov43 & May–Sep44), 6th (Nov43–Sep44), 7th (Sep–Nov43)
25th Sp	Dec 42	Sep 45	Manokwari	7th (Nov43–Jul45), 18th (Oct–Nov43 & Jul–Sep45), 20th (Nov43–May44 & Jan–Sep45), 21st (Jan44–Sep45), 26th & 27th (Jan–Sep45), 29th (May–Sep45)
26th Sp	Nov 43	May 45	Halmahera	18th (Nov43–May44), 19th (Mar–May44), 20th (May44–Jan45), 91st (Apr–May44)
27th Sp	Mar 44	Sep 45	Wewak	90th (Mar–Jul44)
28th	May 44	Jul 45	Biak	18th (May44–Jul45), 19th (May–Sep44), 91st (May–Sep44)
30th	Jan 44	Sep 45	Palau	45th & 46th (Mar44–Sep45)
31st Sp	Jan 42	Feb 43	Manila	
31st Sp	Sep 44	Sep 45	Manila	35th (Nov44–Sep45)
32nd Sp	Pre-war	Sep 45	Davao	33rd (Mar44–Sep45), 36th (May–Jun44)
33rd Sp	Aug 44	Sep 45	Cebu	36th (Aug44–Sep45)
51st	Sep 42	Aug 43	Kuriles	5th (Sep42–Aug43)

Note: dates are official. Units may have been activated before deployment, or kept on the books after destruction.

Three torpedo boats, one minesweeping unit
Signal Section
Repair Section
Medical Section
Intendence Section

Subsequently, a second 8" battery (2 guns) was added, as was a second 8cm battery, and a twin 13mm MG battery. These increased the size of the unit from an original 902 men to 1,122. In May 1943 the Sasebo 7th SNLF (a defensive unit) arrived and it was with this force that the US invasion was met.

The 65 guard units eventually raised (numbered 1 to 110, with gaps) were generally designated in groups. The eight guard units numbered 18th to 29th were subordinated to the 25th Special Base Force, the nine numbered 41st to 49th were subordinated to the 4th Base Force, the eight numbered 61st to 68th were under the 6th Base Force, while the nine numbered 81st to 89th were, most of the time, subordinate to the 8th Base Force.

The only guard units to share a common organization were the 61st to 64th in the Marshalls, each of which started with a strength of 557 men, although these figures climbed slightly through 1942–43 as reinforcements were sent in. As finally configured, the three original naval guard forces to survive the war were armed as follows:

	62nd Gd U	63rd Gd U	64th Gd U
15cm CD guns	3	8	6
12cm CD guns	4	2	2
12.7cm DP twinmounts	6	5	6
7.6cm DP guns	0	0	0
25mm twinmount AA guns	3	3	3
13.2mm AA machine guns	6	21	17
Landing craft	3	4	3
Torpedo boats	0	0	2
Patrol boats	4	10	5

As the Navy advanced south after the outbreak of the war further guard units were formed in the Dutch East Indies and neighboring areas out of the SNLFs that had been used to capture the key ports. Thus the 1st Guard Unit (347 men at Singapore) was formed from the Kure 2nd SNLF, the 2nd Guard Unit (665 men at Tarakan) from the Kure 2nd SNLF, the 3rd Guard Unit (546 men at Makassar) from the Sasebo 2nd SNLF, and the 4th Guard Unit (1,036 men at Ambon) from the Kure 1st SNLF, all activated on 10 March 1942. The 5th Guard Unit followed on 1 July 1942 at Kiska Island out of the Maizuru 3rd SNLF. This new unit had an authorized strength of 711 men (although it actually reached 1,269 before its evacuation in July 1943) and consisted of two batteries of 15cm coastal guns (total 6 guns), one 4-gun battery of 120mm coastal guns, one 4-gun battery of 76mm coastal guns, a field AA battery, a land defense company of three MG platoons, a sea defense unit (13 barges), a 28-man radar platoon, and service elements. The 6th Guard Unit followed in October 1942, and the 7th–9th in 1943, all in or around the East Indies.

In December 1943 the 68th Guard Unit was formed on Eniwetok, only to be destroyed two months later, and in January 1944 the 67th Guard Unit was formed with 22 officers and 383 enlisted on Nauru, plus a detachment of 12 officers and 359 enlisted on Ocean Island. The main body on Nauru, which rose to 1,367 men, manned four 15cm coastal guns, four 8cm coastal guns, four 12.7cm DP twin-mounts, twelve 25mm AA autocannon, and ten 13mm MGs. The Ocean Island detachment, rising to 760 men, manned four 14cm coastal guns, five 8cm AA guns and eight 25mm autocannon.

An example of the defense of a large island group was Truk. Although not a naval base *per se*, it was a key anchorage and trans-shipment point for the Japanese adventures to the south. The original Guard Units, the 41st on Truk itself and the 42nd on Ponape, expanded through the war, but accretions to strength mandated a second headquarters to command additional units and the 43rd Guard Force was formed to control units on Tol Island in the group, plus the 44th on distant Woleai. No further reinforcements arrived, but to rationalize command somewhat three more HQs were formed in March–May 1945, the 47th Guard Unit for units on Moen Island, the 48th for Fefan and Param Islands, and the 49th for Uman Island.

Table 8.4: Naval Guard Forces, Truk, August 1945

	Off	WO	NCO	Other	Heavy Coastal Guns	Heavy AA Guns
41st	61	29	885	795	1x15/40, 2x12/45, 2x8/40	4x12.7 twin
43rd	28	13	499	521	9x15/40	6x12/45
47th	44	5	546	908	4x20/40, 3x15/45, 3x15/40, 2x14/50, 1x12/45, 2x8/40	11x12/45
48th	31	13	412	769	4x15/40, 3x12/45, 3x8/40	12x12/45, 2x8/40
49th	21	9	300	294	3x15.5/60, 2x15/40, 4x14/50	-

As was the case elsewhere, the Guard Units were equipped almost exclusively with obsolete weapons. A single battery of 155mm L/60 weapons in gun shields on concrete pedestals on Uman Island represented the only modern coastal weapons, these having a stated effective range of 27km. For the most part the rest of the coastal defense rested on a battery of old 20cm (8") removed from scrapped cruisers with a range of 18km, and a variety of ancient 15cm L/45 and L/40 guns with ranges of about 10Km.[4] For anti-aircraft defense the modern weapons were four twin-mounts of 12.7cm guns, along with a total of 73 multiple- and 23-single-mount 25mm weapons, and 3 twin- and 89 single-mount 13.2mm machine guns plus, inexplicably, a single quad-mount Hotchkiss original 13.2mm.

The 45th Guard Unit formed in March 1944 on Palau, also subordinate to Truk, had 1,034 men by the Spring of 1944, including two batteries each with four 15cm L/40 Armstrong guns, one battery with two 12cm L40 Armstrongs, two batteries each four 12cm Type 10 DP guns, one battery with twelve 25mm twin-mounts, and radar station with a Type 1 Mark 1, Model 3, along with a harbormaster's office and repair, supply and medical branches. Interestingly, 370 of the personnel, including most of those at the DP batteries, were temporarily assigned crews of sunken ships.

Because the areas to be defended and the probable threats varied widely, no attempt was made to standardize the organization of a naval guard unit. In early 1944, however,

4 Indicative of the age of these weapons, of the 15cm L/40 weapons only two were Japanese-made, the others being Armstrong products, including no fewer than 15 made by the Italian subsidiary Stabilimento Armstrong around 1900–1902.

tables for a standard 25mm AA battery, 12cm DP battery, and 12cm coastal battery were published. The TO&E for a light AA battery called for a 35-man HQ and two 69-man platoons. The battery HQ was provided with two rangefinders and sixteen telephones for distribution as needed, and also included the battery's two medics. Each platoon was made up of a commander, a platoon sergeant, six twin 25mm AA guns with 9-man crews, and a 13-man ammunition squad. The battery also included six Type 92 medium MGs on high-angle mounts, and 8 Type 99 light MGs and 137 rifles for close-in land defense.

The heavy AA battery was made up of a 42-man HQ and a 52-man firing battery. The HQ included a seven-man fire control section, a 4-man ranging section (with two optical rangefinders), an eight-man searchlight section (with two 110–150cm lights), an eight-man generator section, and a six-man fuze setter section. The firing battery was built around four 12cm L/45 guns, each with its 8-man crew, along with 18 ammunition handlers. The battery was also provided with four Type 92 MGs on high-angle mounts, two Type 99 light MGs, 77 rifles and 18 telephones.

The coast defense battery comprised a 19-man HQ and a 34-man firing battery. The latter manned three 12cm coastal guns, each with its 7-man crew. The battery was also provided with two medium and two light MGs, 43 rifles and six telephones.

These efforts at component standardization appear to have had little effect on the guard units. The 45th Guard Unit in the Palau Islands had the following composition in April 1944:

1st Dept (infantry)(86)(4 LMG, 2 MG, 6 small flamethrowers)
2nd Dept (harbormaster)(51)(1 tugboat)
3rd Dept (repair)(50)
4th Dept (medical)(23)
5th Dept (supply)(47)
6th Dept (CD)(67)(1 LMG, 1 MG, 4 15cm/L40 guns, 1 searchlight)
7th Dept (CD)(26)(2 12cm/L40 guns, 1 searchlight)
8th Dept (CD)(65)(1 LMG, 1 MG, 4 15cm/L40 guns, 1 searchlight)
Heavy AA Battery (n/a)(4 12cm/L45 DP guns)
Light AA Battery (166)(25mm twinmount guns)
Radar unit (37)(3 Mk 1 Mod 3 radars)
Others (31)

A few guard units were also extemporized by the local base forces. When the 8th CSNLF and its constituent Kure 6th and Yokosuka 7th SNLFs were disbanded in December 1943 in the Bougainville area the troops were assigned to the 8th Base Force (Rabaul) and used to form the 87th (Buka), 88th (Manus) and 89th (Namatanai) Guard Units, joining the earlier 81st to 86th Guard Units. A portion also remained in the Buin area to help form the Shortlands Guard Unit, also under the 1st Base Force. This guard unit had the following composition:

HQ (13)
1st AA Battery (ex1st AA Unit)(166)
  9 LMG, 2 MG, 8 GL, 4 76mm DP
2nd AA Battery (ex13th AA Unit)(62)

1 MG, 1 GL, 4 13mm MG, 3 twin 25mm AA
3rd AA Battery (ex13th AA Unit)(71)
1 LMG, 2 MG, 1 GL, 4 13mm MG, 1 13mm twin MG, 1 25mm twin AA
4th AA Battery (ex13th AA Unit)(41)
1 LMG, 1 MG, 1 GL, 4 13mm MG)
5th AA Battery (ex13th AA Unit)(83)
2 LMG, 3 MG, 3 13mm MG, 1 13mm twin MG, 1 25mm twin AA
6th AA Battery (ex13th AA Unit)(41)
1 LMG, 3 MG, 1 GL, 1 13mm twin MG, 1 searchlight
1st CD Battery (exKure 6th SNLF)(102)
1 LMG, 4 14cm L50 guns, 1 searchlight
2nd CD Battery (exYokosuka 7th SNLF)(60)
3 LMG, 1 MG, 3 GL, 4 76mm guns, 1 searchlight
3rd CD Battery (exKure 6th SNLF)(91)(lookouts)
5 LMG, 4 GL, 1 flamethrower
4th CD Battery (exYokosuka 7th SNLF)(38)
1 LMG, 1 MG, 2 76mm guns
Land Defense Platoon (exKure 6th SNLF)(33)
1 LMG, 2 MG, 3 GL
Radar Platoon (ex1st Base Force)(18)
2 Mk 1 Mod 2 radars
Engineer Platoon (55)

Another unit formed out the of the 8th CSNLF, as mentioned, was the 88th Guard Unit on the Admiralty Islands. This unit was built around a rifle company, a security unit (essentially a second rifle company), a 12cm coastal battery, a 52-man sea defense section, a 77-man observation unit and service troops for a total of about 750 men.

Air Defense Units

As the Allied drive across the Pacific picked up momentum it became clear that the greatest threat to the Japanese island bases came from the air. The IJN therefore began raising battery-sized "air defense units" (*Bokutai*) to supplement the efforts of the base and guard units. These came in three types designated Types A, B and C. The Type A air defense unit had a strength of 287 men and consisted of a fire control section (with 2 searchlights and rangefinders), three gun platoons (each with two 76mm AA guns) and a machine gun platoon (ten 7.7mm MGs on high-angle mounts).

The Type C air defense unit had a strength of 316 men and was organizationally similar except that each gun platoon had four 25mm machine cannon instead of the 76mm guns. Such a unit had a total of twelve 25mm Type 96 autocannon, ten 7.7mm Type 92 machine guns, one Type 97 1-meter base rangefinder, three Type SU searchlights (75cm or 90cm) and two generators. In practice, the units often reorganized into four near identical platoons. An example of this was the 3rd Platoon of the 11th Naval Air Defense Unit, which consisted of a petty officer commander, five petty officer gun commanders, a cook, a messenger, a medic, two 25mm Type 96 autocannon each with a nine-man crew, and three 7.7mm AA MGs each with a four-man crew.

The Type B unit had a strength of 225 men and was divided into command and observation sections and three machine gun platoons. It was armed with a total of 32 heavy (13.2mm) and ten light (7.7mm) machine guns, along with two searchlights.

The naval air defense units began raising on 5 December 1942 , with the Yokosuka ND responsible for Units 1–10, Kure ND for Units 11–20, Sasebo ND for Units 21–30 and Maizuru ND for Units 31–40 (although 37–40 were never, in fact, raised). This group was destined mainly for the heavy fighting of the south Pacific, with seven of the units going to the Solomons, ten to the Bismarks, eight to New Guinea, and four to the western Dutch East Indies.

The three larger naval districts began raising a second series of AA defense units in August 1943, with the Yokosuka ND responsible for Units 41–50, Kure ND for Units 51–60, and Sasebo ND for Units 61–70. This group was dispersed much more widely in response to the growing pressure being felt on all fronts. Eight of these thirty units went to the Bismarks, two to New Guinea, six to the Carolines, two the Marianas, three to the central East Indies, and three to the Nicobars and Andamans.

A third wave (80-, 90-, 100- and 110-series) was raised in December 1943 to April 1944. These were sent to the eastern East Indies (10 units), the Tateyama gunnery school (5), Palau (3), Philippines (3), Truk (2), New Guinea (2), Burma (2), Borneo and Singapore. A fourth wave of 35 units (numbered 121–139, 141–151, 161–170, 181–188 and 201–207) was raised starting in the spring of 1944. About eight of these went to the Philippines, two to Formosa, one to Iwo Jima, and the rest to the Homeland.

Starting in September 1943 the Navy began the process of disbanding the AA defense units and incorporating their personnel and equipment into existing (or, in a few cases, newly-formed) guard and base units. Thus, of eighteen AA defense units sent to the Bismarks two (12 and 19) were disbanded in November 1943 and no fewer than fourteen in July 1944 (2, 5, 14, 21, 22, 36, 44, 45, 52, 63, 64, 65, 67, and 69). This trend peaked in July 1944 when no fewer than 66 AA defense units were disbanded, including six at Truk (6, 43, 46, 81, 85 and 86), three in the Marianas (82, 83 and 84) and four at Soerabaya (103, 108, 109 and 117).

Emergency Mobilization

The events of early 1944 seem to have spurred the Navy command to undertake a massive, comprehensive mobilization of forces for defense of their bases. On 10 July 1944 they issued a directive for the mobilization of "emergency combat forces" (*Kinkyū Sembi Butai*) that was to take place in two phases, each of two stages. The first phase was to see units activated on 15 July for the first stage, and on 20 July the orders went out for the second stage of units to be activated on1 August. Orders for the initial stage of the second phase were issued on 3 August, with activation dates of 5 September, with units to depart for their destinations in mid-month. The final increment, the second stage of the second phase, activated units for 5 October, with a departure in mid-month.

Reflecting the increasing dominance of US airpower the vast majority of the emergency combat units were anti-aircraft forces, including no less than 48 air defense units of battery-to-battalion size. In addition to those air defense units, it also directed the activation of separate batteries for defense of specified bases from the Dutch East Indies and the Philippines, to the Bonins and Formosa, to Japan proper.

The vast majority of these were, again, anti-aircraft units, the most common by far being batteries configured with either four 12cm DP guns or twelve 25mm autocannon

Table 8.5: Navy Emergency Combat Forces Order Activations				
Activation date	**15-Jul**	**01-Aug**	**05-Sep**	**05-Oct**
Anti-Aircraft Batteries				
2x12.7cm DP	0	3	0	0
6x12cm DP	1	0	0	0
4x12cm DP	15	11	9	35
3x12cm DP	0	0	1	0
2x12cm DP	1	0	0	0
12x25mm AA	29	34	53	67
8x25mm AA	0	0	2	3
6x25mm AA	1	1	12	16
4x25mm AA	0	3	4	11
Coastal Batteries				
6x20cm short	1	0	0	0
4x20cm short	3	0	0	0
2x15.5cm CD	1	0	0	0
3x15cm CD	1	1	0	0
2x15cm CD	2	4	0	1
2x15cm how	1	0	0	0
2x14cm CD	2	7	0	0
3x12cm CD	1	0	2	0
2x12cm CD	1	2	0	0
6x12cm short	1	1	0	1
4x12cm short	1	1	2	0
3x12cm short	0	0	1	0
2x8cm CD	0	1	0	0
Other Units				
6x8cm mortar batteries	7	6	0	0
Radar sections	5	6	18	3
Searchlight units	10	14	14	23
Lookout stations	21	9	15	12

(replaced by 13.2mm machine guns in a minority of units). These were supplemented by lookout stations, fire-control and searchlight-control radar sections, and high-angle and low-angle searchlight units, the latter generally with two lights.

The batteries raised were added to existing command structures, being added to base forces and guard units. Thus, Okinawa received the 149th Air Defense Unit plus two 12cm DP batteries, three 25mm AA batteries, and four coastal batteries, all of which were incorporated into the Okinawa Guard Unit. Iwo Jima received the 125th, 132nd, 141st and 203rd Air Defense Units, which retained their identities, along with two 12cm DP and five 25mm AA batteries that were subordinated to other formations.

Table 8.6: Air Defense Units Activated under Navy Emergency Combat Forces Order of 10 July 1944

Unit	Active	Location	Weapons	Unit	Active	Location	Weapons
89	15-Jul	Cebu	6x12cm DP, 12x25mm AA, 2 SL	162	01-Aug	Shinchiku	36x25mm AA
99	15-Jul	Kendari	6x7cm AA, 12x25mm AA, 2 SL	163	01-Aug	Taihoku	36x25mm AA
116	15-Jul	Nichols Fld	6x12cm DP, 12x25mm AA, 2 SL	164	01-Aug	Amami O Shima	4x12cm DP, 24x25mm AA, 2 SL
117	15-Jul	Nichols Fld	6x12cm DP, 12x25mm AA, 2 SL	165	01-Aug	Ishigaki Jima	4x12cm DP, 24x25mm AA, 2 SL
120	15-Jul	Legaspi	4x7cm AA, 24x25mm AA, 2 SL	166	05-Sep	Ormoc	6x12cm DP
125	15-Jul	Iwo Jima	6x12cm DP, 2 SL	167	05-Sep	Chitose	6x12cm DP
130	01-Aug	Takao	6x12cm DP, 12x25mm AA, 2 SL	168	05-Sep	Bihoro	6x7cm AA
132	15-Jul	Iwo Jima	4x12cm DP, 12x25mm AA, 2 SL	169	05-Sep	Kisarazu	4x12cm DP
133	15-Jul	Cavite	4x12cm DP, 24x25mm AA, 2 SL	170	05-Sep	Tateyama Afd	4x12cm DP
134	05-Sep	Tainan Afd	6x12cm DP	181	01-Aug	Minami Daitu Jima	4x12cm DP, 24x25mm AA, 2 SL
135	05-Sep	Zamboanga	6x12cm DP	182	01-Aug	Miyako Jima	4x12cm, 24x25mm AA, 2 SL
136	05-Sep	Bulan Afd	6x7cm AA	183	05-Oct	Ormoc	6x12cm DP
137	05-Sep	Ishigaki Jima	4x12cm DP, 12x25mm AA	184	05-Oct	Manila	6x12cm DP
138	05-Sep	Kagoshima Afd	4x12cm DP, 12x25mm AA	185	05-Oct	Cebu	6x7cm AA
139	05-Sep	Takao	4x12cm DP	186	05-Oct	Kanoye	4x12cm DP
141	15-Jul	Iwo Jima	6x12cm DP, 2 SL	187	05-Oct	Kasanohara	4x12cm DP
142	15-Jul	Chichi Jima	4x12cm DP, 12x25mm AA, 2 SL	188	05-Oct	Niyasaki	4x12cm DP, 12x25mm AA
143	15-Jul	Nichols Fld	24x25mm AA	201	05-Sep	Katori Afd	4x12cm DP
147	01-Aug	Nii Jima	6x12cm DP, 12x25mm AA, 2 SL	202	05-Oct	Kisarazu	6x12cm DP
148	01-Aug	Kanoya	6x7cm AA, 12x25mm AA, 2 SL	203	05-Oct	Iwo Jima	6x12cm DP
149	01-Aug	Okinawa	36x25mm AA	204	05-Oct	Legaspi	6x75mm AA
150	01-Aug	Kasonohara	36x25mm AA	205	05-Oct	Chitose	4x12cm DP, 12x25mm AA
151	15-Jul	Chichi Jima	4x12cm DP, 24x25mm AA, 2 SL	206	05-Oct	Bihoro	4x12cm DP, 12x25mm AA
161	01-Aug	Shinchiku	6x7cm AA, 12x25mm AA, 2 SL	207	05-Oct	Nichols Fld	4x12cm DP, 12x25mm AA

In January 1945 the four AA defense units on Iwo Jima were disbanded and reorganized into the new Iwo Jima Guard Force. Others were, of course, lost in fighting, including the 60th (type C) on Guam; and the 82nd (type C) and 83rd (type A) on Tinian.

The Navy also organized a small number barrage balloon units later in the war for defense of their homeland and overseas bases. An overseas unit, such as that sent to Manila, had a nominal strength of three officers, 16 NCOs and 160 enlisted and was formed into eight balloon sections. Each section was composed of a commander, 13 balloon handlers, two mooring personnel, and two hydrogen gas generator operators.

Notes on Sources

General Notes

Coincident with the surrender the War Ministry and Navy Ministry directed all their subordinate units and agencies to destroy all their records. Throughout the homeland and across the remains of the empire, bonfires lit up the skies as files were carted out and tossed on the flames. The operation was, unfortunately, hugely successful. Aside from scattered small holdings, almost nothing remained of the documents generated by the IJA and IJN.

There is, thus, no analog to the vast documentary holdings of other militaries, a fact that has stymied much research into Japanese military history. To the extent documentation survives, it falls into three main categories. The first is documents captured by Allied forces during combat operations. Japanese staff work was notoriously lax and indifferently documented, so this, with one exception, yielded only fragmentary pieces of the puzzle. The exception was the capture in 1945 of the headquarters of the 31st Army on Saipan. They had destroyed many of their files before the fall of the island, but what remained was still a treasure trove of information. The two US commands in the Pacific, although nominally joint commands, were actually service commands, with South-West Pacific being firmly under Army control and Central Pacific just as firmly under the Navy. They rarely spoke to each other or traded information and this is reflected in their publications.

The second was the records located after the war by the US occupation authorities. The US had established the Washington Document Center in 1944 as a central repository for captured Japanese documents, although they seem to have had only limited success persuading the services to cooperate until the end of the war. Once the war ended the WDC sent teams into Japan to vacuum up such records as remained, often in competition with other agencies, such as the USSBS. They succeeded in locating several series of records, but their greatest successes came from the fact that the order to destroy files was not always understood to include libraries. The total number of documents seized by the WDC is not clear, but figures of 250,000 to 300,000 are usually mentioned.

Following the dissolution of the WDC the items were handed out in two large lots. Files (groups of individual papers) were given to the National Archives and bound materials (loosely defined to include pamphlets and brochures) were given to the Library of Congress. The National Archives microfilmed some of the larger lots of files and returned them all to Japan in the 1950s and 1960s. The Library of Congress retained their holdings. The LoC WDC military collection remained uncatalogued until 1990 when a researcher from the Japanese Self-Defense Force University arrived and produced a list of 3,400 Army and 2,000 Navy documents that he had located. These holdings are still not incorporated into the LoC catalog, and must be accessed via Hiromi Tanaka's guide.

Finally, there are documents remaining in Japan that escaped the notice of the WDC collectors. These, however, are fairly few and consist for the most part of personal diaries.

In the immediate post-war occupation the US authorities launched several major efforts to document Japan's war efforts using such documents as had survived and the recollections of Japanese officers. The greatest output came from the SCAP program to write a series of Japanese Monographs dealing with the conduct of the war. A total of 185 monographs were written under Allied supervision by Japanese ex-officers who were involved in the relevant operations. Almost all deal with specific operations or campaigns. There is also a separate, much smaller, series dealing with operations in Manchuria.

Another organization that conducted extensive research was the US Strategic Bombing Survey. They prepared reports on the status of bypassed islands in the Pacific and Japanese war industries, along with a wide variety of other specialized reports. Of particular note is their Study No.45 "Japanese Army Ordnance." The report itself, dealing with the production of army weapons, is rather basic, but the supporting materials in their files (now in the National Archives) includes an astounding recreation of production numbers for each item month-by-month through the war.

Finally, both services also launched investigations into Japanese weaponry and equipment in late 1945. The Navy's Naval Technical Mission to Japan turned out rather cursory studies that tended to be dismissive of Japanese efforts, while the Army's studies on Japanese anti-aircraft and coast-defense programs were excellent in-depth studies, albeit limited to the homeland.

There is little in the way of secondary source material in English, the gold standard for such works being Edward Drea.

Sources for Organizational Materials

The Imperial Japanese Army's *Permanent Mobilization Plan,* established in 1926 and updated in 1943, provided the framework for mobilizing the Army. It laid out the principles and procedures for mobilization but did not specify the size or composition of the Army. Instead, an *Annual Mobilization Plan* was to be issued for each year by the Army General Staff after a careful review of force structure requirements. The annual mobilization plans were generally issued late in the year preceeding their effective year and came into effect on 1 April. Thus, the 1936 mobilization plan would have covered the period from 1 April 1936 to 31 March 1937.

The annual mobilization plan laid out in considerable detail the composition of the regular Army for that year. In particular, *Detailed Annex A* of each annual mobilization plan provided very comprehensive tables of equipment summaries of each unit to be activated. The annual mobilization plan, however, was concerned only with the elements of the regular army, not units activated solely for wartime service. Thus, the numerous independent mixed brigades do not show up at all in the annual mobilization plans, nor do they cover many of the garrison and special formations (such as the occupation divisions) raised for the war. An indication of how largely irrelevant these documents had become to the overall war force structure is the fact that no such plans were ever issued for the years 1942, 1943 or 1945.

Units activated for wartime service that did not show up in the annual mobilization plan were authorized by *Temporary Mobilization Plans* or *Temporary Mobilization Orders* issued by the General Staff. This vehicle was also used to reorganize units from the structure shown in the Annual Mobilization Plan to new configurations as required for the war. As with the Annual Mobilization Plans, these also included a detailed annex that laid out in considerable detail the organization the unit was to adopt.

Making the historian's task difficult is the fact that the IJA was notoriously poor at record-keeping. The central repository for order-of-battle information was the General Staff's *Unit Organization Tables* (*Hengō Butai Gaikanhyō*), a series of printed forms filled out by hand, one record for each division- and brigade-size unit and one record for groups of small units (e.g. heavy artillery regiments). A record showed the components of that unit (down only one level, however), the date of activation or assignment to the unit, and the unit's subordination. The records turned over to the Allies have a start date of August 1944, so units disbanded before then are not included.

A second file source is the *Unit Strength Tables*, filled out for each separately-numbered unit in the Army and containing a chronological record of that unit's TO&E and authorized strength. These were filed by subordination. Thus, when a unit was transferred from, say, the 3rd Army to the 31st Army its Unit Strength Table would be physically moved from the 3rd Army file to the 31st Army file. When a unit was disbanded or deactivated its Unit Strength Table was moved to a "disbanded unit file." Although probably practical at the time the system is cumbersome for historians, for a unit's exact place in the chain of command as of the surrender (when the records were frozen) must be known in order to locate its Unit Strength Table.

A third file source is the Unit Historical Tables (*Shuyō Butai Ryakureki Hyō*). The use of arabic numerals exclusively and the lack of cross-outs in reorganized units strongly suggests that these were not wartime-generated documents, but rather created in response to SCAP demands for information immediately following the surrender. A table exists for each division- and brigade-sized unit in the IJA showing its constituent units and each change in authorized strength from mobilization to the end of the war, along with footnotes explaining the strength changes. The strengths given, however, only rarely match those shown in the wartime Unit Strength Tables. In addition, the tables for some units are manifestly incomplete, suggesting that the compilers did not have access to the full set of IJA records.

The two annual mobilization plans relevant to the war were captured by the United States in the Pacific, translated and issued as printed intelligence guides. These form the basis for any study of the IJA during the war. They are:

IJA General Staff, *Japanese Army 1941 Mobilization Plan Order, Detailed Annex (A)*, issued on 10 September 1940; translated as CINCPAC-CINCPOA Items No. 9683A (26 September 1944), 9683B (nd) and 9683C (28 October 1944).

IJA General Staff, *Japanese Army 1944 Mobilization Plan Order, Detailed Annex (A)*, issued on 30 August 1944; translated as CINCPAC-CINCPOA Item No. 9175, 12 September 1944.

Copies of both of these documents are held by the Operational Archives Branch of the Naval Historical Center at the Washington Navy Yard, Washington, DC. Portions of Item No. 9683A/B/C, disassembled and refiled in different order, can also be found in the Public Records Office (WO 208/1256) in London.

Another main source is the strength and order of battle files maintained or generated by the Army Section of IGHQ:

IJA General Staff, Untitled [Unit Strength Tables]; Files G-1 to G-97, Reels 21 to 32 (11,780 frames), held in RG-242 Captured Documents Section of National Archives, Washington, DC.

IJA General Staff, *Hengō Butai Gaiken Hyō* [Unit Organization Tables, August 1944 to

August 1945]; File F-6, Reel 33 (486 frames), held in RG-242 Captured Documents Section of National Archives, Washington, DC.

IJA General Staff, *Shuyō Butai Ryakureki Hyō* [Unit Historical Tables]; Files G-16 and G-17 on Reel 32 (356 frames), held in RG-242 Captured Documents Section of National Archives, Washington, DC.

GHQ, Supreme Commander Allied Powers; *Final Report on Progress of Demobilization of the Japanese Armed Forces*; 31 December 1946; copy held by the US National Archives (RG 407, Box 1352) at Suitland, MD.

IJN General Staff, 1st Section, *Teikoku Kaigun Senji Hensei Kaitei* [Revisions of Imperial Naval Wartime Organization]. This is a chronological listing of the activations, transfers and deactivations of naval forces for the period May 1942 to December 1944. It is not clear if this is a wartime document or one generated after the war for SCAP. The US gave it the file number HS-49A, under which it can be found on Reel 19 (131 frames) of the National Archives RG-242 Japanese Records microfilm collection.

The Japanese-language documents were returned to Japan after microfilming, and many are now available on the JACAR (Japan Center for Asian Historical Records) website.

The third major group of source documents are the orders and directives issued by the Army Section of the Imperial General Headquarters during the war. During the occupation of Japan the historical section of FEC collected and translated these and published them locally. The Army Directives are primarily administrative in nature and are only occasionally of historical interest. The Army Orders, on the other hand, are command documents, ordering the activation of units, changes in command structure, movement of units, etc. The collection is about 98 percent complete and represents probably the best English-language source collection on the IJA during the war.

Headquarters Far East Command, Military History Section, *Imperial General Headquarters Army Directives*, Vol I (Directives 698-1269/23 July 1940 - 14 September 1942), Vol II (Directives 1275-2225/18 September 1942 - 14 October 1944), n.d.

Headquarters Far East Command, Military History Section, *Imperial General Headquarters Army Orders*, Vol I (Orders 401-660/15 December 1939 - 14 July 1942), Vol II (Orders 661-1131/20 July 1942 - 18 September 1944), Vol III (Orders 1132-1388/20 September 1944 - 22 August 1945), n.d.

Copies of both of these works are held by, and available on microfilm from, the US Army's Office of the Chief of Military History.

The above documents establish the framework for a study of the Japanese ground forces but provide little data on units activated or reorganized under Temporary Mobilization Orders during the war. They also do not cover non-Army units, particularly the naval ground units.

For such information the researcher outside of Japan must turn to the primary sources within the US military intelligence files of 1941–45. The published products they produced included a number of inaccuracies, most of which are corrected with the benefit of hindsight through the use of the documents given above, but the files of the various translation services are a gold mine of primary source material.

Especially useful are the translations done on Saipan by the US Pacific Fleet and

Pacific Ocean Areas, which simply bear the impretur of CINCPAC-CINCPOA. The IJN Administrative Orders Manual was essentially a filing system for orders related to the manning of ships and ground units. As such it provided excellent information on the various ground forces of the Navy. The set captured on Saipan was current through April 1944 and extracts, grouped by type of unit, were translated and published by CINCPAC. The other treasure trove found on Saipan were order of battle lists for some divisional districts in Japan, showing the field units they were responsible for. The most important were:

CINCPAC-CINCPOA Item No. 10309, Special Translation No.10; *Data on Special Naval Landing Forces*; December 1944. This represents a translation of the portions of the IJN Administrative Orders Manuals collection captured on Saipan dealing with SNLFs, including activation orders, administrative directives and TO&Es through May 1944, totaling 164 pages.

CINCPAC-CINCPOA Item No. 10804, *Data on Naval Guard Units and Specially Established Guard Units*; October 1944. This is derived from the same source as No. 10309 above, but extracts the information dealing with guard units.

CINCPAC-CINCPOA Item No. 10805, *Data on Base Forces and Special Base Forces*; December 1944. This is derived from the same source as No. 10309 above, but extracts the information dealing with base forces.

CINCPAC-CINCPOA Item No. 10982, Special Translation No. 37; *Data on AA Defense Units*; January 1945. This is derived from the same source as No. 10309 above, but extracts the information dealing with naval AA defense units.

CINCPAC-CINCPOA Special Translation No.75, Bulletin No. 163-45; *Japanese Order of Battle*; July 1945. This is a 186-page translation of tables showing the complete order of battle, mobilization dates and orders, and TO&E strengths of all the field units raised by the 2nd Guards, 51st and 52nd Depot Divisions. A detailed look at a small but representative slice of the IJA.

CINCPAC-CINCPOA Special Translation No.75, Supplement 1, Bulletin No. 163-45; *Japanese Order of Battle;* July 1945. A 39-page follow-on to the above detailing the field units supported by the 53rd and 54th Depot Divisions

Most of the CINCPAC-CINCPOA Bulletins and Special Translations are held by, and available on microfilm from, the Operational Archives Division of the US Naval Historical Center.

In addition to the above organization, which serviced the central Pacific, the Allies fielded a number of other translation units, most notably ATIS (Allied Translation and Interpretation Service) for SW Pacific under MacArthur, SEATIC (South East Asia Translation and Interrogation Center) and SINTIC (Sino Translation and Interrogation Center) for the China Command. The bulk of their translations of Japanese documents are collected together in their own record groups at the National Archives at College Park, MD, but others are scattered about in various operational command records. In the case of SINTIC, a good deal of these documents are post-war translations of documents supplied by the Japanese during demobilization. Hundreds of these translations were consulted and the most important are listed below:

ATIS, Current Translations No.154; Regulations Concerning the Reorganization of 23

Div (Provisional); 19 Oct 44. Provides organizational data on 23rd Division for move to the Philippines.

ATIS, Enemy Publication No.5; Note-Book Containing Tables of Organization of Certain Japanese Army Units (Giruwa, 22 Jan 43); excellent overview of army organization, corroborated by the Mobilization Plans.

ATIS, Enemy Publication No.387; 1st Division Mobilization Plan Orders; 4 Jan 44. Provides tables of organization for the components of the 1st Division as reorganized in early 1944.

ATIS, Item No.1339; Regulations and Charts Kept by 54th Infantry Regiment; 29 May 1943. Provides tables of organization for the components of the 17th Division in 1943.

ATIS, Item No.1372; Regulations for the Reorganization of the Force; 1 July 1941. Provides tables of equipment for 17th Division in 1941.

CINCPAC-CINCPOA, Item No.8964; 31st Army Organization Order No.45 - Regulations for the Temporary Organization (Reorganization) of Forces under the command of the 31st Army; orig. date 30 May 1944. Orders and defines the reorganization of forces in the Central Pacific into independent mixed brigades and regiments.

CINCPAC-CINCPOA, Item No.8965; Information on the Allocation of Organized Forces; n.d. Further details on the reorganization of 31st Army as explained in the above document (8964).

CINCPAC-CINCPOA, Item No.9085; 135th Regiment T/O; nd.

CINCPAC-CINCPOA, Item No.9116; Principle Ordnance of Northern Marianas Group as of 30 May 1944, orig date 30 May 1944. A comprehensive table showing the weapons available to each unit in that island chain.

CINCPAC-CINCPOA, Item No.9487; 14th Division OpOrd A No.69, Annexed Volume, Essentials of Organization; nd. Provides TO&Es for the 53rd IMB.

CINCPAC-CINCPOA, Item No.9510; Tsukuba Butai Order, Annex No.1; orig date 31 Dec 1943. Shows organization and personnel strength of units subordinate to 18th Army.

CINCPAC-CINCPOA, Item No.9579A&B; Extracts from Yokasuka Naval District Secret Regulations No.4-358; 1 April 1938 corrected to June 1943. Contains table of organization for Yokasuka CSNLF.

CINCPAC-CINCPOA, Item No.9614; War Ministry Order A-106 - Summary of the Reorganization (Temporary Organization A) of the 3rd, 5th and 13th Divisions, the Independent Mixed Brigades, and the South Seas Detachments, and of the 250th Return (Demobilization); 16 Nov 43. Provides detailed tables of organization for the island defense divisions, the 24th-30th IMBs, 31st-34th IMRs, and the amphibious brigades.

CINCPAC-CINCPOA, Item No.9660; Reports on Supplies and Ammunition, Charts of Personnel and Material Added to and Subtracted from Various Units; nd. Tables showing modifications made to standard TO&Es for the expeditionary units.

CINCPAC-CINCPOA, Item No.9691; Chichi Jima Fortress HQ: Table of Military Equipment on Hand, as of April 1944. An extremely detailed listing of all types of military equipment held by the fortress, broken down by unit.

CINCPAC-CINCPOA, Item No.9859; T/O for an Infantry Regiment, undated, handwritten copy.

CINCPAC-CINCPOA, Item No.9882; letter from Commanding General Chichi Jima to Commanding General 31st Army; orig date 5 May 44. Lists all units and officers assigned to the Chichi Jima Fortress.

CINCPAC-CINCPOA, Item No.11004; Weapons Available for Various Islands; n.d. A comprehensive survey of the ordnance available to the subordinate units of 31st Army.

CINCPAC-CINCPOA, Item No.9893; Personnel and Horses in the Ogasawara Group at Present; orig date 25 Mar 44. A strength breakdown of the Chichima Fortress by unit.

CINCPAC-CINCPOA, Item No.9962; Temporary Organization (Reorganization) and Return (Demobilization) of CPA Forces; nd, c.May 1944. Gives complete tables of organization for the 109th Division, the IMBs and IMRs as reorganized by 31st Army in 1944 in the Central Pacific Area.

CINCPAC-CINCPOA, Item No.10119; Detailed Summary of Temporary Organization (Reorganization) of the 278th Return (Demobilization); nd. Provides detailed tables of equipment, complementing the organization tables in Item No.9962 above.

CINCPAC-CINCPOA, Item No.10952; Extract From a Table of Equipment for the 53rd Independent Mixed Brigade; nd. Supplements CINCPAC-CINCPOA Item No.9487 above.

PACMIRS, Document DB-26; Tables of Allocation of Ordnance, Ammunition, Materiel, Swords and Pistols, and Powders and Firing Accessories for 136th Infantry Regiment; June 1944. Provides Tables of Equipment in use on Saipan. Held by Public Record Office (WO 208/1256) in London.

PACMIRS, Document 1220; Regulations for Organization and Dispatch of Transferred Units, June 1944. Organization and reorganization tables for the expeditionary units.

SEATIC, Intelligence Bulletin No.218; Table of Organization of Units in 29 Army Staff of 29 Army; 28 Nov 45.

SINTIC, Document Translation No.157; Order of Battle and Strength of Japanese Army in China; 27 October 1945.

SINTIC, Document Translation No.246; Organizational Charts of Various Units, China Expeditionary Forces of Japanese Army (excluding 13th Flight Division); 18 February 1946. An excellent summary of all the IJA combat units in China at the end of the war.

SINTIC, Document Translation No.331; Provisions Regarding the Temporary Activation of Divisions, Independent Mixed Brigades, Independent Garrison Units, etc., and the 322nd Return of Troops (Demobilization) etc.; orig date 20 Mar 45. The organization order, included appended TO&Es, for the 86th IMB.

SINTIC, Weekly Intelligence Notes No.14, Annex III; Status of Disarmament of Japanese Troops in China, North French Indochina, and Taiwan; 29 November 1945.

Unknown Provenance: Order of Battle of Japanese Forces in the Dutch East Indies, supplied by the Japanese for demobilization purposes. Mostly dated mid-September 1945. Held by Public Records Office (WO 208/1698) in London.

Other, smaller, items are scattered throughout the files of the intelligence sections that operated against the Japanese, as well as in the rest of the SINTIC, ATIS, CINCPAC-CINCPOA and SEATIC files. Particularly useful items in the former category can be found in the files of the 10th US Army for late in the war and during its role in the occupation of Japan, when it supervised the demobilization of the IJA.

Another source of information, invaluable although generally less specific, is the series of Japanese monographs prepared by former IJA officers under the supervision of the Military History Section of GHQ FEC during the immediate post-war period. All were consulted, but those most relevent to a study of IJA organization were as follows:

Headquarters, Army Forces Far East, Military History Section; Japanese Monograph No.17, *Homeland Operations Record* (Revised Edition, including Nos 17, 18, 19 & 20).
Japanese Monograph No.45, *History of Imperial General Headquarters Army Section*.
Ibid; Japanese Monograph No. 138, *Japanese Preparations for Operations in Manchuria, Jan 43-Aug 45*.
Ibid; Japanese Studies on Manchuria Vol V; *Infantry Operations*.
Ibid; Japanese Studies on Manchuria Vol VI; *Armor Operations*.
Ibid; Japanese Studies on Manchuria Vol VII; *Supporting Arms and Services: Artillery and Antiaircraft Artillery Operations*.

Immediately after the war the US Army Forces Pacific also commissioned two narrow studies that produced exceptionally thorough and detailed reports on homeland defenses at the end of the war through interviews, surrendered documentation and personal inspection.

GHQ USAFPAC, Seacoast Artillery Research Board, *Survey of Japanese Coast Artillery*, 1 February 1946, in NARA RG-407/98-GHQ2-15.5
GHQ USAFPAC, AAA Research Board, *Survey of Japanese Antiaircraft Artillery*, 1 February 1946, in NARA RG-407/98-GHQ2-15.0

Every effort has been made to avoid the use of the processed intelligence reports issued by the Allies during the war, as opposed to the translations of captured documents listed above. A perusal of War Department TM-E30-480 *Handbook on Japanese Military Forces* dated 1944 will show the distortions that creep into even the best of the intelligence processes. It does contain useful subjective assessments of Japanese tactics and procedures, however. The TM was reprinted by a commercial firm, The Military Press, in 1970, but without 31 of the 34 pages in Chapter III dealing with army tactical organization. The entries in Chapter III, however, were so general and often misleading that SWPA was preparing to issue its own handbook on IJA organization when the war ended. A draft copy of *Organization of the Japanese Ground Forces* by G-2, GHQ, SWPA up to 15 August 1945, however, is held by the National Archives at Suitland, MD. This work is much more comprehensive and accurate than the War Department manual but never saw the light of day. It is still limited, however, by the lack of knowledge of events in the Japanese homeland and Manchuria, and, to a lesser extent, in SE Asia.

A related source is the US Army's *Order of Battle of the Japanese Armed Forces* (5th Ed., 1945), which has also been reprinted commercially by Game Marketing Company (Allentown, PA, 1981). Omissions are many but mistakes few in this work, which thus provides a handy, albeit limited, reference for the historian.

For a general overview of the Japanese Army's role in the war there are two main English-language sources. The most well-known is the Hayashi book, widely available in the US. The MacArthur book was well written by a combination of former IJA staff officers and the intelligence section of MacArthur's HQ during the post-war period and was beautifully produced with lavish color maps and facsimilies, but covers only the Japanese response to MacArthur's campaigns and the defense of the homeland. The Hattori book is available in English only as copies of a typescript translation, with a copy being held by the Pentagon Library.

Hayashi, Saburo; *Kogun, the Japanese Army in the Pacific War*; (Greenwood Press, 1959, Westport, CT).

Reports of General MacArthur, Vol II (Japanese Operations in the Southwest Pacific Area); (GPO, 1966, Washington, DC)

The definitive English-language study of the IJA is the one by Drea:

Drea, Edward J.; *Japan's Imperial Army, It's Rise and Fall, 1853-1945*, (Univ Press of Kansas, 2009, Lawrence, KS).

Other titles published by Helion & Company

Thunder at Prokhorovka. A Combat History of Operation Citadel, Kursk, July 1943
David Schranck
ISBN 978-1-909384-54-5 Hardback

D-Day - The Last of the Liberators: Some of The Last Veterans of the Normandy Landings Retrace their Steps Seventy Years Later
Robin Savage
ISBN 978-1-909982-31-4 Hardback

Für Volk and Führer. The Memoir of a Veteran of the 1st SS Panzer Division Leibstandarte SS Adolf Hitler
Erwin Bartmann
ISBN 978-1-909384-53-8 Hardback

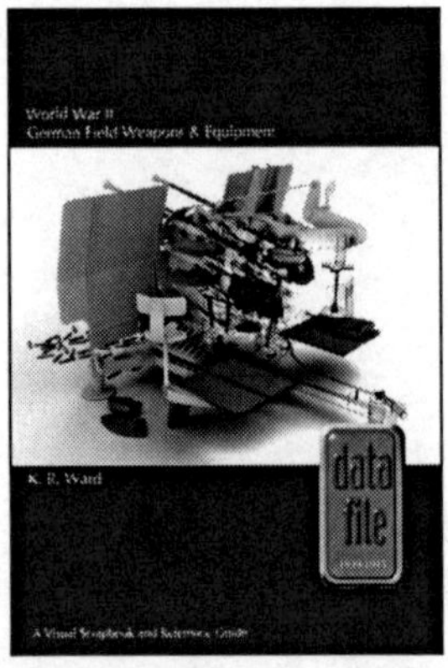

World War II German Field Weapons & Equipment. A Visual Reference Guide
Keith Ward
ISBN 978-1-909384-44-6 Paperback

HELION & COMPANY
26 Willow Road, Solihull, West Midlands B91 1UE, England
Telephone 0121 705 3393 Fax 0121 711 4075
Website: http://www.helion.co.uk
Twitter: @helionbooks | Visit our blog http://blog.helion.co.uk

Lightning Source UK Ltd.
Milton Keynes UK
UKOW03f1326060814

236455UK00004B/46/P